Mechanics of Composite Materials

Mechanics of Composite Materials

Contributors

Kyung-In Jang, Ha Uk Chung et al.

AURIS
Reference

www.aurisreference.com

Mechanics of Composite Materials

Contributors: Kyung-In Jang, Ha Uk Chung et al.

Published by Auris Reference Limited
www.aurisreference.com

United Kingdom

Copyright 2016
Printed in 2017 for Sale in the Indian Subcontinent

The information in this book has been obtained from highly regarded resources. The copyrights for individual articles remain with the authors, as indicated. All chapters are distributed under the terms of the Creative Commons Attribution License, which permit unrestricted use, distribution, and reproduction in any medium, provided the original author and source are credited.

Notice

Contributors, whose names have been given on the book cover, are not associated with the Publisher. The editors and the Publisher have attempted to trace the copyright holders of all material reproduced in this publication and apologise to copyright holders if permission has not been obtained. If any copyright holder has not been acknowledged, please write to us so we may rectify.

Reasonable efforts have been made to publish reliable data. The views articulated in the chapters are those of the individual contributors, and not necessarily those of the editors or the Publisher. Editors and/or the Publisher are not responsible for the accuracy of the information in the published chapters or consequences from their use. The Publisher accepts no responsibility for any damage or grievance to individual(s) or property arising out of the use of any material(s), instruction(s), methods or thoughts in the book.

Mechanics of Composite Materials

ISBN: 978-1-78154-944-5

British Library Cataloguing in Publication Data
A CIP record for this book is available from the British Library

Printed in the United Kingdom

Exclusively distributed by CBS Publishers & Distributors Pvt. Ltd.

Sales & Distribution Rights only for India, Pakistan, Bangladesh, Sri Lanka, Nepal and Bhutan. This book is not to be sold outside these territories.

Contents

List of Abbreviations vii
List of Contributors... ix
Preface.. xvii

Chapter 1 Soft Network Composite Materials with Deterministic and Bio-Inspired Designs 1
Kyung-In Jang , Ha Uk Chung , Sheng Xu , Chi Hwan Lee, Haiwen Luan, Jaewoong Jeong, Huanyu Cheng, Gwang-Tae Kim , Sang Youn Han, Jung Woo Lee, Jeonghyun Kim, Moongee Cho , Fuxing Miao, Yiyuan Yang, Han Na Jung, Matthew Flavin , Howard Liu , Gil Woo Kong , Ki Jun Yu , Sang Il Rhee , Jeahoon Chung , Byunggik Kim , Jean Won Kwak , Myoung Hee Yun, Jin Young Kim, Young Min Song, Ungyu Paik, Yihui Zhang, Yonggang Huang & John A. Rogers

Chapter 2 Biaxial Tensile Strength Characterization of Textile Composite Materials.. 25
David Alejandro Arellano Escárpita, Diego Cárdenas, Hugo Elizalde, Ricardo Ramirez and Oliver Probst

Chapter 3 Molecular Simulations on Interfacial Sliding of Carbon Nanotube Reinforced Alumina Composites 55
Yuan Li, Sen Liu, Ning Hu, Weifeng Yuan and Bin Gu

Chapter 4 Mechanical Coating Technique for Composite Films and Composite Photocatalyst Films 79
Yun Lu, Liang Hao and Hiroyuki Yoshida

Chapter 5 Generation of R-Curve from 4ENF Specimens: An Experimental Study... 115
V. Alfred Franklin and T. Christopher

Chapter 6 On the Impact of Manufacturing Uncertainty in Structural Health Monitoring of Composite Structures: A Signal to Noise Weighted Neural Network Process 137
Hessamodin Teimouri, Abbas S. Milani, Rudolf Seethaler, Amir Heidarzadeh

Chapter 7 Enhancement in the Electrical and Thermal Properties of Ethylene Vinyl Acetate (EVA) Co-Polymer by Zinc Oxide Nanoparticles 155
Jose Sebastian, Eby T. Thachil, Jobin Job Mathen, Joseph Madhavan, Prince Thomas, Jacob Philip, M. S. Jayalakshmy, Shahrom Mahmud, Ginson P. Joseph

Chapter 8	**Graphene-Boron Nitride Composite: A Material with Advanced Functionalities** 175	
	Sumanta Bhandary and Biplab Sanyal	
Chapter 9	**Properties of MWNT-Containing Polymer Composite Materials Depending on Their Structure** 193	
	Ilya Mazov, Vladimir Kuznetsov, Anatoly Romanenko and Valentin Suslyaev	
Chapter 10	**Carbon Nanotube Reinforced Alumina Composite Materials** 221	
	Go Yamamoto and Toshiyuki Hashida	
Chapter 11	**Manufacturing and Properties of Quartz (SiO_2) Particulate Reinforced Al-11.8%Si Matrix Composites** 243	
	M. Sayuti, S. Sulaiman, B.T.H.T. Baharudin, M.K.A Arifin and T.R. Vijayaram	
	Citations 277	
	Index 279	

List of Abbreviations

ANN	Artificial Neural Networks
CNTs	Carbon nanotubes
CNF	Central-notched flexure
CMCs	Ceramic matrix composites
CVD	Chemical vapor deposition
DSD	Damage signature database
DAQ	Data acquisition
EMI	Electromagnetic irradiation
FE	Finite element
h-BNC	Hexagonal BNC
INF	Internal-notched flexure
MCT	Mechanical coating technique
MMC	Metal matrix composites
MMCs	Metal matrix composites
MWNTs	Multiwall carbon nanotubes
PVD	Physical vapor deposition
PMCs	Polymer matrix composites
SEM	Scanning electron microscope
SN	Signal-to-noise
SWNTs	Single-wall nanotubes
SWCNT	Single-walled carbon nanotube
SHM	Structural Health Monitoring
T-T	Tension-tension
TC	Textile composites
TEM	Transmission electron microscope
UDC	Unidirectional composites
WWFE	World Wide Failure Exercises
ZGNRs	Zigzag nanoribbons

List of Contributors

Kyung-In Jang
Department of Materials Science and Engineering, Frederick Seitz Materials Research Laboratory, University of Illinois, Urbana-Champaign, Urbana, Illinois 61801, USA

Ha Uk Chung
Department of Materials Science and Engineering, Frederick Seitz Materials Research Laboratory, University of Illinois, Urbana-Champaign, Urbana, Illinois 61801, USA

Sheng Xu
Department of Materials Science and Engineering, Frederick Seitz Materials Research Laboratory, University of Illinois, Urbana-Champaign, Urbana, Illinois 61801, USA

Chi Hwan Lee
Department of Materials Science and Engineering, Frederick Seitz Materials Research Laboratory, University of Illinois, Urbana-Champaign, Urbana, Illinois 61801, USA

Haiwen Luan
Department of Civil and Environmental Engineering, Department of Mechanical Engineering, Center for Engineering and Health and Skin Disease Research Center, Northwestern University, Evanston, Illinois 60208, USA

Jaewoong Jeong
Department of Electrical, Computer and Energy Engineering, University of Colorado, Boulder, Colorado 80309, USA

Huanyu Cheng
Department of Civil and Environmental Engineering, Department of Mechanical Engineering, Center for Engineering and Health and Skin Disease Research Center, Northwestern University, Evanston, Illinois 60208, USA

Gwang-Tae Kim
Department of Materials Science and Engineering, Frederick Seitz Materials Research Laboratory, University of Illinois, Urbana-Champaign, Urbana, Illinois 61801, USA

Sang Youn Han
Department of Materials Science and Engineering, Frederick Seitz Materials Research Laboratory, University of Illinois, Urbana-Champaign, Urbana, Illinois 61801, USA
Samsung Display Co. Display R&D Center, Yongin-city, Gyeongki-do 446–711, Republic of Korea

Jung Woo Lee1
Department of Material Science and Engineering, Department of Energy Engineering, Hanyang University, Seoul 133-791, Republic of Korea

Jeonghyun Kim
Department of Materials Science and Engineering, Frederick Seitz Materials Research Laboratory, University of Illinois, Urbana-Champaign, Urbana, Illinois 61801, USA
Department of Material Science and Engineering, Department of Energy Engineering, Hanyang University, Seoul 133-791, Republic of Korea

Moongee Cho
Department of Materials Science and Engineering, Frederick Seitz Materials Research Laboratory, University of Illinois, Urbana-Champaign, Urbana, Illinois 61801, USA

Fuxing Miao
Department of Civil and Environmental Engineering, Department of Mechanical Engineering, Center for Engineering and Health and Skin Disease Research Center, Northwestern University, Evanston, Illinois 60208, USA
Department of Mechanical Engineering and Mechanics, Ningbo University, Ningbo 315211, China

Yiyuan Yang
Department of Materials Science and Engineering, Frederick Seitz Materials Research Laboratory, University of Illinois, Urbana-Champaign, Urbana, Illinois 61801, USA

Han Na Jung
Department of Materials Science and Engineering, Frederick Seitz Materials Research Laboratory, University of Illinois, Urbana-Champaign, Urbana, Illinois 61801, USA

Matthew Flavin
Department of Materials Science and Engineering, Frederick Seitz Materials Research Laboratory, University of Illinois, Urbana-Champaign, Urbana, Illinois 61801, USA

Howard Liu
Department of Materials Science and Engineering, Frederick Seitz Materials Research Laboratory, University of Illinois, Urbana-Champaign, Urbana, Illinois 61801, USA

Gil Woo Kong
Department of Materials Science and Engineering, Frederick Seitz Materials Research Laboratory, University of Illinois, Urbana-Champaign, Urbana, Illinois 61801, USA

Ki Jun Yu
Department of Materials Science and Engineering, Frederick Seitz Materials Research Laboratory, University of Illinois, Urbana-Champaign, Urbana, Illinois 61801, USA

Sang Il Rhee
Department of Materials Science and Engineering, Frederick Seitz Materials Research Laboratory, University of Illinois, Urbana-Champaign, Urbana, Illinois 61801, USA

Jeahoon Chung
Department of Materials Science and Engineering, Frederick Seitz Materials Research Laboratory, University of Illinois, Urbana-Champaign, Urbana, Illinois 61801, USA

Byunggik Kim
Department of Materials Science and Engineering, Frederick Seitz Materials Research Laboratory, University of Illinois, Urbana-Champaign, Urbana, Illinois 61801, USA

Jean Won Kwak
Department of Materials Science and Engineering, Frederick Seitz Materials Research Laboratory, University of Illinois, Urbana-Champaign, Urbana, Illinois 61801, USA

Myoung Hee Yun
Department of Materials Science and Engineering, Frederick Seitz Materials Research Laboratory, University of Illinois, Urbana-Champaign, Urbana, Illinois 61801, USA
School of Energy and Chemical Engineering, Ulsan National Institute Science and Technology (UNIST), Ulsan 689-798, Republic of Korea

Jin Young Kim
School of Energy and Chemical Engineering, Ulsan National Institute Science and Technology (UNIST), Ulsan 689-798, Republic of Korea

Young Min Song
Department of Electronic Engineering, Biomedical Research Institute, Pusan National University, Geumjeong-gu, Busan 609-735, Republic of Korea

Ungyu Paik
Department of Material Science and Engineering, Department of Energy Engineering, Hanyang University, Seoul 133-791, Republic of Korea

Yihui Zhang
Department of Civil and Environmental Engineering, Department of Mechanical Engineering, Center for Engineering and Health and Skin Disease Research Center, Northwestern University, Evanston, Illinois 60208, USA
Center for Mechanics and Materials, Tsinghua University, Beijing 100084, China.
Correspondence and requests for materials should be addressed to Y.Z. or to Y.H

Yonggang Huang
Department of Civil and Environmental Engineering, Department of Mechanical Engineering, Center for Engineering and Health and Skin Disease Research Center, Northwestern University, Evanston, Illinois 60208, USA

John A. Rogers
Department of Materials Science and Engineering, Frederick Seitz Materials Research Laboratory, University of Illinois, Urbana-Champaign, Urbana, Illinois 61801, USA

David Alejandro Arellano Escárpita
Mechatronics Engineering Department, Instituto Tecnológico y de Estudios Superiores de Monterrey, Campus Ciudad de México, Col. Ejidos de Huipulco, Tlalpan, México D.F., Mexico

Diego Cárdenas
Mechatronics Engineering Department, Instituto Tecnológico y de Estudios Superiores de Monterrey, Campus Ciudad de México, Col. Ejidos de Huipulco, Tlalpan, México D.F., Mexico

Hugo Elizalde
Mechatronics Engineering Department, Instituto Tecnológico y de Estudios Superiores de Monterrey, Campus Ciudad de México, Col. Ejidos de Huipulco, Tlalpan, México D.F., Mexico

Ricardo Ramirez
Mechatronics Engineering Department, Instituto Tecnológico y de Estudios Superiores de Monterrey, Campus Ciudad de México, Col. Ejidos de Huipulco, Tlalpan, México D.F., Mexico

Oliver Probst
Physics Department, Instituto Tecnológico y de Estudios Superiores de Monterrey, Campus Monterrey, Monterrey, N.L., Mexico

Yuan Li
Department of Nanomechanics, Tohoku University, Aramaki-Aza-Aoba, Aoba-ku, Sendai,, Japan

Sen Liu
Department of Mechanical Engineering, Chiba University, Yayoi-cho, Inage-ku, Chiba,, Japan

Ning Hu
Department of Mechanical Engineering, Chiba University, Yayoi-cho, Inage-ku, Chiba,, Japan

Weifeng Yuan
School of Manufacturing Science and Engineering, Southwest University of Science and Technology, Mianyang, P.R.China

Bin Gu
School of Manufacturing Science and Engineering, Southwest University of Science and Technology, Mianyang, P.R.China

Yun Lu
Graduate School & Faculty of Engineering, Chiba University, Japan

Liang Hao
Graduate School, Chiba University, Japan

Hiroyuki Yoshida
Chiba Industrial Technology Research Institute, Japan

V. Alfred Franklin
Faculty of Mechanical Engineering, Sardar Raja College of Engineering, Alangulam, Tirunelveli 627808, India

T. Christopher
Faculty of Mechanical Engineering, Government College of Engineering, Tirunelveli 627007, India

Hessamodin Teimouri
School of Engineering, University of British Columbia, Kelowna, Canada

Abbas S. Milani
School of Engineering, University of British Columbia, Kelowna, Canada

Rudolf Seethaler
School of Engineering, University of British Columbia, Kelowna, Canada

Amir Heidarzadeh
Department of Aerospace Engineering, Sharif University of Technology, Tehran, Iran

Jose Sebastian
Department of Polymer Science and Rubber Technology, Cochin University of Science and Technology, Kerala, India

Eby T. Thachil
Department of Polymer Science and Rubber Technology, Cochin University of Science and Technology, Kerala, India

Jobin Job Mathen
Department of Physics, Loyola College, Chennai, India

Joseph Madhavan
Department of Physics, Loyola College, Chennai, India

Prince Thomas
Department of Physics, St. Thomas College, Palai, India

Jacob Philip
Department of Instrumentation, Cochin University of Science and Technology, Kerala, India

M. S. Jayalakshmy
Department of Instrumentation, Cochin University of Science and Technology, Kerala, India

Shahrom Mahmud
School of Physics, Universiti Sains Malaysia, Minden, Malaysia

Ginson P. Joseph
Department of Physics, St. Thomas College, Palai, India

Sumanta Bhandary
Department of Physics and Astronomy, Uppsala University, Uppsala, Sweden

Biplab Sanyal
Department of Physics and Astronomy, Uppsala University, Uppsala, Sweden

Ilya Mazov
Boreskov Institute of Catalysis, Novosibirsk, RussiaNational Research Technical University "MISIS", Moscow,, Russia

Vladimir Kuznetsov
Boreskov Institute of Catalysis, Novosibirsk,, Russia

Anatoly Romanenko
Nikolaev Institute of Inorganic Chemistry, Novosibirsk,, Russia

Valentin Suslyaev
National Research Tomsk State University, Tomsk,, Russia

Go Yamamoto
Fracture and Reliability Research Institute (FRRI), Tohoku University, Japan

Toshiyuki Hashida
Fracture and Reliability Research Institute (FRRI), Tohoku University, Japan

M. Sayuti
Department of Mechanical and Manufacturing Engineering, Faculty of Engineering, Universiti Putra Malaysia, Serdang, Selango, Malaysia
Department of Industrial Engineering, Faculty of Engineering, Malikussaleh University, Lhokseumawe, Aceh, Indonesia

S. Sulaiman
Department of Mechanical and Manufacturing Engineering, Faculty of Engineering, Universiti Putra Malaysia, Serdang, Selango, Malaysia

B.T.H.T. Baharudin
Department of Mechanical and Manufacturing Engineering, Faculty of Engineering, Universiti Putra Malaysia, Serdang, Selango, Malaysia

M.K.A Arifin
Department of Mechanical and Manufacturing Engineering, Faculty of Engineering, Universiti Putra Malaysia, Serdang, Selango, Malaysia

T.R. Vijayaram
Faculty of Engineering and Technology (FET) Multimedia University, Jalan Ayer Keroh Lama, Bukit Beruang, Melaka, Malaysia

Preface

A composite material is a material made from two or more constituent materials with significantly different physical or chemical properties that, when combined, produce a material with characteristics different from the individual components. The text *Mechanics of Composite Materials* introduces the basic concepts of mechanical behavior of composite materials. In first chapter, we introduce a type of bio-inspired, soft deterministic composite that can quantitatively reproduce the mechanics of non-mineralized biological materials, including the precise non-linear stress/strain response of human skin and its subtle spatial variations across different locations on the body. Biaxial tensile strength characterization of textile composite materials has been discussed in second chapter. In third chapter, a series of pull-out simulations of either SWCNT or MWCNT from alumina matrix are carried out based on molecular mechanics (MM) to investigate the corresponding interfacial sliding behaviors in CNT/alumina composites. Mechanical coating technique for composite films and composite photocatalyst films has been presented in fourth chapter. Fifth chapter examines the fracture energy of four-point end-notched flexure (4ENF) composite specimens made of carbon/epoxy and glass/epoxy. The aim of sixth chapter is to conduct an investigation into the development of a robust structural health monitoring (SHM) via a weighted artificial neural network (ANN), which can be immune against potential manufacturing errors in the structure. Seventh chapter deals with the effect of ZnO nanopowder on the electrical, optical, mechanical properties EVA polymer matrix. Eighth chapter focuses on graphene-boron nitride composite, a material with advanced functionalities. In ninth chapter, we describe preparation of MWNT-containing composite materials using coagulation precipitation technique and polymethylmethacrylate (PMMA) and polystyrene (PS) as matrices. Tenth chapter presents the novel processing approach based on the precursor method. The failure mechanism of the MWCNTs during crack opening in a MWCNT/alumina composite is also investigated through transmission electron microscope (TEM) observations and single nanotube pullout tests. Manufacturing and properties of quartz (SiO_2) particulate reinforced Al-11.8% SI matrix composites are discussed in last chapter.

Chapter 1

SOFT NETWORK COMPOSITE MATERIALS WITH DETERMINISTIC AND BIO-INSPIRED DESIGNS

Kyung-In Jang[1], Ha Uk Chung[1], Sheng Xu[1], Chi Hwan Lee[1], Haiwen Luan[2], Jaewoong Jeong[3], Huanyu Cheng[2], Gwang-Tae Kim[1], Sang Youn Han[1,4], Jung Woo Lee1,[5], Jeonghyun Kim[1,5], Moongee Cho[1], Fuxing Miao[2,6], Yiyuan Yang[1], Han Na Jung[1], Matthew Flavin[1], Howard Liu[1], Gil Woo Kong[1], Ki Jun Yu[1], Sang Il Rhee[1], Jeahoon Chung[1], Byunggik Kim[1], Jean Won Kwak[1], Myoung Hee Yun[1,7], Jin Young Kim[7], Young Min Song[8], Ungyu Paik[5], Yihui Zhang[2,9], Yonggang Huang[2] & John A. Rogers[1]

[1]Department of Materials Science and Engineering, Frederick Seitz Materials Research Laboratory, University of Illinois, Urbana-Champaign, Urbana, Illinois 61801, USA.

[2]Department of Civil and Environmental Engineering, Department of Mechanical Engineering, Center for Engineering and Health and Skin Disease Research Center, Northwestern University, Evanston, Illinois 60208, USA.

[3]Department of Electrical, Computer and Energy Engineering, University of Colorado, Boulder, Colorado 80309, USA.

[4] Samsung Display Co. Display R&D Center, Yongin-city, Gyeongki-do 446–711, Republic of Korea. [5]Department of Material Science and Engineering, Department of Energy Engineering, Hanyang University, Seoul 133-791, Republic of Korea.

[6]Department of Mechanical Engineering and Mechanics, Ningbo University, Ningbo 315211, China.

[7] School of Energy and Chemical Engineering, Ulsan National Institute Science and Technology (UNIST), Ulsan 689-798, Republic of Korea.

[8]Department of Electronic Engineering, Biomedical Research Institute, Pusan National University, Geumjeong-gu, Busan 609-735, Republic of Korea.

[9] Center for Mechanics and Materials, Tsinghua University, Beijing 100084, China. Correspondence and requests for materials should be addressed to Y.Z. or to Y.H.

ABSTRACT

Hard and soft structural composites found in biology provide inspiration for the design of advanced synthetic materials. Many examples of bio-inspired hard materials can be found in the literature; far less attention has been devoted to soft systems. Here we introduce deterministic routes to low-modulus thin film materials with stress/strain responses that can be tailored precisely to match the non-linear properties of biological tissues, with application opportunities that range from soft biomedical devices to constructs for tissue engineering. The approach combines a low-modulus matrix with an open, stretchable network as a structural reinforcement that can yield classes of composites with a wide range of desired mechanical responses, including anisotropic, spatially heterogeneous, hierarchical and self-similar designs. Demonstrative application examples in thin, skin-mounted electrophysiological sensors with mechanics precisely matched to the human epidermis and in soft, hydrogel-based vehicles for triggered drug release suggest their broad potential uses in biomedical devices.

INTRODUCTION

Concepts in materials science that draw inspiration from the natural world have yielded an impressive collection of important advances in recent years[1, 2, 3, 4, 5, 6, 7, 8, 9, 10]. Structural materials are of particular interest, due to their essential roles in nearly every engineered system. Biology provides examples of two general classes of such materials: (1) hierarchically assembled composites that combine hard (~GPa) inorganic minerals such as calcium carbonate or hydroxyapatite with organic polymer additives and (2) non-mineralized, soft (~MPa) materials that embed sparse networks of wavy, fibrous materials such as collagen, elastin or keratin in extracellular matrices[11, 12]. The first offers linear elastic response for strains up to a fraction of a per cent, with a 'brick-and-mortar' arrangement of organic and inorganic constituents to impart levels of fracture toughness that are essential to seashell nacre, dentine and bone. Many sophisticated examples of synthetic materials that exploit these design concepts can be found in the literature[12, 13, 14, 15, 16, 17]. The second involves tangled networks of coiled fiberous polymers, typically in a ground substance that includes interstitial fluid, cell adhesion proteins and proteoglycans[18]. Tensile loads cause these fibres to unfurl, straighten, buckle, twist and stretch in a manner that imparts a low-modulus response for relative small strains (for example, ligament: ~0–2%, epidermis: less than ~10%) with a sharp transition to a high modulus regime for larger strains (for example, ligament: ~5%, epidermis: ~30%)[19, 20, 21, 22]. This 'J-shaped' stress–strain response combines soft, compliant mechanics and large levels of stretchability, with a natural 'strain-limiting' mechanism that protects biological tissues from excessive strain[23, 24]. Although such soft, non-mineralized biological structures offer great potential in areas ranging from artificial tissue constructs to bio-integrated devices, they have received far less attention compared with their mineralized counterparts[25, 26].

Here we introduce a type of bio-inspired, soft deterministic composite that can quantitatively reproduce the mechanics of non-mineralized biological materials, including the precise non-linear stress/strain response of human skin and its subtle spatial variations across different locations on the body. The concepts use planar, lithographically defined networks related to those found in lightweight, impact resistant, loading bearing structures[27, 28, 29] with serpentine microstructures originally developed for interconnects in stretchable electronics. A low-modulus elastomer or hydrogel provides a supporting matrix[30, 31]. When formed with network geometries optimized using tools of computational mechanics, such composites can yield a wide range of desired mechanical properties, including isotropic and anisotropic responses and spatially heterogeneous characteristics. Successful experimental demonstrations of these soft network composites in combination with electrophysiological sensors and drug-release vehicles indicate their potential for practical applications in biomedical devices.

RESULT

Bio-Inspired Soft and Thin Film Composites

Figure 1 presents a schematic illustration of the strategy in the context of an artificial skin construct. Here a two-dimensional (2D) network of photolithographically defined polyimide filaments[28, 31] (HD-4110, HD Microsystems, USA) resides in the middle of a soft, 'skin-phantom' matrix that is vapour permeable ($\sim 10\,g\,h^{-1}\,m^{-2}$ at a thickness of 100 μm), ultra soft (E~3 kPa), highly elastic (up to ~250% tensile strain), biocompatible and adherent to biological materials such as skin (~2 kPa) (Silbione RT Gel 4717A/B, Bluestar Silicones, USA). See Fig. 1a–d andSupplementary Figs 1 and 2. The deterministic architecture of the former component, here in a uniform triangular lattice configuration of repeating, filamentary building block units with 'horseshoe' geometries (as shown in the inset of Fig. 1d, which consists of two identical circular arcs, each with an arc angle of , radius of R and width of w), defines the mechanical properties, through a role that is analogous to that of collagen and elastin in biological systems. As in biology, this composite exhibits tensile responses to mechanical loading that consist of three phases, as reflected in the experimental data of Fig. 1e. The first phase ('toe' region) involves large-scale, bending-dominated deformations of the constituent filaments, to yield an ultralow effective modulus (approximately few or few tens of kPa). In the second phase ('heel' region), continued stretching causes the filaments to rotate, twist and align to the direction of the applied stress, with a corresponding increase in modulus. Complete extension defines a transition point of the J-shaped stress–strain curve into the third phase, 'linear', region, where stretching of the filaments themselves dominates the response. The modulus in this phase can be several orders of magnitude higher than that in the initial phase. Deformation finally continues until the point of ultimate tensile strength, where plastic yielding and rupture of the network defines a fourth region of behaviour.

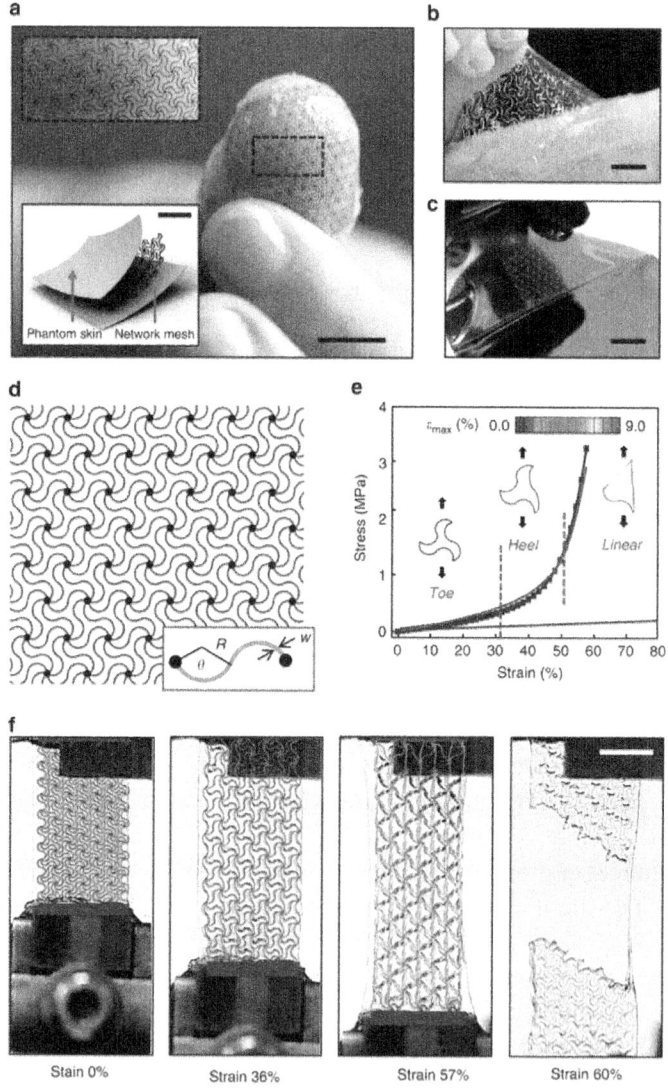

Figure 1: Soft, deterministic network composites in designs inspired by non-mineralized biological materials.

(a) Optical images and an exploded view schematic illustration (lower left inset) of a skin-like composite that consists of a lithographically defined wavy filamentary network of polyimide, analogous to a collagen/elastin structure, embedded in a soft breathable elastomer, analogous to a biological ground substance. The image shows this material wrapped onto the tip of the thumb. (b) Optical image of a similar material partially peeled away from the skin of the forearm. (c) Optical image of the polyimide network during removal from a PMMA-coated silicon wafer. (d) Schematic illustration of a wavy network constructed from a collection of 'horseshoe' building blocks configured into a triangular array; the inset at the bottom right provides the key geometrical parameters of the building block. (e) Experimental (denoted by line) and computational (FEA; denoted by line plus square symbol) results for the stress–strain response of this type of network (triangular lattice geometry with $=180°$, $R=400$ μm, $w^*=0.15$ and thickness$=55$ μm). Blue and green colours represent the soft materials with and without network mesh, respectively. The three regimes of behavior are analogous to those that occur in biological materials. The network responds primarily by bending and stretching in regimes of small (toe) and large (linear) strain, respectively. The intermediate regime (heel) marks the transition. (f) Optical images with overlaid FEA results for the composite shown in (a) evaluated at different tensile strains. The polyimide network is in a lattice geometry as in (d). The colour in e,f denotes the magnitude of maximum principal strain. All scale bars are 1 cm.

The local slope of stress–strain curve (that is, the tangent modulus) increases slowly at low strains (for example, <40%), where bending motions dominate the deformation of the network, as in Fig. 1f (~36% strain). As the horseshoe shapes begin to reach full extension (~57% strain), the slope of the stress–strain curve increases rapidly due to the transition into a stretching-dominated deformation mode (as shown in Fig. 1f). With further stretching, the strain in the constituent network material (that is, polyimide) rapidly increases, finally terminating with rupture at the ultimate tensile strength (~3 MPa). Dilatation of the triangular shaped unit cell at low strains leads to a negative Poisson effect in this region (Fig. 1f and Supplementary Fig. 3), with a disappearance of this behaviour as the horseshoe shapes reach full extension. The experimental and computational (finite element analysis, FEA, see Methods section for details) results exhibit quantitative agreement in both the nature of the physical deformations and the stress–strain curves, throughout the entire range of stretching. The net effect is a compliant artificial structure with non-linear properties, that is, ~30-fold increase in the tangent modulus (that is, local slope of stress–strain curve) with strain, of potential value in active or passive devices that integrate intimately with the human body, as illustrated in conformal wrapping on flat and curved regions of the skin (Fig. 1a,b and Supplementary Fig. 4)[32, 33].

Deterministically Defined Non-Linear Mechanical Responses

The mechanical properties can be adjusted to match, precisely, the properties of the skin or other organs. This tunability follows from the ability, via a simple lithographic process, to render the networks into nearly any 2D configuration[29, 34, 35, 36]. Here theoretical descriptions of the mechanics represent essential tools for optimized selection of key design parameters, including the material type, the network topology, the filament dimensions and the microstructure geometry, to meet requirements of interest. Spatially homogeneous or heterogeneous mechanical properties are possible, with isotropic or anisotropic responses. In all cases, the design is inherently scalable in terms of a limited set of non-dimensional parameters that define the microstructure geometry.

Figure 2 summarizes a collection of theoretical and experimental results on network topologies corresponding to triangular, Kagome and honeycomb lattices. Because of their six-fold symmetry, these networks each offer isotropic elastic properties at small strain. Diamond and square networks represent examples of topologies that provide anisotropic elastic responses (Supplementary Fig. 5). The building blocks for all such cases (Fig. 1d and 2a andSupplementary Fig. 5) can be represented by three non-dimensional parameters that characterize the horseshoe shape, that is, the arc angle , normalized width $w^*=w/R$ and normalized thickness $t^*=t/R$. The relative density $(\bar{\rho})$, defined by the ratio of the mass density of the network to that of a corresponding solid film, is approximately linearly proportional to the normalized cell width, as given by

Biaxial Tensile Strength Characterization of Textile Composite Materials 7

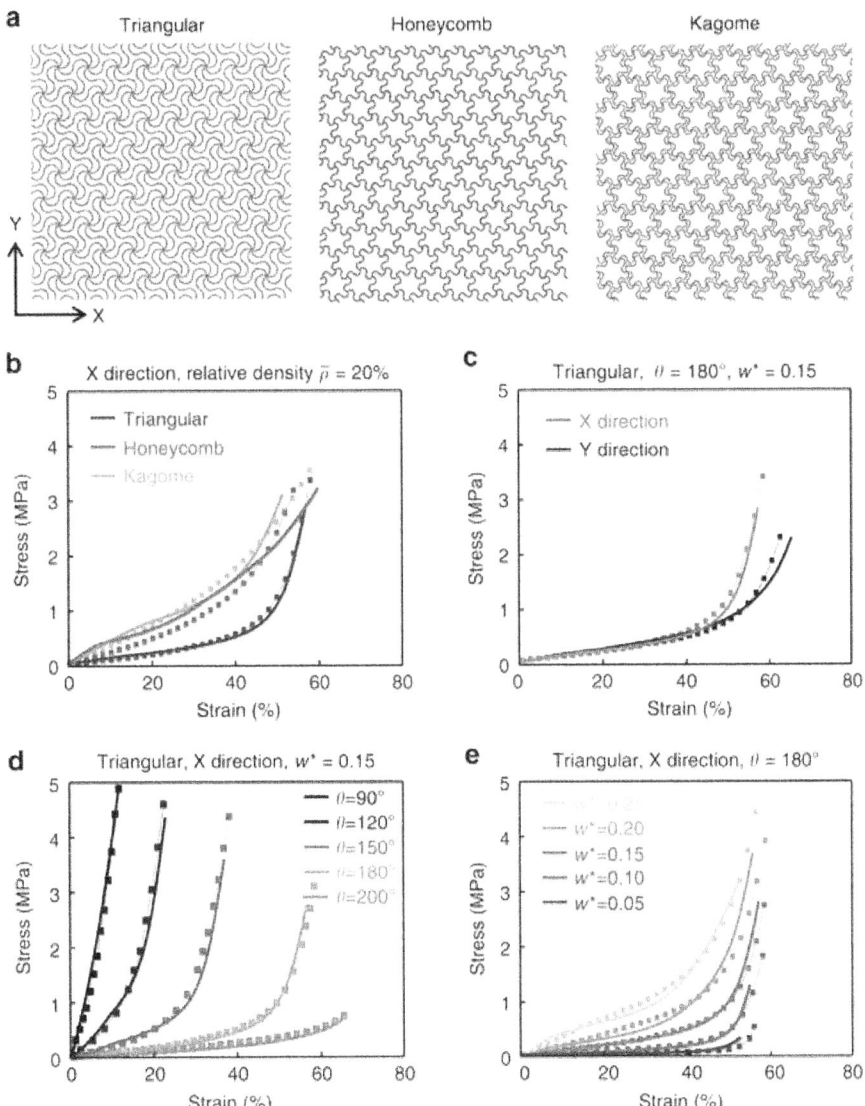

Figure 2: Wavy network architectures and design rules for tailored, non-linear stress–strain responses.

(a) Schematic illustrations of three different wavy network architectures, in which the node connection between the unit cells forms triangular (left), honeycomb (centre) and Kagome (right) lattices. Key parameters of these networks define the non-linear mechanical responses: lattice topology, direction of applied stress and arc angle ( ) and the normalized width (w^*) of the horseshoe building blocks, as illustrated in frames b–e, respectively. In b–e, the experimental and FEA results are denoted by line, and line plus square symbol, respectively. The triangular lattice exhibits the most pronounced transition from low to high tangent modulus. Results of parametric studies of this type of mesh appear in d and e. The filament thicknesses are 55 μm in all cases.

$$\bar{\rho}_{triangular} = \frac{\sqrt{3}}{4} \frac{w^* \theta}{\sin^2(\theta/2)},$$

$$\bar{\rho}_{Kagome} = \frac{\sqrt{3}}{8} \frac{w^* \theta}{\sin^2(\theta/2)},$$

$$\bar{\rho}_{honeycomb} = \frac{\sqrt{3}}{12} \frac{w^* \theta}{\sin^2(\theta/2)}. \quad (1)$$

These results show that the triangular and honeycomb networks are the most and least densely distributed, respectively, in accordance with the total number of connected filaments per node (Z), that is, $Z_{triangular}$=6, $Z_{honeycomb}$=3, Z_{Kagome}=4 (ref. 26). Mechanical evaluation of a complete design set reveals the influence of the key parameters on the stress/strain behaviour, as in Fig. 2b–e. The data indicate that the triangular network exhibits the most prominent strain-limiting behaviour for a given relative density. Studies of other design parameters (direction, arc angle and arc width) show that moderate anisotropic mechanical responses arise from different strains ($\varepsilon_{trans}^x = \theta/[2\sin(\theta/2)]^{-1}$ and $\varepsilon_{trans}^y = \sqrt{3}\theta/[3\sin(\theta/2)]^{-1}$)

(See Supplementary Note I and Supplementary Fig. 6 for details) needed to fully align the horseshoe microstructures along the x and y directions. Fig. 2d,e illustrate that the arc angle controls the transition from low to high tangent modulus (that is, the transition strain) and the normalized width defines the sharpness in this transition. The quantitative agreement between FEA predictions and experimental measurement in all of these cases further establishes the computational approaches as reliable design tools.

The underlying nature of the deformations in the networks that lead to these different effective properties is important to understand. These motions consist, in general, of a combination of twisting, translations and in- and out-of-plane bending. Fig. 3 demonstrates that the geometry of the building block microstructures defines the extent of out-of-plane deformations induced by buckling[30, 31]. Here the cross-sectional aspect ratio (that is, w/t) plays a prominent role because it determines the ratio of the stiffness ($\propto wt^3$) for out-of-plane deformations (that is, twisting and bending) to that ($\propto w^3t$) for in-plane bending. Figure 3a–d shows buckling and non-buckling deformations in networks with $w/t \approx 1.82$ and 0.73, respectively. The FEA results are consistent with observations from scanning electron microscope (SEM) images. Generally, out-of-plane buckling can be constrained by embedding the network structure in a solid elastomer. The modulus and thickness of the elastomer determine the extent of this constraint. For the ultralow modulus (~3 kPa) elastomer and the thickness (100 μm) used in the examples of Fig. 1, the resulting reduction in the out-of-plane displacement of buckling is ~4% (relative) for stretching of ~40%, for the network material shown in Fig. 3a,b (with t=27.5 μm, w=50 μm). Since buckling usually leads to a softening in the overall mechanics (Fig. 3e,f), reductions in thickness lead to increases in the slope of the stress–strain curve across the transition strain, that is, they enhance the sharpness of the transition. Compared with the parameters of Fig. 2, the effect of thickness is relatively minor. With increasing applied strain, the tangent modulus (Fig. 3f) increases slowly and then more sharply until it reaches a maximum at ε_{peak}≈60%, after which it decreases. This final softening occurs in a regime of behaviour where the network material dominates; here the tangent modulus decreases at high strain levels (as shown in Supplementary Fig. 7), for stresses calculated by the reaction force divided by the initial area, due to a reduction in the cross-sectional area that follows from the Poisson effect. For any given network, a critical thickness exists below which buckling will occur on stretching to $_{peak}$. This critical thickness (see Methods section) appears as function of width and arc angle in Fig. 3g,h, indicating that large widths and/or arc angles facilitate buckling.

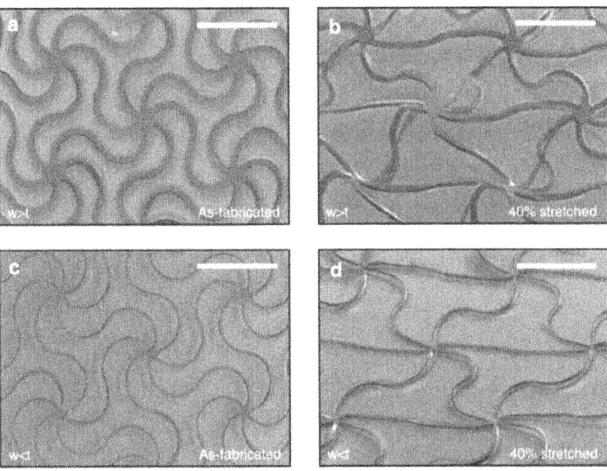

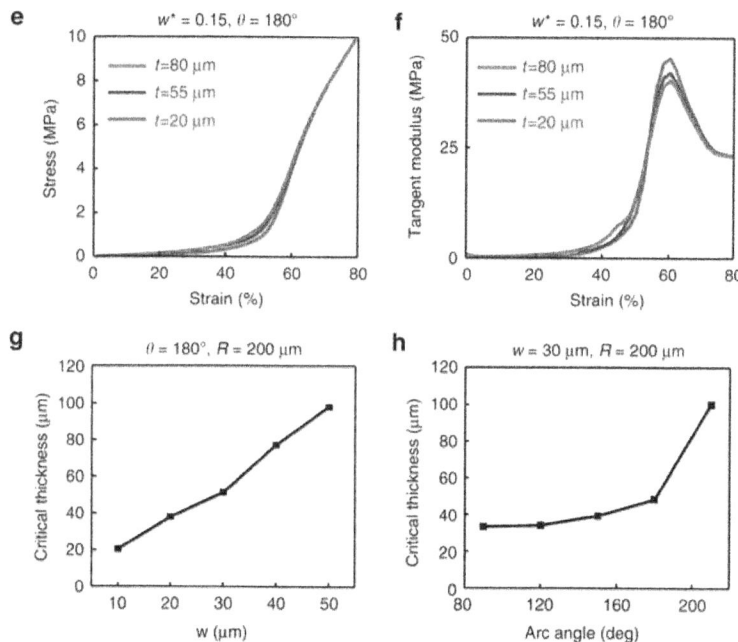

Figure 3: Buckling mechanics of triangular network architectures under uniaxial tensile loading.

(a–d) Colourized scanning electron microscope (SEM) images and overlaid FEA results of two different polyimide network structures uniaxially stretched to 40%. When $w>t$ the structures exhibit significant out-of-plane buckling. All scale bars are 2 mm. FEA results on (e) stress–strain responses and (f) corresponding tangent moduli for networks with three different thicknesses, with =180° and w^*=0.15. (g) Critical thickness as a function of filament width for =180° and R=200 μm. (h) Critical thickness as a function of arc angle for w=30 μm and R=200 μm.

Hierarchical and Self-Similar Network Configurations

Hierarchical layouts occur frequently in biological tissues where they provide additional levels of control over the key properties[10, 23, 37]. Similar strategies can be exploited as extensions to the network configurations of Figs 1, 2, 3. The building block for the example in Fig. 4a adopts a self-similar geometry formed by connecting horseshoe microstructures in layouts that reproduce the same overall geometry, but at a larger scale (as illustrated in Supplementary Fig. 8). Figure 4bshows that FEA predictions of stress–strain curves agree remarkably well with experimental results, even for these complex cases (see Supplementary Fig. 9 for detailed dimensions). The results indicate that this type of hierarchical design (that is, 2nd order) offers a much higher stretchability (Fig. 4b) than the corresponding non-hierarchical design (that is, 1st order) because of the increased lengths of the constituent filaments and associated reduced levels of strain in these materials. Figure 4c,d show that this system exhibits two transition points due to a deformation mechanism that involves sequential unravelling of the 1st order and then the 2nd order microstructure, illustrated in Fig. 4e. Below a strain of ~57%, the large scale, 1st order structure unravels, with little change in the 2nd order; beyond ~57%, the 2nd order unravels until the ultimate strength is reached at full extension. Related mechanisms occur in recently reported classes of electrical interconnects in stretchable electronics[38, 39]. These concepts can be extended to higher order designs, thereby expanding the range of mechanical responses that can be realized.

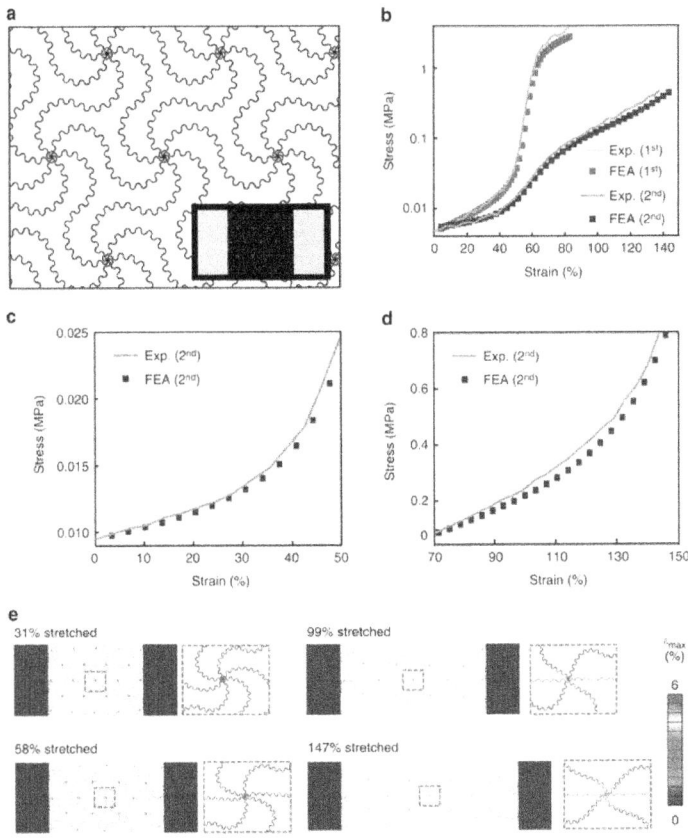

Figure 4: Self-similar network designs for multistage non-linear mechanical response and extreme stretchability.

(a) Schematic illustration of a wavy network with a 2nd order self-similar architecture consisting of horseshoe building blocks. (b) Stress–strain responses (with the stress in logarithmic scale) for wavy networks constructed using 1st and 2nd order designs. For strains between 0–60% and 70–145%, results appear in linear scale in c and d for the 2nd order design. (e) Deformation configurations at four different levels of stretching with associated maximum principle strain.

Materials That Reproduce the Mechanics of Human Skin

Figure 5 and Supplementary Fig. 4 present results of designs that exploit the physics of these deterministic network composites to achieve stress/strain responses that precisely match those of skin, while also providing spatial control over the characteristics, with relevance to tissue engineering, bio-integrated electronics and other applications. We note that the Poisson's ratio is another, related parameter that can be considered. Here we focus only on the stress–strain behaviours. The sample in Fig. 5a is a soft and breathable sheet consisting of thin elastomer (~100 μm thickness) with an embedded network structure of polyimide (PI) (20~50 μm thickness). As shown by results in Fig. 5b, the stress–strain curves of real human skin (extracted from refs40, 41) at different areas across the body can be accurately reproduced. Here an iterative process that uses FEA as a design tool determines the necessary parameters of the building block microstructures (mesh material: PI for back I, back, II and abdomen; filling material:*Silbione* RT 4717A/B for back I and abdomen, *Silbione* HS FIRM LV A/B for back II; lattice pattern: triangular/w^*=0.15/ =110°/R=400 μm for back I, triangular/w^*=0.11/ =150°/ R=400 μm for back II and triangular/w^*=0.12/ =200°/R=400 μm for abdomen).

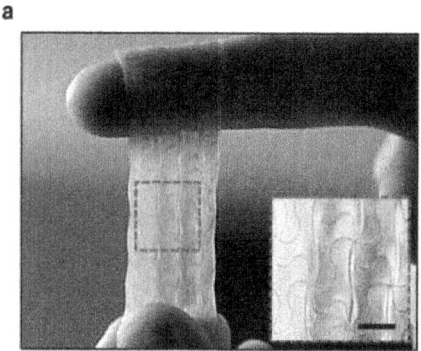
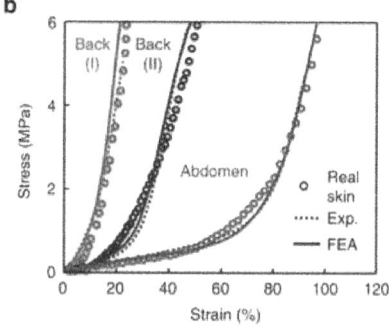

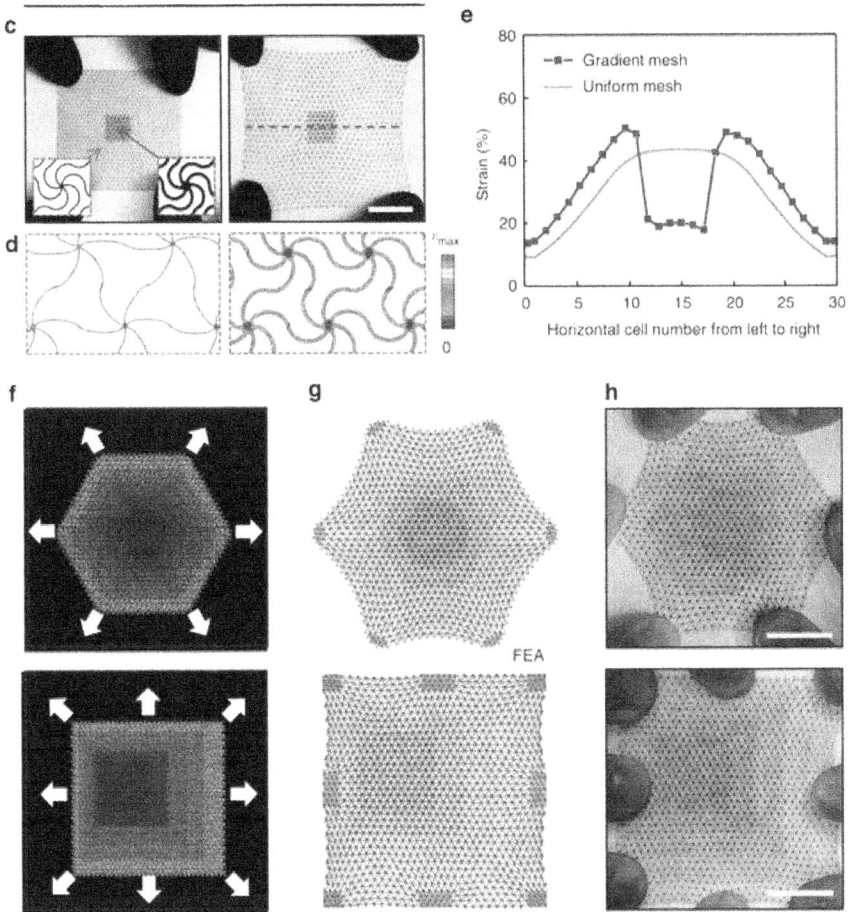

Figure 5: Skin-like mechanics in a network composite material and spatially heterogeneous designs.

(a) Image of uniaxially stretched composite with mechanics tailored to match the non-linear stress–strain characteristics of human skin. The scale bar in the inset image is 5 mm. This material consists of three layers: (top)—a soft silicone elastomer (Silbione HS Firm Gel LV; E=3~5 kPa; t=200 μm); (middle)—a polyimide wavy network structure in a triangular lattice (PI; E=2.5 GPa; w^*=0.15; =150°; t=55 μm); and (bottom) another silicone elastomer identical to that on top. (b) Stress–strain responses of human skin (circle) and engineered skin-like composite (dot line), with FEA modeling results for the latter (solid line), for various locations on different individuals (red: back area of person I, blue: back area of person II, green: abdomen area of person III). (c) Images and (d) FEA results of bi-axial stretching (to 35%)

of a spatially heterogeneous wavy network structure in an overall triangular lattice of horseshoe building blocks (polyimide, $t=55$ μm, $=180°$) in which the centre and outer regions have $w^*=0.10$ and $w^*=0.25$, respectively. The colour in d denotes the magnitude of the maximum principal strain, with the maximum values of 1.5% and 1.8% at the centre and outer regions, respectively. Nominal strains are 16.6% and 48.1% at the centre and outer regions, respectively. Scale bar is 2 cm. (e) Distribution of strain measured based on the length change of each horseshoe microstructure along the purple dash line in right panel of (c). Schematic illustrations of two gradient designs (f), illustrations of the deformed configurations determined by FEA (g) and experimental images (h), under bi-axial stretching.

The lithographic processes for creating the networks afford access to gradient forms of this type of artificial skin, in which spatially varying values of the widths of the horseshoe microstructures yield corresponding variations in the effective mechanical properties (Fig. 5a,b). Fig. 5c presents a simple example that incorporates a microstructure with enhanced stiffness (that is, 100 μm versus 40 μm in width) in the central region; the resulting mechanics leads to reductions in the levels of strain in this region (by a factor of ~2.3, as compared to the case of uniform microstructures) on overall stretching of the skin, as shown in Fig. 5d,e. The deformations predicted by FEA match those observed in experiment (Supplementary Fig. 10). Figure 5f–h shows advanced designs that incorporate isotropic and anisotropic gradients in properties, respectively, and the corresponding deformations under bi-axial stretching. Comparisons of the resulting distributions in strain to those of uniform microstructures (Supplementary Figs 11 and 12) illustrate the capability of such layouts to achieve nearly any desired spatial variation in strain, where FEA can guide the selection of designs to match requirements (Supplementary Figs 12 and 13). These ideas are fully compatible with existing chemistries and materials approaches in tissue engineering, in the sense that embedded lattice structures can provide the necessary mechanical response without altering the physicochemical and biochemical properties of the matrix, as shown for the example in Fig. 1 (ref. 42).

Substrates for Bio-Integrated Electronics

These types of bio-inspired soft composites represent ideal platforms for stretchable electronic systems that intimately integrate with the human body[43, 44, 45, 46, 47]. Figure 6a–c and Supplementary Fig. 14 shows an example that consist of a thin (2 μm), filamentary metal mesh that rests on a layer of silicone (~60 μm) with an embedded network structure (described in Methods section) designed to match the mechanical properties of the epidermis. The result is a skin-mounted sensor for electrocardiography (ECG) that has sufficiently small thickness and low modulus (at low strain) to maintain conformal contact with the skin, but with skin-like physical toughness (Fig. 6b) to allow multiple cycles of application and removal without damage to the device or the skin (Fig. 6c). Fig. 6d–h and Supplementary Fig. 15 presents a different type of system in which a similar platform acts as a support for a responsive hydrogel[48, 49] that can be activated wirelessly by exposing an integrated dipole antenna (30 mm for each branch, for operation at ~2.4 GHz without hydrogel, and at 1.9 GHz with hydrogel as shown in Fig. 6f) with filamentary mesh layout (Cu traces, 3-μm thick and 10-mm wide, encapsulated above and below with polyimide) to radio frequency radiation. The collected energy creates oscillating current in a connected Joule heating element (Au/Cr 50/5 nm thick) to increase locally the temperature of the hydrogel (inset infrared image in Fig. 4d). Above the low critical solution temperature (Fig. 6g), the hydrogel undergoes a change from a swollen (transparent) to shrunken (white colour) state, thereby releasing its contents (that is, water-soluble drug) to the surroundings (that is, skin), as shown in Fig. 6h. These simple devices, along with other examples that appear in theSupplementary Figs 16 and 17, including active semiconductor devices such as transistors and light-emitting diodes, provide evidence for the utility of bio-inspired soft composite materials of the type introduced here.

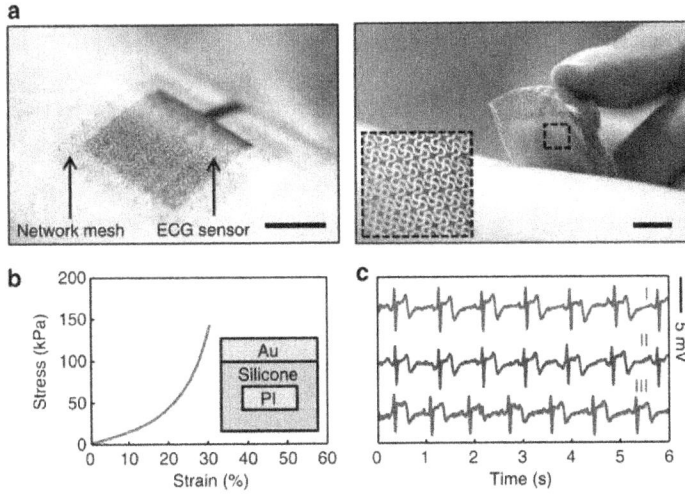

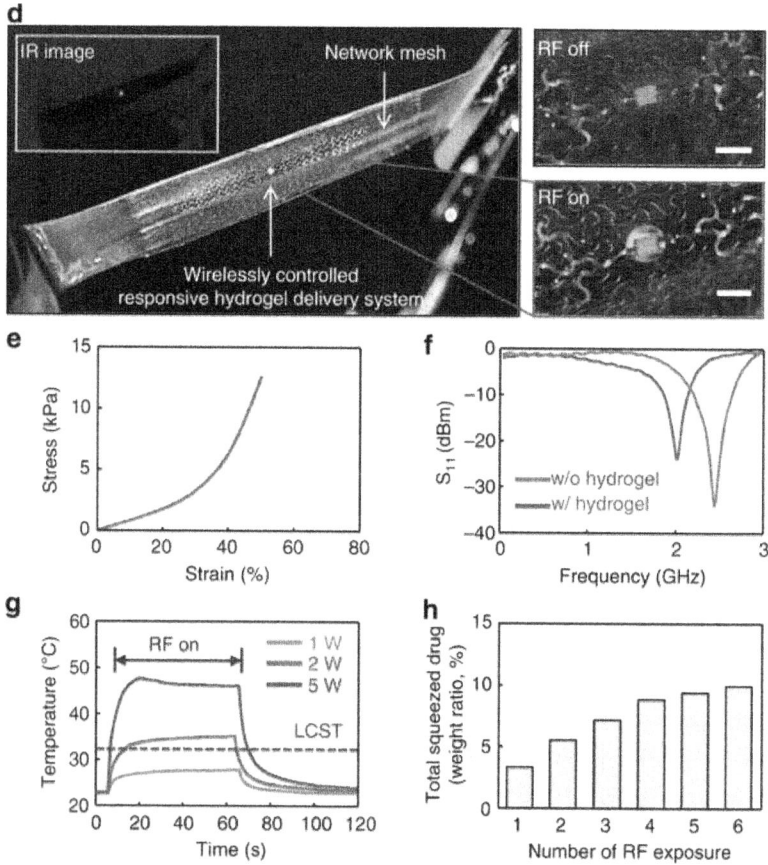

Figure 6: Deterministic soft composite materials as substrates for skin-mounted electronics and wirelessly controlled responsive hydrogels.

(**a**) Lamination and delamination of a soft, skin-like ECG sensor onto the forearm. The magnified view on the right illustrates the filamentary serpentine metal mesh structures that define the electrodes. All scale bars are 1 cm. (**b**) Stress–strain measurements on this device and schematic cross-sectional illustration. The wavy polyimide network used in the composite substrate adopts a triangular lattice of horseshoe building blocks, with =120°, w=40 μm, t=55 μm. (**c**) ECG signals measured using devices without (I, blue) and with (II, purple) the soft composite substrate. The result (III, red) corresponds to a measurement performed after applying and removing the device with composite substrate 20 times. All signals show expected PQRST features in the waveforms. (**d**) Optical and infrared (IR) images of a wirelessly controlled responsive hydrogel delivery system. All scale bars are 1 mm. This system consists of three functional layers: a thermally responsive hydrogel membrane, a stretchable radio frequency antenna with Joule heating element and a composite substrate. As shown in inset IR image, the wirelessly activated heater locally increases the temperature of the hydrogel. As demonstrated in right two magnified images, when the temperature exceeds the low critical solution temperature (LCST) of the hydrogel, the material changes in phase from a swollen (transparent) to a shrunken (white) state, corresponding to a large volume contraction. This process induces release of the aqueous contents of the hydrogel (that is, water-soluble drugs) to the surroundings. The wavy polyimide network used in the composite substrate adopts a triangular lattice of horseshoe building blocks, with $=150°$, $w=40\,\mu m$, $t=55\,\mu m$. The stress–strain response appears in (**e**). (**f**) S_{11} coefficient measured from the wireless heating element, evaluated with and without the hydrogel. (**g**) Transient control of temperature of the hydrogel on the skin using the wireless heating element, and measured using an IR camera. The temporal behaviour during heating and cooling defines the phase of the hydrogel and the resulting delivery mode. (**h**) Total expelled water (weight ratio, %) as a function of number

DISCUSSION

The materials approach, fabrication strategies and mechanical design methods reported here provide immediate access to soft composites with deterministic tailored, non-linear mechanical properties. These concepts are applicable to a wide range of constituent materials for both the matrices and the networks. Many application opportunities exist in tissue engineering and biomedical devices. Integrating active functionality into the networks and extending their coverage into three-dimensional (3D) spaces represent some directions that might be interesting to explore.

METHODS

Finite Element Analysis

Three-dimensional FEA enabled analysis of the full deformation mechanics and computation of small and large strain responses under uniaxial and bi-axial loads. Experimentally measured non-linear stress–strain curves of the constitutive materials (Supplementary Fig. 7) served as inputs. Eight-node 3D solid elements and four-node shell elements were used for the cases of $w<2t$ and $w>2t$, respectively, and refined meshes were adopted to ensure the accuracy. Linear buckling analyses determined the critical buckling strains and corresponding buckling modes. These results served as initial geometric imperfections for post-buckling simulations. The critical thicknesses (in Fig.

3g,h) were determined by comparing the critical buckling strain with the peak strain to reach the peak tangent modulus. A sufficiently large number of unit cells was adopted to avoid edge effects (Supplementary Fig. 18)[50].

Fabrication of Polyimide Networks

Copper (50 nm) deposited on a glass slide (75 × 50 × 1.0 mm³) served as a sacrificial layer to facilitate release. Spin casting on top of this substrate yielded a film of photodefinable polyimide (PI; 55 μm in thickness, HD Microsystems, USA). Photolithographic patterning of this PI followed by thermal curing (2 h. at 250 °C in a vacuum oven) defined the desired mesh structure. Wet etching eliminated the copper to allowed release for subsequently integration with a soft matrix material.

Fabrication of The Skin-Like Composite

Spin casting (30 s. at 1,000 r.p.m.) and thermal curing (5 min. at 70 °C) yielded a tacky (adhesion ~1.8 kPa), breathable (penetration: 170 mm/10, DIN ISO 2137), and ultra soft (E ~3 kPa) elastomer (Silbione RT 4717A/B, Bluestar silicones, France) membrane on a water-soluble tape composed of sodium carboxymethyl cellulose and wood pulp (Aquasol, USA). Transfer printing a polyimide network onto the surface of this elastomer and uniformly coating it with a layer of the same material completed the fabrication.

Fabrication of The Electrophysiological (Ep) Sensor

Spin casting and baking (3 min at 180 °C) formed a layer of poly(methylmethacrylate) (PMMA; 0.8 μm in thickness, Microchem, USA) on a glass substrate. A spin cast and thermally cured (2 h. at 250 °C) layer of PI served as an overcoat. Electron beam evaporation yielded metal bilayers of Cr (7 nm)/Au (100 nm). Photolithography and wet chemical etching defined the open mesh structure for the EP sensor. Reactive ion etching (20 sccm O_2, 300 mTorr, 200 W) removed the PI layers in regions not protected by the patterned metal traces. Immersing the glass substrate in acetone dissolved the PMMA and allowed retrieval of the sensor onto a water-soluble tape (3 M, USA) for delivery to a substrate composed of a polyimide network embedded in a soft silicone matrix (Ecoflex, USA).

Measurements of electrocardiogram (ECG) signals

The experiments used a custom LABVIEW interface. EP sensors with and without composite substrates were placed on the proximal left and right forearm for detection of ECG signal with a common ground electrode on the human subject's left hip. Voltage differences between bipolar pairs of electrodes were amplified and digitized with data acquisition (DAQ) system at 1,000 Hz with a 0.1–100 Hz online band-pass filter to remove slow drifts and high-frequency non-physiological noise, and a 60-Hz notch filter to attenuate electrical line noise. All experiments were conducted under approval from Institutional Review Board at the University of Illinois at Urbana-Champaign (protocol number: 13229). The eight subjects (age: 21~35 years, all males) were all coauthors. All work involved informed consent from the subjects.

Fabrication of the wireless responsive hydrogel system

Spin casting and baking (3 min at 180 °C) formed a layer of poly(methylmethacrylate) (PMMA; 0.8 μm in thickness, Microchem, USA) on a glass substrate. A spin cast and thermally cured (2 h. at 250 °C) layer of PI served as an overcoat. The wireless heater used photolithographically patterned multilayers of Cr (5 nm)/Au (50 nm)/Cu (3 μm) deposited by electron beam evaporation. A spin cast layer of PI (2,000 r.p.m.) passivated and isolated the devices. Reactive ion etching (20 s.c.c.m. O_2, 200 W, 200 mTorr) through a photolithographically patterned hard mask (Cu, 100 nm thick) removed the PI in the regions between the devices. A film of water-soluble tape (3 M, USA) allowed retrieval of the wireless heat unit from the glass substrate and delivery to a composite substrate consisting of a polyimide network embedded in a silicon matrix (Ecoflex) and coated with a thin layer of a silicone adhesive (Silbione). Spin casting and polymerizing a precursor to a responsive hydrogel yielded a thin (100~200 μm) coating on top of the wireless antenna structure and Joule heating element. Immersion in water dissolved the backing tape to complete the fabrication.

Characterization of Wireless Heater Module And Responsive Hydrogel

A network analyzer (E5602, Agilent technologies, USA) with calibration kit (85033E, Agilent technologies, USA) enabled measurement of the return loss (S_{11}) and the resonance frequency of the RF antennas. An analogue signal generator (N5181A, Agilent technologies, USA), an amplifier (1119, EMPOWER RF system, USA) and a DC power supply (U8031A, Agilent technologies, USA) provided a source of RF power. A directional antenna with 10.5 dBi gain (204411, Wilson electronics, USA) and an RF power meter (43, Bird technologies, USA) allowed controlled exposure of a wireless heater encapsulated in a hydrogel membrane to RF radiation. An infrared (IR) camera (A655SC, FLIR) revealed the resulting temperature distributions. Amounts of water expelled from the hydrogel were evaluated by weighting.

Measurements of Stress–Strain Responses

Mechanical responses of all samples were measured with a dynamic mechanical analyzer (DMA; TA instruments, Q800). Characterizing the applied force versus the displacement under uniaxial tensile loading at room temperature yielded data for determination of the mechanical modulus. Each of the reported results corresponds to an average of measurements on four samples.

ACKNOWLEDGEMENTS

This work was supported by the US Department of Energy, Office of Science, Basic Energy Sciences under Award # DE-FG02-07ER46471 and used facilities in the Frederick Seitz Materials Research Laboratory and the Center for Microanalysis of Materials at the University of Illinois at Urbana-Champaign. Y.H. acknowledges the support from NSF (CMMI-1400169). K.-I.J. acknowledges support from a Basic Science Research Program through the National Research Foundation of Korea (NRF) funded by the Ministry of Education (D00008). S.Y.H. acknowledges support from Samsung Display Co. through a visiting research scholar program. H.C. is a Howard Hughes Medical Institute International Student Research fellow. J.W.L, J.K. and U.P. acknowledge the support from Global Research Laboratory (GRL) Program (K20704000003TA050000310) through the National Research Foundation of Korea (NRF) funded by the Ministry of Science.

REFERENCES

1. Ortiz, C. & Boyce, M. C. Bioinspired structural materials. *Science* 319, 1053–1054 (2008).
2. Sanchez, C., Arribart, H. & Guille, M. M. G. Biomimetism and bioinspiration as tools for the design of innovative materials and system. *Nat. Mater.* 4, 227–288 (2005).
3. Wong, T.-S. *et al*. Bioinspired self-reparing slippery surfaces with pressure-stable omniphobicity. *Nature* 477, 443–447 (2011).
4. Capadona, J. R., Shanmuganathan, K., Tyler, D. J., Rowan, S. J. & Weder, C. Stimuli-responsive polymer nanocomposites inspired by the sea cucumber demis. *Science* 319, 370–374 (2008).
5. Aizenberg, J. *et al*. Skeleton of Euplectella sp.: Sturctural hierarchy from the nanoscale to the macroscale. *Science* 309, 275–278 (2005).
6. Pokroy, B., Kang, S. H., Mahadevan, L. & Aizenberg, J. Self-organization of a mesoscale bristle into ordered, hierarchical helical assemblies. *Science* 323, 237–240 (2009).

7. Ma, M., Guo, L., Anderson, D. G. & Langer, R. Bio-inspired polymer composite actuator and generator driven by water gradient. *Science* 339, 186–339 (2013).
8. Morin, S. A. *et al.* Camouflage and display for soft machines. *Science* 337, 828–832 (2012).
9. Kim, S., Laschi, C. & Trimmer, B. Soft robotics: a bioinspired evolution in robotics. *Trends Biotechnol.* 31, 287–294 (2013).
10. Cranford, S. W., Tarakanova, A., Pugno, N. M. & Buehler, M. J. Nonlinear material behavior of spider silk yields robust webs. *Nature* 482, 72–76 (2012).
11. Aikawa, E. *Calcific Aortic Valve Disease* Ch. 1, Intech (2013).
12. Munch, E. *et al.* Tough, bio-inspired hybrid materials. *Science* 322, 516–520 (2008).
13. Bonderer, L. J., Studart, A. R. & Gauckler, L. J. Bioinspired design and assembly of platelet reinforced polymer films. *Science* 319, 1069–1073 (2008).
14. Mayer, G. Rigid biological systems as models for synthetic composites. *Science* 310, 1144–1147 (2005).
15. Tang, Z., Kotov, N. A., Magonov, S. & Ozturk, B. Nanostructured artificial nacre. *Nat. Mater.* 2, 413–418 (2003).
16. Bouville, F. *et al.* Strong, tough and stiff bioinspired ceramics from brittle constituents. *Nat. Mater.* 13, 508–514 (2014).
17. Weiner, S. & Addadi, L. Design strategies in mineralized biological materials. *J. Mater. Chem.* 7, 689–702 (1997).
18. Marieb, E. N. & Hoehn, K. *Human anatomy & physiology* Ch. 4 Pearson-Benjamin Cummings: Boston, (2013).
19. Meyers, M. A., McKittrick, J. & Chen, P. Y. Structural biological materials: critical mechanics-materials connections. *Science* 339, 773–779 (2013).
20. Komatsu, K. Mechanical strength and viscoelastic response of the periodontal ligament in relation to structure. *J. Dent. Biomech* 2010, 502318 (2010).
21. Provenzano, P. P., Heisey, D., Hayashi, K., Lakes, R. & Vanderby, R. Subfailure damage in ligament: a structural and cellular evaluation. *J. Appl. Physiol.* 92, 362–371 (2002).
22. Čretnik, A. *Achilles Tendon* InTech (2012).

23. Pritchard, R. H., Huang, Y. Y. S. & Terentjev, E. M. Mechanics of biological networks: from the cell cytoskeleton to connective tissue. *Soft Matter* 10, 1864–1884 (2014).

24. Onck, P. R., Koeman, T., van Dillen, T. & van der Giessen, E. Alternative explanation of stiffening in cross-linked semiflexible networks. *Phys. Rev. Lett.* 95, 178102 (2005).

25. Naik, N., Caves, J., Chaikof, E. L. & Allen, M. G. Generation of spatially aligned collagen fiber networks through microtransfer molding. *Adv. Healthc. Mater.* 3, 367–374 (2014).

26. Hong, Y. *et al*. Mechanical properties and in vivo behavior of a biodegradable synthetic polymer microfiber-extracellular matrix hydrogel biohybrid scaffold. *Biomaterials* 32, 3387–3394 (2011).

27. Evans, A. G., Hutchinson, J. W., Fleck, N. A., Ashby, M. F. & Wadley, H. N. G. The topological design of multifunctional cellular metals. *Prog. Mater. Sci.* 46, 309–327 (2001).

28. Deshpande, V. S., Ashby, M. F. & Fleck, N. A. Foam topology: bending versus stretching dominated architectures. *Acta Mater.* 49, 1035–1040 (2001).

29. Fleck, N. A., Deshpande, V. S. & Ashby, M. F. Micro-architectured materials: past, present and future. *Proc. R. Soc. A* 466, 2495–2516 (2010).

30. Li, T. & Suo, Z. Compliant thin film patterns of stiff materials as platforms for stretchable electronics. *J. Mater. Res.* 20, 3274–3277 (2005).

31. Kim, D.-H. *et al*. Materials and noncoplanar mesh designs for integrated circuits with linear elastic responses to extreme mechanical deformations. *Proc. Natl Acad. Sci. USA* 105, 18675–18680 (2008).

32. Motte, S. & Kaufman, L. J. Strain stiffening in collagen|networks. *Biopolymers* 99, 35–46 (2013).

33. Dorfmann, L. The mechanics of soft biological systems. *Math. Mech. Solids* 18, 559–560 (2013).

34. Lakes, R. Materials with structural hierarchy. *Nature* 361, 511–515 (1993).

35. Ashby, M. F. Overview No. 92: Materials and shape. *Acta Metall. Mater.* 39, 1025–1039 (1991).

36. Ashby, M. F. Overview No. 80: On the engineering properites of materials. *Acta Metall.* 37, 1273–1293 (1989).

37. Satoh, K. *et al*. Cyclophilin A enhances vascular oxidative stress and the development of angiotensin II–induced aortic aneurysms. *Nat. Med.* 15, 649–656 (2009).

38. Xu, S. *et al*. Stretchable batteries with self-similar serpentine interconnects and integrated wireless recharging systems. *Nat. Commun.* 4, 1543 (2013).

39. Zhang, Y. *et al*. A hierarchical computational model for stretchable interconnects with fractal-inspired designs. *J. Mech. Phys. Solids* 72, 115–130 (2014).

40. Shergold, O. A., Fleck, N. A. & Radford, D. The uniaxial stress versus strain response of pig skin and silicone rubber at low and high strain rates. *Int. J. Impact Eng.* 32, 1384–1402(2006).

41. Annaidh, A. N., Bruyère, K., Destrade, M., Gilchrist, M. D. & Otténio, M. Characterization of the anisotropic mechanical properties of excised human skin. *J. Mech. Behav. Biomed. Mater.* 5, 139–148 (2012).

42. Yannas, I. V. & Burke, J. F. Design of an artificial skin. I Basic design principles. *J. Biomed. Mater. Res.* 14, 65–81 (1980).

43. Someya, T. *Stretchable electronics* Wiley-VCH Verlag and Co.: Weinheim, (2013).

44. Rogers, J. A., Someya, T. & Huang, Y. Materials and mechanics for stretchable electronics. *Science* 327, 1603–1607 (2010).

45. Kaltenbrunner, M. *et al*. An ultra-lightweight design for imperceptible plastic electronics.*Nature* 499, 458–463 (2013).

46. Xu, S. *et al*. Soft Microfluidic Assemblies of Sensors, Circuits, and Radios for the Skin.*Science* 344, 70–74 (2014).

47. Jang, K.-I. *et al*. Rugged and breathable forms of stretchable electronics with adherent composite substrates for transcutaneous monitoring. *Nat. Commun.* 5, 4779 (2014).

48. Xia, L.-W. *et al*. Nano-structured smart hydrogels with rapid response and high elasticity.*Nat. Commun.* 4, 2226 (2013).

49. Son, D. *et al*. Multifunctional wearable devices for diagnosis and therapy of movement disorder. *Nat. Nanotech.* 9, 397–404 (2014).

50. Onck, P. R., Andrews, E. W. & Gibson, L. J. Size effects in ductile cellular solids. Part I: modeling. *Int. J. Mech. Sci.* 43, 681–699 (2001).

CITATION

Jang, K.-I. *et al*. Soft network composite materials with deterministic and bio-inspired designs. *Nat. Commun.* 6:6566 doi: 10.1038/ncomms7566 (2015).

Chapter 2

BIAXIAL TENSILE STRENGTH CHARACTERIZATION OF TEXTILE COMPOSITE MATERIALS

David Alejandro Arellano Escárpita[1], Diego Cárdenas[2], Hugo Elizalde[2], Ricardo Ramirez[2] and Oliver Probst[3]

[1] Mechatronics Engineering Department, Instituto Tecnológico y de Estudios Superiores de Monterrey, Campus Ciudad de México, Col. Ejidos de Huipulco, Tlalpan, México D.F.,, Mexico

[2] Mechatronics Engineering Department, Instituto Tecnológico y de Estudios Superiores de Monterrey, Campus Ciudad de México, Col. Ejidos de Huipulco, Tlalpan, México D.F.,, Mexico

[3] Physics Department, Instituto Tecnológico y de Estudios Superiores de Monterrey, Campus Monterrey, Monterrey, N.L.,, Mexico

INTRODUCTION

Woven architecture confers textile composites (TC) multidirectional reinforcement while the undulating nature of fibres also provides a certain degree of out-of-plane reinforcement and good impact absorption; furthermore, fibre entanglement provides cohesion to the fabric and makes mould placement an easy task, which is advantageous for reducing production times [1]. These features make TC an attractive alternative for the manufacture of high-performance, lightweight structural components. Another interesting feature of TC is that they can be entangled on a variety of patterns, depending of the specific applications intended. Despite the wide interest of textile composites for industry and structural applications, most of the research efforts for strength characterization has focused on unidirectional composites (UDC), resulting in a large number of failure theories developed for UDC (around 20, as inferred from the conclusions of the World-Wide Failure Excercise (WWFE) [2]); some of the most popular failure models are used indistinctly for UDC and TC: most designers use Maximum Strain, Maximum Stress, Tsai-Hill and Tsai-Wu both for UDC or TC as stated in reference [3], despite the fact that none of the aforementioned failure criteria has been developed specifically for TC, which has led to use of high safety factors in critical structural applications to overcome associated uncertainties [4]. The most successful approaches to predict TC strength are based on phenomenological modeling of interactions between constituents at different scales (matrix-yarn-fiber), providing new insight into TC failure mechanisms. However, the implementation of phenomenological models as design tools is considerably more complex than that of traditional failure criteria, while still exhibiting significant deviation from the scarce experimental data [5] available. This scarcity of experimental data to validate or reject failure theories continues to be a major obstacle for improving TC models. Recent investigations reporting biaxial tensile strength tests in 2D-triaxial TC employing tubular specimens suggested that the failure envelope predicted by the maximum strain criterion fits the experimental data in the tension-tension (T-T) quadrant [6] fairly well. Other tests performed on cruciform specimens indicated that the maximum stress criterion is more adequate [5]; however, the authors of ref. [5] expressed some concerns about the generality of the experimental methodology for the case of non quasi-isotropic lay-up configurations, such as the one studied in their work. In view of the lack of consensus for accurate TC strength prediction [7],[8], and as stated by researchers who participated in the World Wide Failure Exercises (WWFE) [2] more experimental data, better testing methods and properly designed specimens are needed to generate reliable biaxial strength models.

BIAXIAL TESTING REVIEW

Combined multi-axial strength characterization of composites is far from straightforward, as three basic elements are required: i) An apparatus capable of applying multi-axial loads, ii) a specimen capable of generating a homogeneous stress and strain field in a predefined gauge zone, producing failure inside this zone for correct strength characterization, and iii) a measurement system capable of acquiring the applied loads and resulting specimen strains. Although the general procedure is similar to that for uni-axial testing, significant complications arise due to the requirements outlined above; moreover, the required equipment is costly and available generally only at large specialized research centres. Regarding the specimens, the ability of generating a homogeneous multi-axial strain field inside a pre-specified gauge zone is not straightforward mainly due to geometric stress concentrations. Finally, the data acquisition system requires a free surface in order to perform direct measurements. In practice these factors limit the number of combined loads that can be applied to a single specimen to only two, although some researchers have proposed apparatuses designed to apply tri-axial loads, albeit at the expense of limiting the access for full field strain measurements. Efforts on multi-axial testing have been disperse and rather unsuccessful in defining adequate testing methodologies, as evidenced by the lack of standardization by international organisms which have otherwise generated well-known standards for uni-axial characterization of composites, such the ASTM D3039 (standard testing procedures for obtaining tensile properties of polymer matrix composites), British Standard: BS 2782: Part 3: Method 320A-F: Method for obtaining mechanical properties of plastics, BS EN ISO 527 Part 5: Plastics. Determination of tensile properties and test conditions for unidirectional fibre-reinforced plastic composites, CRAG (Composite Research Advisory Group) Test Methods for the Measurement of the Engineering Properties of Fibre Reinforced Plastics, Standard ASTM D6856: Testing procedures for textile composite materials, Japanese Industrial Standard JIS K7054: Tensile Test Method for Plastics Reinforced by Glass Fibre, Russian Standards GOST 25.601-80: Design Calculation and Strength Testing Methods of mechanical testing of polymeric composite materials. In brief, there exist at least seven standards for tensile uni-axial characterization, while none specific standard for bi-axial testing. This demonstrates the need for developing biaxial testing methodologies. In this chapter a review is presented of the state of the art of multi-axial testing with emphasis on biaxial tensile specimens, testing machines and data measurement systems. The reasons for concentrating on biaxial loads are: i) The complexity of testing systems increases considerably with the the number of independent applied. ii) Most structural applications of composites uses thin skins, resulting in shell structures, in which the thickness of the laminates is significantly smaller than the other dimensions. One characteristic of shell structures is that buckling failure modes are the limiting factors in the case compressive loads [9]; consequently, the structural strength depends little on the materials strength and mostly on the geometry and stiffness. On the other hand, when tensile loads are applied to shell, the structures tend to be stable, and the final failure does depend on the materials strength. Evidently, given these fundamentally different failure modes in the cases of compressive and tensile loads, respectively, a combination of biaxial load conditions (compressive-compressive, compressive-tensile, tensile-compressive, and tensile-tensile) can lead to a quite complex behaviour and the need for developing predictive failure models that can account for this complexity.

Biaxial and Multiaxial Specimens

To generate useful strength data, a biaxial specimen must be capable of meeting a set of requirements [10],[11],[12],[13]: i) A sufficiently wide homogeneous biaxially-stressed zone must be generated for strain measurements, ii) Failure must occur within this zone. iii) No spurious loads (other than tension/

compression) should be acting on the gauge specimen. iv) The specimen should accept arbitrary biaxial load ratios. The very design of specimens that recreate biaxially loaded components has become a constantly evolving field, aiming to provide optimal geometry, manufacture and general arrangement for a valid and reliable test [14]. Specimens designed for biaxial testing can be classified into three main groups: i) tubes, ii) thin plates and iii) cruciforms. A review of these groups and their main features is given below.

Tubular Specimens

Multi-axial stress states were formerly created with thin-walled tubes subjected to internal pressure, torsion and axial loads [10],[11],[15]. These specimens allow simultaneous application of tensile and compressive longitudinal loads, as well as tangential and shear loads, therefore representing a versatile scheme for the conduction of multi-axial characterization (Figure 1).

However, the existence of stress gradients across the tubular wall makes this method less accurate than setups based on flat plates, which are also more representative of common industrial applications than the tubular geometry. Some studies also reveal high stress concentrations on the gripping ends. A further disadvantage is a pressure leakage after the onset of matrix failure, although some correction can be provided by internal linings [15].

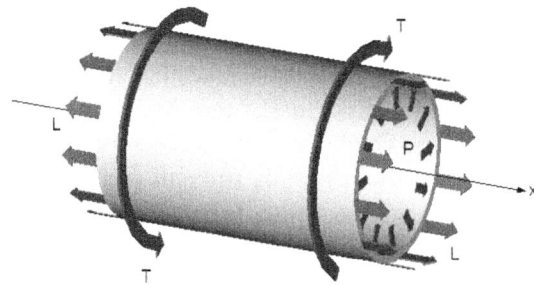

Figure 1: Thin-walled tube specimen.

Thin Plates

Round or elliptical flat sheets subject to pressure in the hydraulic bulge test [16], as shown in Figure 2, can develop a biaxial stress state, although the technique has several disadvantages, for example, non-homogeneous stress distributions induced by gripping of the edges [17]. Also, just like the rhomboidal plate case, the loading ratio is shape-dependant [18] and can therefore not be varied during the test to obtain a full characterization.

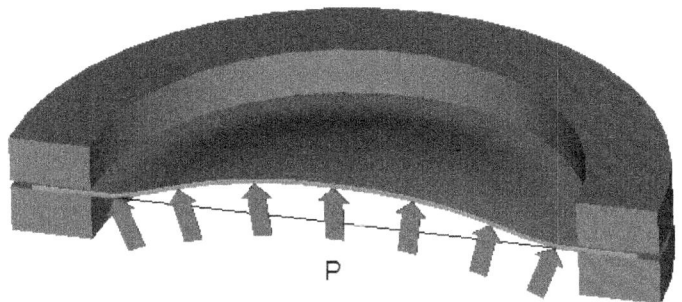

Figure 2: Elliptical flat sheet used in the bulge test.

Cruciform Specimens

Testing biaxially-loaded cruciform specimens represent a more direct approach for obtaining true biaxial stress states, and consequently this method has gained wide acceptance [7],[8],[10],[11],[15]. As suggested by many researchers in the field [7],[10],[13], an ideal cruciform specimen should accomplish the following features: *i*) It should be capable of generating a sufficiently wide and homogenous biaxial stress/strain field in the gauge area, *ii*) failure must occur in the predefined gauge zone, *iii*) the cruciform should accept arbitrary biaxial load ratios for generating a complete failure envelope (within a desired range), *iv*) both the tested and the reinforcement layers should be of the same material, *v*) the transition between the gauge zone and the reinforced regions should be gradual enough as to avoid undesirable high stress concentrations, *vi*) the cruciform fillet radius should be as small as possible in order to reduce stress coupling effects, and *vii*) stress measurements in the test area should be comparable to nominal values obtained by dividing each applied load by its corresponding cross-sectional area. Although various cruciform geometries containing a central-square thinned gauge zone have been proposed in the literature, none can claim full satisfaction of the above requirements due to difficulties inherent to biaxial tests [10]. A cruciform with a thinned central region and a series of limbs separated by slots is presented in Figure 3a. [19]; the slotted configuration allows greater deformations to occur in the thinned section, thus enforcing failure there. Nevertheless, thickness-change can induce undesirable stress concentrations that usually lead to premature failure outside the gauge zone. Also, the extensive machining required for thinning is an undesirable feature.

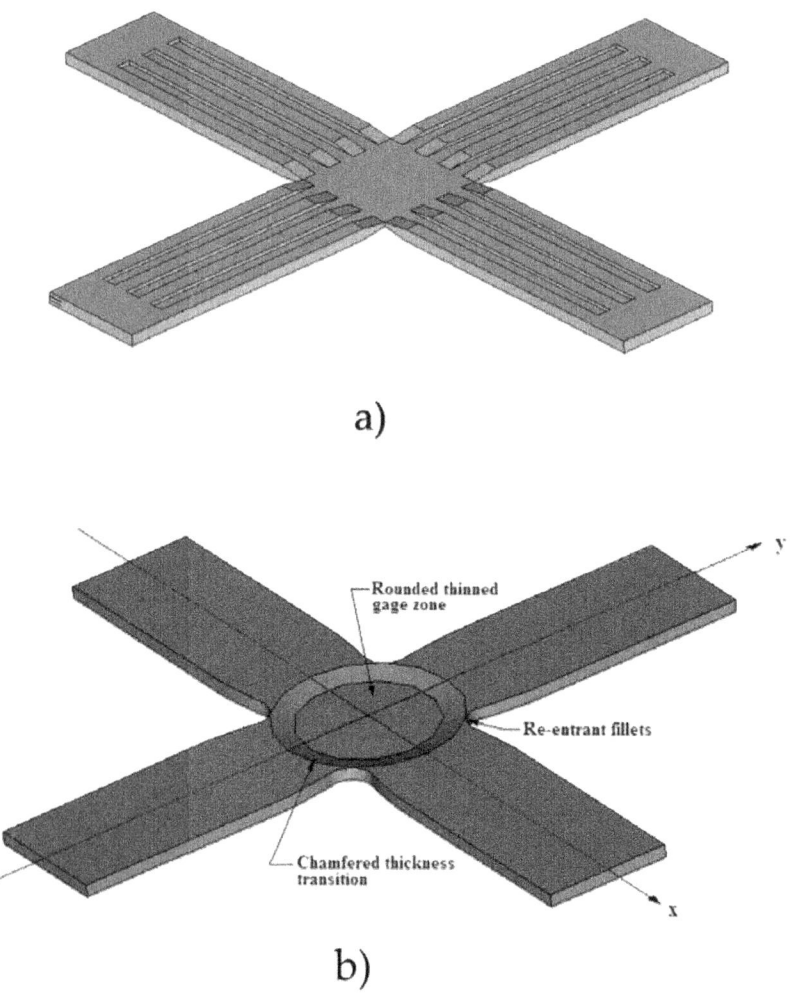

Figure 3: a). Slotted configuration[19] b). Thinned circular zone in the gauge zone [13].

Another cruciform, shown in Figure 3b, with a thinned circular zone in the gauge area [13] exhibits failure outside it, mainly because manufacturing defects caused unexpected higher strength in one axis. The implementation of a rhomboidal shaped test zone is suggested in [20], although, to the authors knowledge, no results with this geometry have been reported so far. Some experiments concluded that loading must be orthogonal to the fibre orientation to produce failure in the test zone [12]. The main difficulty in obtaining an optimal configuration is eliminating stress concentrations in the arms joints. To solve this, an iterative optimization process (numerical/experimental) yielded optimum geometric parameters of the specimen [21]. Results from this study led to a configuration characterized by a thinned square test zone and filleted corners between arms. Given that failure is prone to occur in the arms, reference [23] presented a design where a small cruciform slot is placed in the centre to cause load transfer from the arms to this region (figure 4a).

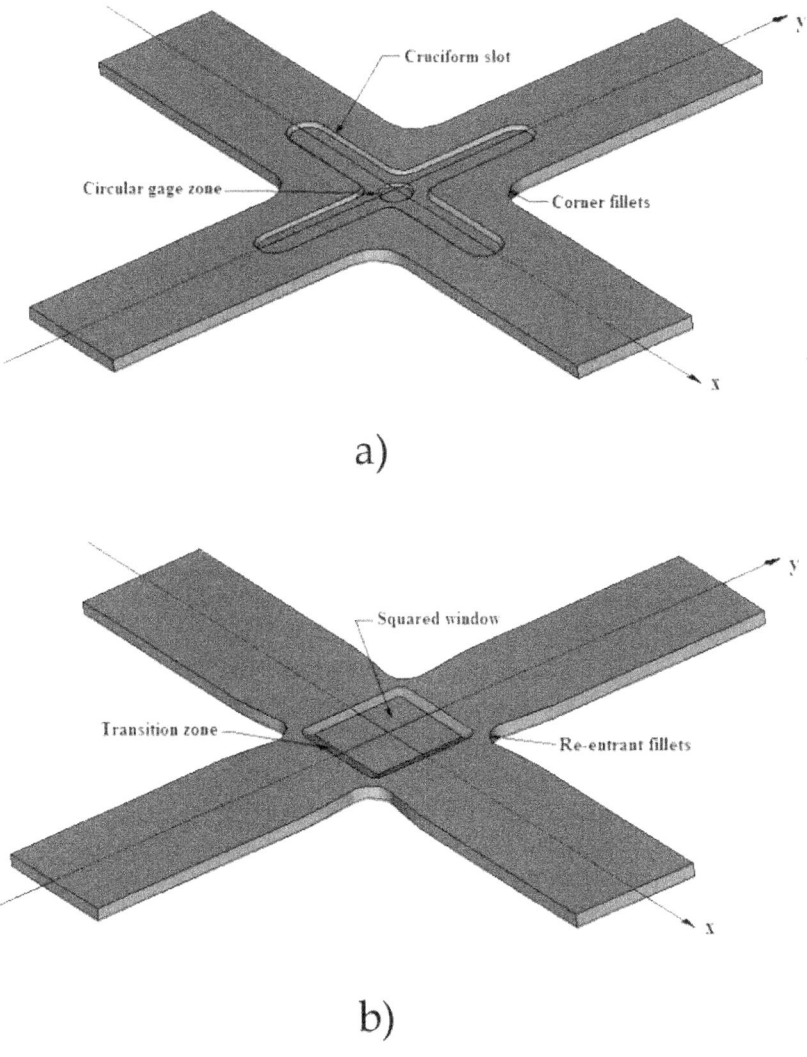

Figure 4: a). Inner cruciform slot [23]. b). Cruciform with thinned rounded square gauge zone and filleted corners [11].

Nevertheless the gauge zone is much reduced, and this makes this specimen useless for TC characterization. In the cruciform proposed by Ebrahim et al [10] failure in the gauge zone is achieved. The design is characterized by a thinned rounded square gauge zone and considers a gradual thickness reduction in the biaxially loaded zone, and also filleted corners as shown in Figure 4b. Results were satisfactory, but it was found that the top and bottom edges of the depression presented high strain gradients. Based on the aforementioned references, a comprehensive study was conducted by the authors to obtain an improved cruciform design. A main feature of this new design is a rhomboid-shaped gauge zone which led to a much more homogeneous strain/strain distribution because of the alleviation of stress concentrations which occur in other designs due to the short distance between the gauge zone and the corners of the arms. Additionally, the corners are filleted to avoid another zone of stress concentration. The specimen is comprised of different layers where the inner layer is under study, whereas the outer ones (equal quantity on each side) are only for reinforcement.

Enhanced Rhomboid-Windowed Cruciform Specimen

In order to avoid premature failure due stress concentrations, a modified cruciform was proposed by considering this design concepts: i) Given that fillets are prime examples of stress concentrators, both the cruciform and gauge zone fillets should be as far apart as possible from each other, thus favouring a rhomboid-windowed gauge zone. This modification also intends to minimize regions of stress interactions, which cause lack of homogeneity in the strain field and even premature failure, as reported for some square-windowed specimens [22],[24]. Traditional (instead of re-entrant) fillets were preferred to maintain this stress concentrator as separated as possible from the gauge zone. ii) Since the focus of this research are textile composites (TC), the proposed specimen also features wider arms and a larger gauge zone, seeking to reduce the textile unit cell vs. gauge zone length ratio. This modification is in tune with ASTM standards on testing procedures for textile composites [25]. iii) To avoid polluting the obtained strength data with in-situ effects, adhesion between adjacent layers and other multilayer-related uncertainties, characterization is performed for a single-layer central gauge zone, while a number of reinforcement layers are added outside the gauge zone to enforce failure inside it. The resulting rhomboid windowed cruciform shape was similar to other specimens employed for fatigue characterization of ABS plastic, which report a smooth biaxial strain field at the gauge zone [26]. Basic dimensions were selected from a specimen reported in literature [27]: arm width $w =$ 50mm and cruciform fillets $R=25$mm. The rhomboid window length l was set identical to the arm width, $l=50$mm while the window's fillet radius r was set as 10% of l; the geometry is sketched in figure 5. Finite-element (FE) analysis demonstrated that this geometry generates a more uniform strain distribution, while the maximum shear strain in the cruciform fillet is relatively slow.

Once the suitability of a rhomboid windowed cruciform specimen for creating a biaxial strain state was established, a geometrical optimization process based on the experiment design methodology was conducted. Suitable objective functions were defined in order to homogenize the ε_x and ε_y strain fields inside the rhomboid gauge zone while maintaining shear strain γ_{xy} field close to zero. Details of the optimization process can be found in reference [29]

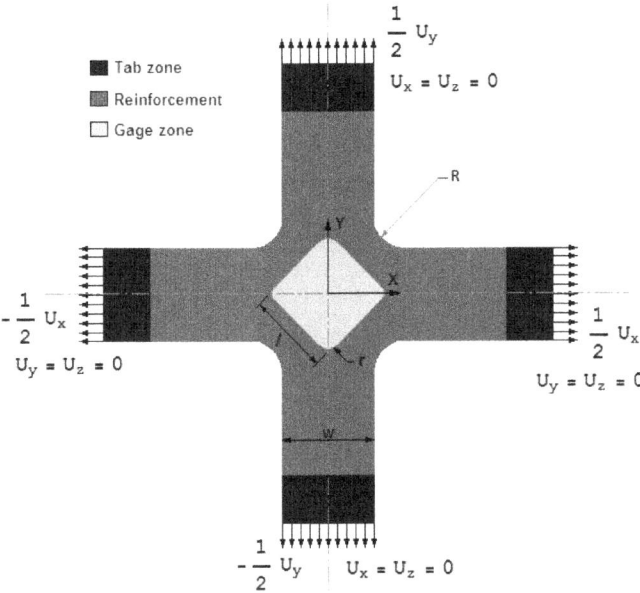

Figure 5: Geometry of the proposed cruciform specimen. Lay up for the reinforcement region is $[0]_5$, while for the gauge zone is [0] (that is, a single layer). All dimensions are given in mm.

Table 1: In-plane measured mechanical properties for a TC conformed of: Epoxy West System 105/206 reinforced with fibreglass cloth style #7520, bidirectional plain weave 8.5 oz./sq. yd, with 18L x 18W threads per inch count

E_{11}	[GPa]	25.0
E_{22}	[GPa]	25.0
ν_{12}	[-]	0.2
G_{12}	[GPa]	4.0

Evaluation of specimen using finite element analysis was carried by applying boundary conditions as defined in figure 5, with U_x and U_y chosen to produce a maximum strain (ε_x or ε_y) of 2% inside the gauge zone, corresponding to typical failure strain values reported for glass-epoxy TC [5]. The materials properties correspond to a generic plain weave bidirectional textile, as presented in Table 1.

The optimized geometry is defined in Figure 6, while the results of the FE analysis are shown in Figure 7, which splits the geometry into top and bottom sections for simultaneously illustrating the ε_x and γ_x strain fields, respectively, in a single graph; due to full symmetry, the ε_y strain field is identical to the ε_x field when rotated by 90°. The resulting geometry generates a very homogeneous strain field in the gauge zone and keeps shear strains near zero, while keeping shear strains in the fillet regions below the failure value. These results are believed to represent a great improvement if compared with other specimens reported in the literature.

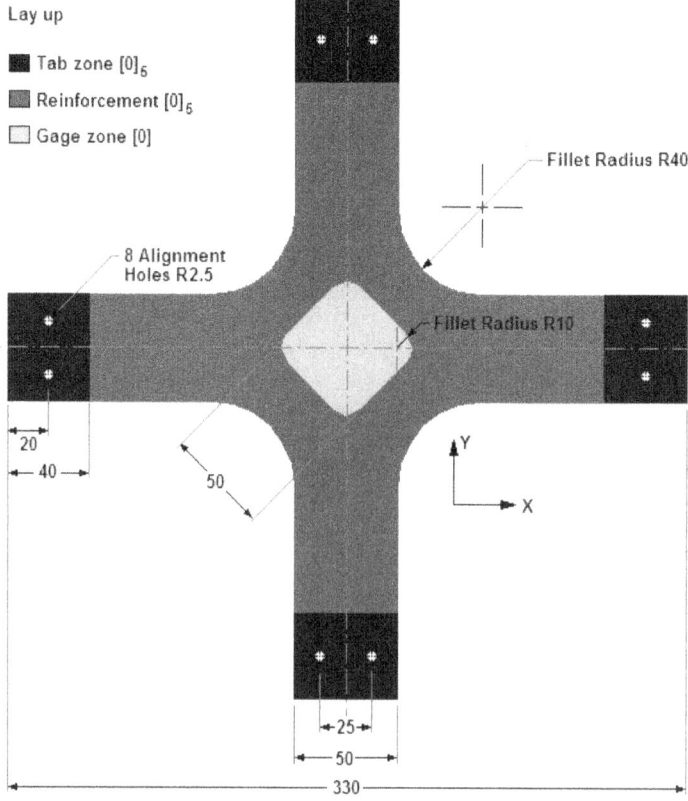

Figure 6: Optimized specimen specifications Dimensions are in mm.

Biaxial Testing Machines

To apply biaxial loads on cruciform specimens a specific device is required, which can meet the following requirements [28],[30],[12],[31]: i) The loads applied to a cruciform specimen must be strictly in tension or compression, avoiding spurious shear or bending loads. ii)The restriction previously stated implies that orthogonality among load axes must be guaranteed at all times during the test, and, consequently iii) the centre of the specimen must remain either still or the load axes must displace with it. An efficient method to ensure the previous condition is to apply equal displacements in the loaded axis. These requirements can be accomplished by using an active control system, or by passive mechanical methods, such that the one described later. A review of the most common biaxial testing systems is presented next.

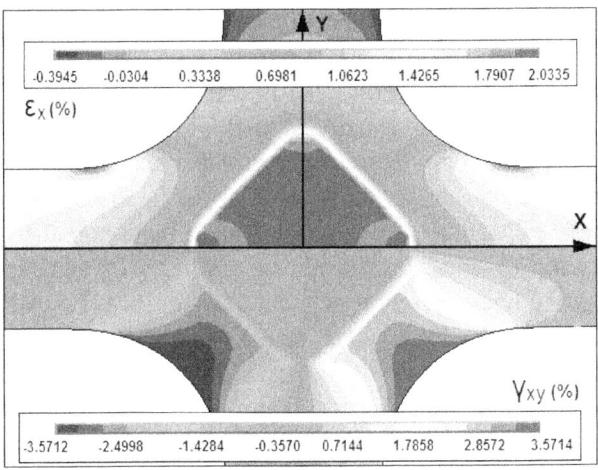

Figure 7: Linear strain field (upper part of the graph) and shear strain field (lower part) within and near the gauge zone of the optimized specimen.

Hydraulic Systems

Hydraulic systems rely on hydraulic actuators to apply loads to the specimen; they typically use double-acting pistons with a closed-loop servo control system which sense displacements and/or loads as feedback, as implemented in the design by Pascoe and de Villiers [32]. This configuration (sketched onfigure 8) which comprises the use of independent actuator for each applied load, allows the centre of the specimen to move during the test, which is an undesirable condition; this adverse feature can be avoided by implementing a control system that ensures synchronization of opposite actuators [33],[18] thereby avoiding motion of the centre of the specimen.

This configuration also allows the load ratios to be varied in order to obtain a full failure envelope. None of the systems mentioned could ensure equal displacement in both extremes of each axis, even the one using synchronization control, therefore allowing the centre of the specimen to move. If systems are implemented to correct this problem, the design and manufacturing costs inevitably increase. Fessler [34] proposed a machine in which motion is allowed only in one direction at one arm for each cruciform axis. This is the most common basic configuration found in the literature related to biaxial characterization of composites [35],[33]. In an attempt to simplify the previous concept while maintaining symmetric load conditions, some modifications have been proposed; for example, each loading axis, consisting of a pair of opposite hydraulic actuators, can be connected to a common hydraulic line so the force exerted by each side is the same and thus movements of the centre of the specimen are eliminated. Although the common hydraulic line ensures equal force in both extremes of one axis this does not ensure equal displacements. Another variation to hydraulic systems is described in the US Patent No. 5279166, which describes a biaxial testing machine consisting of two independently orthogonal loading axes capable of applying tension and/or compression loads; two ends of the specimen are gripped to fixed ends while the complementary ends are fixed to grips attached to actuators that apply the load, made in an attempt to reduce the complexity and hence the costs of biaxial testing machines (fig. 9). This configuration results in significant displacements of the centre of the specimen, although it is stated that the machine has a mechanism that helps maintain the centre of the specimen and ensure that the loads are always orthogonal. In spite of these features, under large displacements the mechanism used is not capable of maintaining the orthogonally of the loads as shown by a quick finite-element evaluation, whose results are shown Figure 8; moreover, the resulting displacement field is completely asymmetric, a condition which generates undesirable shear stress. While most of the biaxial testing hydraulic machines are original developments, a commercial biaxial testing machine has been developed by the company MTS in conjunction with NASA. It uses four independent hydraulic actuators, each with a load cell and hydraulic grippers, and an active alignment system for the specimen. While solving most of the problems mentioned above, the cost of this system is too high for entry-level composites development laboratories.

Biaxial Tensile Strength Characterization of Textile Composite Materials 37

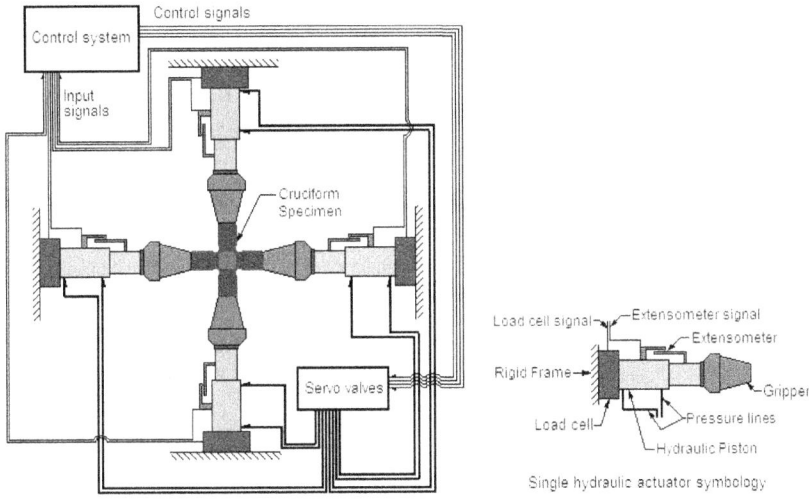

Figure 8: Use of independent actuator per load applied.

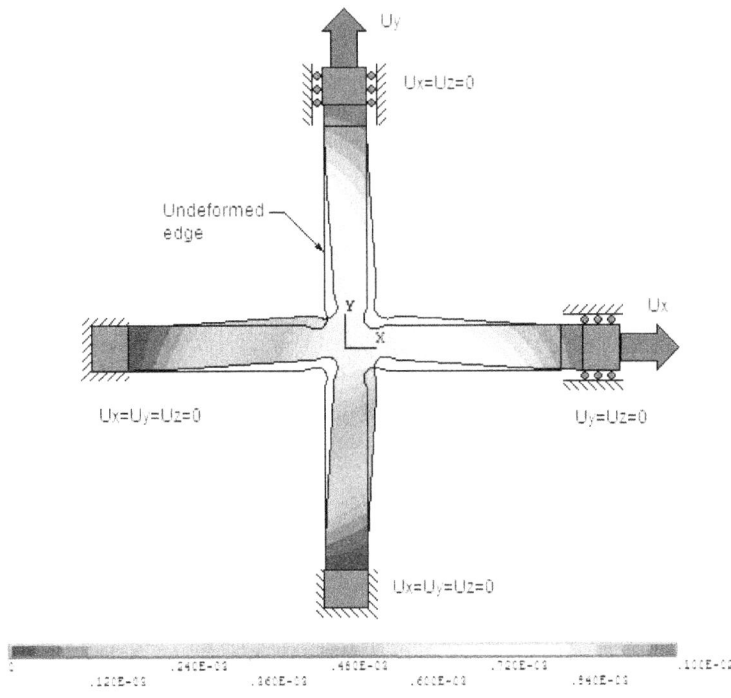

Figure 9: Contour plots of the magnitude of the displacement vector for the case of a configuration where one end of each axis is fixed and the other is displaced.

Mechanical Systems

Mechanical systems owe their name to the fact that they are based on the kinematics of their mechanisms to maintain load symmetry, no matter if the actuators are hydraulic or mechanic; even the application of deadweight to the specimen through systems of ropes, pulleys, levers and bearings has been considered, as presented by Hayhurst et al [36]. In practice, the mechanical systems proposed for the characterization of composite materials are mainly test rigs designed to be adapted to conventional uniaxial testing machines; basically, they are mechanisms consisting of coupled jointed-arms capable of applying in-plane biaxial loads to cruciform specimens. The load ratio is dependent on the geometrical configuration of the device [31] and can therefore be varied only by changing the length of one element, an impractical solution. Similar devices are found in French Patent No. 2579327 [37] and US Patent No. 7204160. A simpler mechanism is presented in US Patent No. 5905205 [30] which uses a four-bar rhomboid-shaped mechanism on which the loading ratios are changed before the test by certain variations in the assembly of the members. One of most practical mechanical systems found consist of four arms, joined at one side to a common block fixed via revolute joints to an universal test machine actuator through a load cell [26] which permits monitoring the applied force, while the other sides are linked, also with revolute joints, to a sliding block each; those blocks slide over a flat plate, fixed to the universal machine's frame. The sliding blocks assemble the grippers which hold the specimen.

A Novel Biaxial Testing Apparatus

After reviewing the existing machines and mechanisms on which biaxial tests can be carried out, some conclusions can be drawn; in the case of some of them, the lack of a mechanism that automatically corrects any load difference that could lead to the displacement of the centre of the specimen makes them unsuitable for reliable tests ; in those case where such mechanism does exist, it is controlled by means of an active system that increases design complexity and costs.

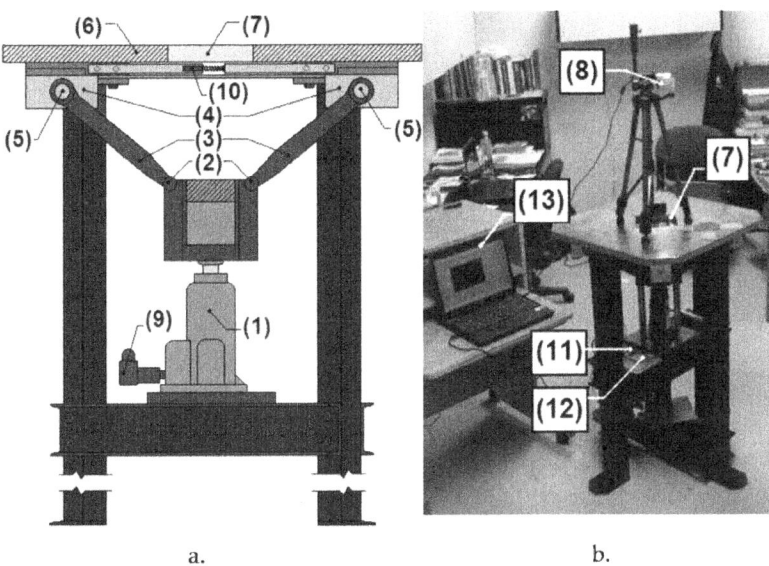

Figure 10: a). Sketch of the biaxial testing machine, showing one load/displacement axis sketch. b). System general arrangement.

The design proposed by the authors considers these drawbacks, as well as the testing requirements previously stated; in addition, construction costs for the novel proposed design are considerably lower compared to other systems. The resulting apparatus is sketched in Figure 10a and a photograph of the completed machine in Figure 10b. The operational principles are described in the following (numbers in parentheses refer to the components identified in the corresponding figures): The loads are applied through a symmetrical slider-crank-slider mechanism meeting the following requirements: The hydraulic piston (1), which is the first slider, is attached on its base to the machine frame and provides the load drive, while its piston is linked by a revolute joint (2) to a pair of arms (3) arranged symmetrically, which in turn are connected by cylindrical joints (5) to the blocks where the grips holding the tab zone of the specimen are installed (4); these cylindrical joints allow to absorb small misalignment in the loads, as established in *iv*.

The grip blocks are lubricated and slide on the lower side of a flat plate (6) featuring a rectangular window (7) allowing a full-field view from the top of the machine, where a high-definition digital camera (HDDC) was installed (8), thereby satisfying the requirement *iii*. A similar arrangement was installed at a right angle with respect the first one, ensuring the indepence of the load axes as required by *i*. Data acquisition is conducted by measuring the pressure in the hydraulic cylinders (9) and correcting this information by considering the geometry of the mechanism, while the displacements are measured directly at the grips through resistive displacement sensors (10); all sensors are powered by a power board to provide a common voltage reference (11), and the signals are acquired through a National Instruments 8-channel analogical data acquisition board (12). The information was stored and processed on a laptop (13) by using a Lab View routine.

Data Acquisition Techniques

Unlike uniaxial tensile tests in which ultimate failure stress and strains data can be straightforwardly obtained from the collected load and displacement data, in the case of biaxial tests the strength values cannot be calculated directly in this way because the stress and strain fields are not necessarily homogeneous along the specimen and generally depend on the load in a non-trivial way due the complex geometry. For this reason biaxial testing requires a method capable of measuring the full strain field in the biaxially loaded zone of the specimen. Given that strain cannot be measured directly it is necessary to measure the displacement field, from which the strain field can be easily calculated. Using the strain field and a constitutive model the stress field can also be calculated. However, full-field measurement techniques are not standard data acquisition methods and in order to identify the most suitable technique for this research a survey was realized.

Full Field Strain Measurement Methods

The first method considered was reflective photo-elasticity; it is based on birefringence, a physical property which consists of the change of the refraction index of a material when shear stresses are applied. It has been used since decade of the 1950s [38], so is a well characterized technique. However, some limiting factors have been identified for the purposes of the curent project: 1) The preparation of the samples is extremely laborious and requires the application of a layer of birefringent material on the surface to be observed, with a thickness of a few millimetres [38].If compared with the thickness of the composite layer under study which is of the orderof about 0.2-0.3mm it is clear that the application of the measurement layer significantly affects the test results. Another technique considered was Moiré interferometry. This technique requires the printing of a pattern of lines on a transparent medium, which is then illuminated by a LASER source, generating an interference pattern which depends on the deformation of the specimen [39]. However, this method has the disadvantage that the data reduction process is tedious and complex [40], and the results heavily depend on the analyst's experience. After considering these options, a technique called digital image correlation

(CDI) was identified from biaxial testing literature [10], [11], [41]. The basic concept consists of obtaining digital images of studied geometry on it initial, non-deformed state and after being subjected to a deformation. The surface of the part under study is pre-printed with a random speckle pattern, so that the displacements between corresponding points on photographs of the non-deformed and deformed states, respectively, can be identified by a computer algorithm. This method has some advantage over the ones mentioned above [42]:

1)The experimental setup and specimen preparation are relatively simple; only one fixed CCD camera is needed to record the digital images of the test specimen surface before and after deformation. ii) Low requirements as to the measurement environment: 2D DIC does not require a laser source. A white light source or natural light can be used for illumination during loading. Thus, it is suitable for both laboratory and field applications. iii) Wide range of measurement sensitivity and resolution: Since the 2D DIC method deals with digital images, the digital images recorded by various high spatial-resolution digital image acquisition devices can be directly processed by the 2D DIC method. For the reasons stated above the 2D DIC method is currently one of the most actively used optical measurement techniques and demonstrates increasingly broad application prospects. Nevertheless, the 2D DIC method also has some disadvantages: i) The surface of the planar test object must have a random grey intensity distribution. ii)The measurements depend heavily on the quality of the imaging system. iii) At present, the strain measurement accuracy of the 2D DIC method is lower than that of interferometric techniques, and is not recommended for the measurement of very small and non-homogeneous deformations. Despite these restrictions the low cost associated with equipment and the low specimen preparation requirements makes Digital Image Correlation the preferred technique for the purposes of this study. The drawbacks can be largely avoided by using the highest-definition camcorder commercially available, using an established Digital Image Correlation program and using a specimen that generates a relatively homogeneous strain field. It was shown above that by proper design and optimization a very homogeneous strain field can indeed be obtained in the gauge zone, so this restriction of the DIC technique was of no concern to this project. Finally, the expected strain values were large enough to be safely detected by the DIC technique.

EXPERIMENTAL SETUP

Specimen Manufacture

As stated by recent research [11] the milling process typically employed to thin the gauge zone produces undesirable damage and stress concentrations in unidirectional (UD) composites; for textile composites (TC), milling would exacerbate this problem due to its more complex 3D structure, making milling an unacceptable choice. The main concern is to preserve the integrity of the textile structure, especially when characterizing a single lamina. To generate a damage-free cruciform specimen with a single-layered gauge zone, a novel manufacturing process was developed by the authors, explained below: 1). Non-impregnated fabric sheets were fixed to a 6mm thick plywood base to ensure dimensional stability, with a printed grid to help proper fibre alignment of each cloth. The whole arrangement was cut into a square pre-form using a water jet, also cutting away the rhomboidal window corresponding to the gauge zone, as shown in the Figure 11. Afterwards, the material was oven-dried at 60°C during 12 hours to eliminate moisture.

2) The following numeric values inside brackets refer to indications given in Figure 12. Two reinforcement layers (1) corresponding to the bottom side of the cruciform specimen were placed in a lamination frame, consisting of a flat surface (2) surrounded by a square border (3) with a side length equal to that of the specimen. A pre-formed 2-layer rhomboid step (4) was located at the centre, corresponding to the location of the gauge zone, to ensure planarity of the central layer (5). The reinforcement layers were manually resin-impregnated and, immediately after this, the central layer (5) was placed and impregnated. Finally, the process was repeated for the last two reinforcement layers (6), as shown. Room environment was controlled during the lamination process at a temperature of 80±2°C and 50-60% relative humidity. Immediately after the impregnation process was completed, the laminate was placed in a vacuum bag consisting of a peel ply (7), perforated film (8), bleeder cloth (9) and the bag itself (10), using sealing tape to ensure vacuum seal (11). 0.8 bar vacuum pressure was applied through a valve located at a corner (12), sufficiently away from where the final shape would be cut. The whole arrangement was cured during 4 hours inside a pre-heated oven at 80±2°C, as measured by a thermocouple (13) located at the gauge zone, as shown in Figure 12. 3) After curing, the final cruciform geometry was obtained through water jet cutting. Nine specimens were prepared meeting the dimensional specifications in Figure 6.

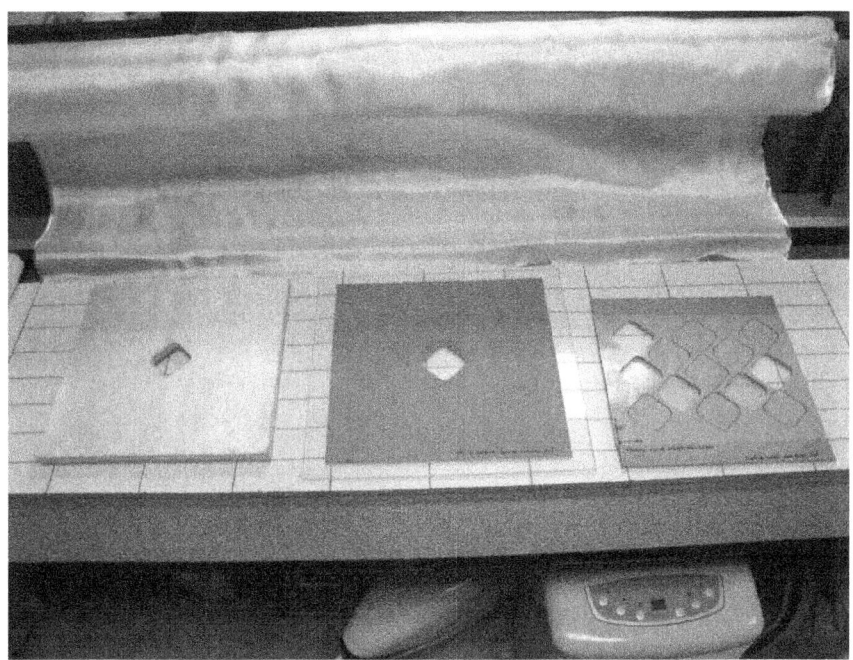

Figure 11: Rhomboid window cutted on the reinforcement layers and other auxiliary tools.

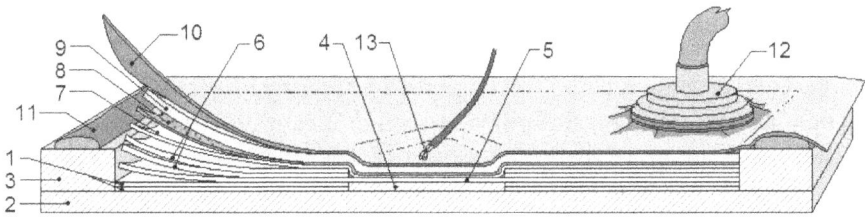

Figure 12: Arrangement for specimen manufacture.

To provide visual reference for the digital image correlation (DIC) strain field measurement [24][41], specimens were painted with a black-dot random speckle pattern over a white-mate primer, as shown inFigure 13. This technique was preferred over the spraying technique reported in [41], as it might result in an inadequate control of the dot size distribution, leading to uncertainty in the DIC measurements. Additionally, five uniaxial, $[0]_5$ layup specimens were prepared in order to perform uniaxial tests to provide precise input data for the development of failure criteria.

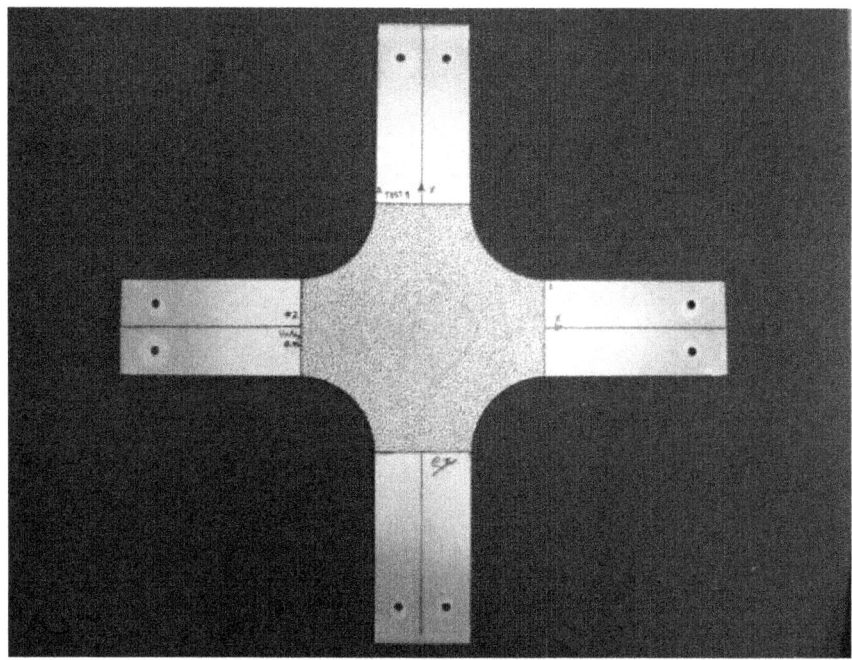

Figure 13: Finished specimen exhibiting its speckle pattern used with the digital image correlation technique

Biaxial Testing

Experimental validation of the optimized cruciform was conducted with the biaxial testing apparatus described in section 2.2.3 as follows: after mounting the specimen in the grippers, a pre-load of 500N was applied to each axis prior to tightening the mounting bolts. (Figure 14) Then, preload and alignment bolts were removed, setting the measured displacements and loads to zero. A high-quality video of the specimen was recorded with a high definition cam coder with adjustable focus and exposure parameters functions for subsequent DIC analyses.

A chronometer synchronized with the computer clock was placed near the specimen and inside the camera vision field, to ensure its inclusion in the captured images; this provided a time reference to relate each video frame with correspondent load data. After starting video recording and the data capture routine, biaxial displacements were applied at a rate of 1mm/min until final failure. This load rate was selected based on the ASTM 3039 standard [43], which recommends a displacement speed such that failure occurs 1 to 10 minutes after the start of the test.

Data acquisition and reduction was conducted as follows: two video frames were taken from the recorded video sequence, one corresponding to the beginning of the test and another just prior to final failure, as shown in figure 15. Both images were fed into the open access software DIC2D (developed by Dr. Wang's team at the Catholic University of America) to obtain the full strain field (ε_x, ε_y and γ_{xy}). The three tests performed covered a range of biaxial ratio BR values in the vicinity of the critical condition BR=1: BR=1.5 (Test #1), BR=1.25 (Test #3), and BR=1 (Test #5). Figure 15 shows the final failure sequence representative of the tests conducted. It should be noted that the failure occurred well within the gauge zone as expected from the FE-predicted strain fields.

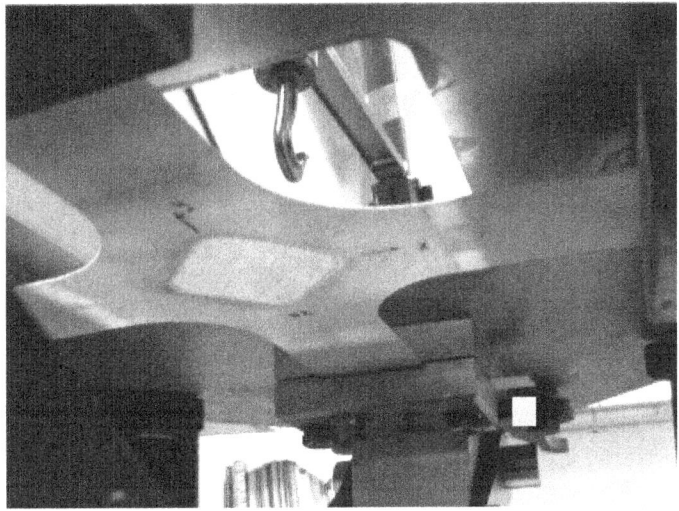

Figure 14: Cruciform specimen mounted in the biaxial testing machine.

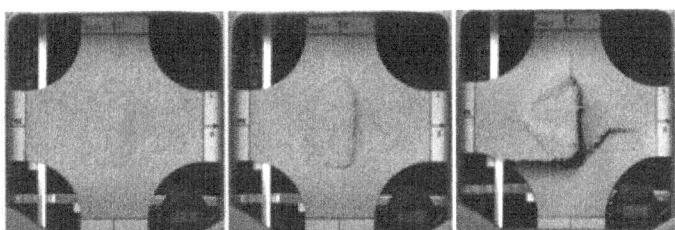

Figure 15: Final failure sequence recorded at 30 frames per second.

The final failure is clearly fibre-dominated, due to its catastrophic nature; it is possible to identify the final failure onset region inside the rhomboidal gauge zone, as required for a successful test. Regarding the strain field, it can be seen from Figure 16 that the agreement between the experimentally results (obtained from DIC) and the FE prediction is remarkably good. The DIC and the FE images show the same symmetry of the experimental shear strain pattern and similar homogeneity and smoothness, and the absolute strain values cover a similar range. This can be considered an additional indicator of the success of the experimental procedure presented in this work.

The same procedure used to characterize the ultimate strain can be used to obtain the matrix onset failure envelope (as opposed to fibre failure), but due to the fact that this phenomenon cannot be deduced visually a different approach was used for this purpose.

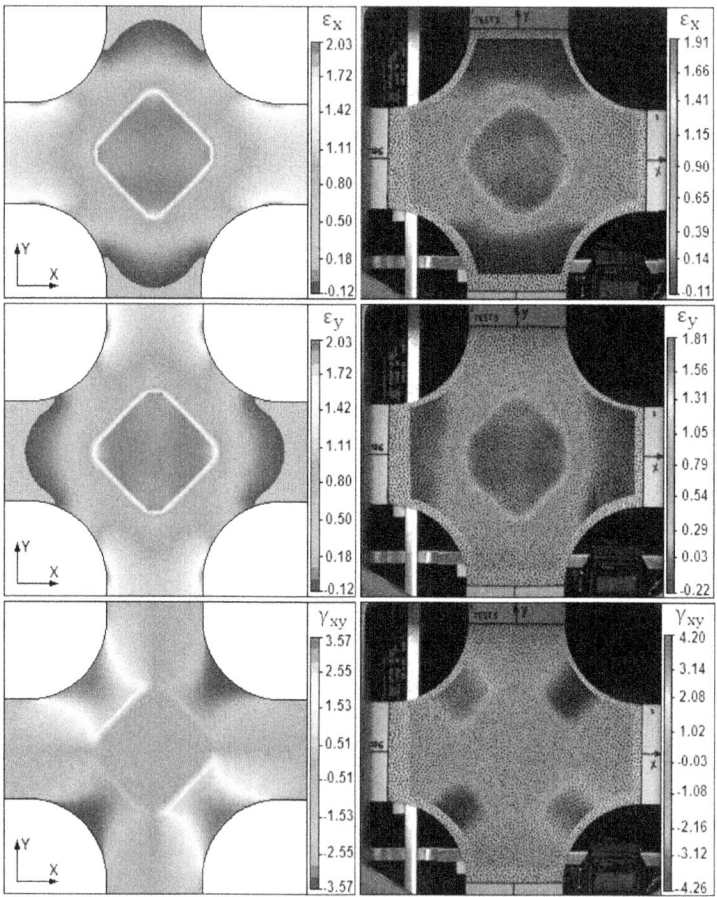

Figure 16: FE vs DIC strain field comparison for Test #5. The first column corresponds to the FE results, while the second column exhibits the results of the digital image correlation (DIC) process. The first and second row show the linear strain field, while the last row exhibits the shear strain stress field.

The load vs. displacement plots were used to identify the change in the slope which evidences matrix damage, as shown in the Figure 17. This method is proposed as an extrapolation of the method employed for uniaxial tests defined by the ASTM 3039 standard for the uniaxial tensile characterization of composites [43]. Linear fits were obtained for every linear segment of curves corresponding to every perpendicular axis, and the intersections were calculated solving the resulting equations, which allowed to quantify the strain values corresponding to the onset of matrix damage, considering that the latter occurs at the first observed slope change. Once the displacement and strain were identified, the digital image corresponding were used to perform a DIC analysis and to get the full field strain in the same fashion described previously.

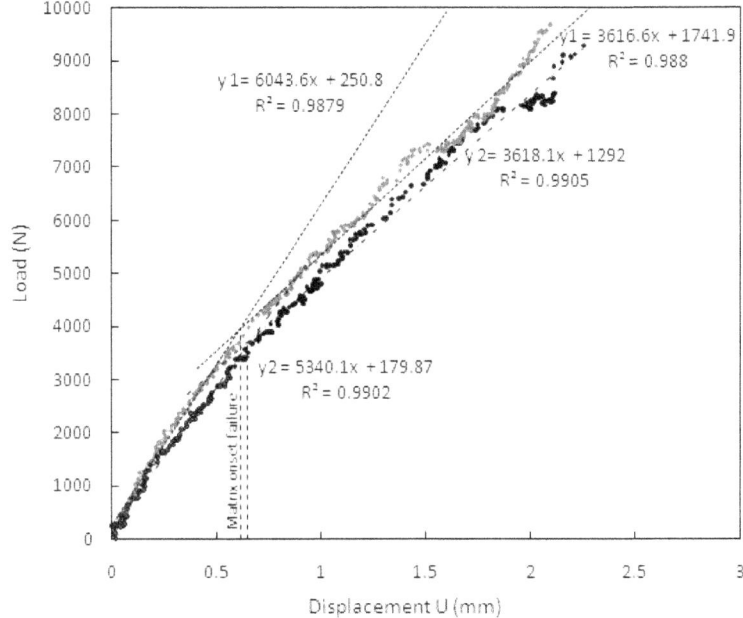

Figure 17: Load vs diplacement for biaxial test #3. The location where the change of slope occurs is interpreted as the onset of matrix failure.

Is important to remark that the use of the slope change in the load vs. displacements curves can be significantly influenced by geometrical effects and materials non-linearity, and other auxiliary techniques such as sonic emission or in-situ x-ray scanning should be employed to verify that this change can be effectively used as a matrix damage onset indication.

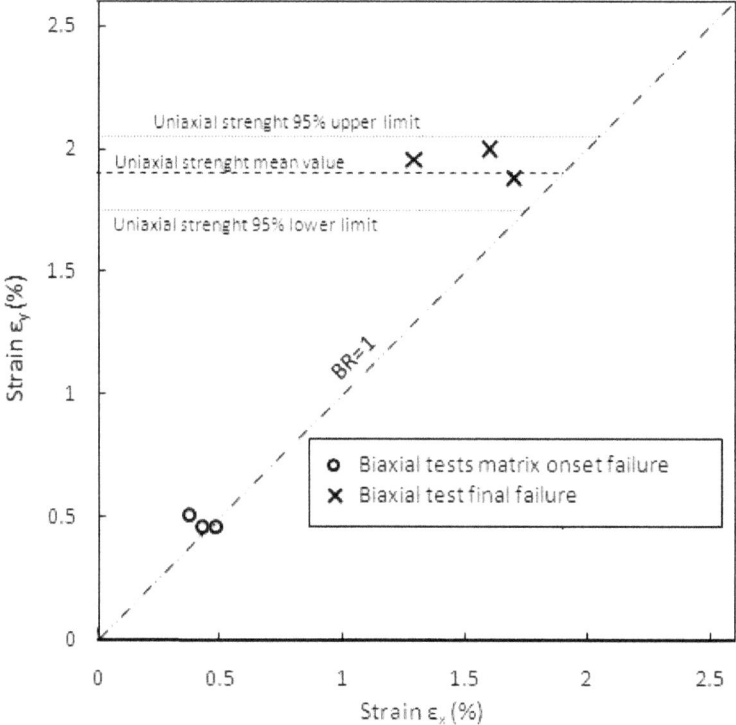

Figure 18: Failure envelope data obtained from the experimental data.

Experimental strength data for single layer biaxial strength obtained from the experimental program are presented in Figure 18 as well as data from uniaxial test performed on five layer specimens. Confidence intervals calculated for the failure strains observed on uniaxial tests are presented in the figure. It should be noted that single-layer strength data fall inside the 95% confidence limits which suggest that interactions between ε_1 and ε_2 strains are significant for single layer laminates. This finding should not be used as design criterion before more experimental data are obtained, but it gives a good indication of the feasibility of the methodology presented for the purposes of failure analysis.

CONCLUSIONS

Improvement over existing cruciform specimens for biaxial testing was achieved by proposing a specimen with rhomboidal thinned gauge zone, based on conclusions from a qualitative stress concentration analysis. An optimization based on the experiment design methodology was performed to achieve a highly homogeneous strain distribution withinu the rhomboidal gage zone while shear strains in the cruciform fillets were kept well below the failure values in order to avoid the premature failure typically affecting this kind of specimens. The resulting geometry generates very homogeneous strain field within the gauge zone and keeps shear strains near zero, while keeping shear strains in fillets below the failure value; this is believed to represent a great improvement over other specimens reported in the literature. In addition to meeting the requirements for equi-biaxial tests, the specimen was evaluated under various biaxial ratios, demonstrating that is practically insensitive to biaxial ratio, and hence can be used without any modification to obtain the full tension-tension failure envelope.

A manufacturing process which avoids machining operations normally required to generate the thinned gauge zone was developed, in an attempt to preserve the textile architecture from machining micro-damage. It consists in cutting the rhomboidal windows from the reinforcement layers prior to its matrix impregnation by using a water jet cutting machine. Despite the highly manual work involved in the specimens manufacturing process, it measured specifications were according to those extrapolated from ASTM 3039 for composite materials unidirectional samples.

Validation of the specimen's geometry and manufacturing technique was made through experimental testing, which were conducted on the in-house-developed biaxial machine. The cruciform's full strain field was measured via digital image correlation; the results demonstrate, in close agreement with the results obtained from finite-element(FE) simulations, that the specimen generates a significantly more homogeneous biaxial load state in the gauge zone than others reported in literature, and failure occurs, for all the tests, inside the gauge zone, as intended.

REFERENCES

1. J. J. Crookston, A. C. Long, I. A. A. Jones, review. summary, mechanical. of, prediction. properties, for. methods, reinforced. textile, composites. polymer, Proceedings of the Institution of Mechanical Engineers, Part L: Journal of Materials: Design and Applications 2005219291109

2. Soden PD, Kaddour AS, Hinton M.J.Recommendations for designers and researchers resulting from the world-wide failure exercise. Composites Science and Technology 2004

3. Kaddour AS, Hinton MJ.Instructions To Contributors Of The Second World-Wide Failure Exercise (Wwfe-II): Part(A).

4. Lomov SV, Huysmans G, Luo Y, Parmas RS, Prodromou A, Verpoest I, Phelan FR. Textile composites: modelling strategies. Composites: Part A 2001; 32(10):1379-1394.

5. Welsh JS, Mayes JS, Key CT, McLaughlin RN. Comparison of MCT failure prediction techniques and experimental verification for biaxially loaded glass fabric-reinforced composite laminates.Journal of Composite Materials 2004382421652181

6. Swanson SR, Smith LV.Comparison of the biaxial strength properties of braided and laminated carbon fibre composites. Composites: Part B 19962717177

7. Welsh JS, Adams DF.An experimental investigation of the biaxial strength of IM6/3501-6 carbon/epoxy cross-ply laminates using cruciform specimens. Composites: Part A 2002336829839

8. J. S. Welsh, J. S. Mayes, A. . Biskner, D. biaxial, testing, predictions. failure, I. of, M7/977, quasi-isotropic. carbon/epoxy, laminates, Composite Structures 2006

9. R. K. Ng, A. Yousefpour, M. Uyema, M. N. Ghasemi, Analysis. Design, Manufacture, of. Test, Water. Shallow, Vessels. Pressure, E. Using-Glass, Woven. Epoxy, Material. Composite, a. for, Underwater. Semi-Autonomous, Vehicle, Journal of Composite Materials 2002362124432478

10. L. Ebrahim, W. Van Paepegem, J. Degrieck, C. Ramault, A. Makris, D. Van Hemelrijck, Strain distribution in cruciform specimens subjected to biaxial loading conditions. Part 1: Two-dimensional versus three-dimensional finite element model. Polymer Testing 2010291713

11. L. Ebrahim, W. Van Paepegem, J. Degrieck, C. Ramault, A. Makris, D. Van Hemelrijck, Strain distribution in cruciform specimens subjected to biaxial loading conditions. Part 2: Influence of geometrical discontinuities. Polymer Testing 20102911 32138

12. Z. Fawaz, analytique. Étude, et. numérique, portant. expérimentale, rupture. sur la, fatigue. et la, des. biaxiales, renforcées. lamelles, de fibres, PhD Thesis, Sherbrooke: Université de Sherbrooke, 1992

13. Gupte AA.Optimization of Cruciform Biaxial Composite Specimen. In: Master on Science Thesis. South Dakota State University, 2003

14. Sepúlveda CG.Biaxial Testing of Composite Materials: Technical Specifications and Experimental Set-up Maestría en Ciencias con Especialidad en Sistemas de Manufactura México Mayo 2009

15. Soden PD, Hinton MJ, Kaddour AS.Biaxial test results for strength and deformation of a range of E-glass and carbon fibre reinforced composite

laminates: failure exercise benchmark data. In: Failure Criteria in Fibre Reinforced Polymer Composites: the World-Wide Failure Exercise. Oxford: Elsevier Science LTD, 20045296

16. J. E. Bird, J. Duncan, Strain hardening at high strain in aluminum alloys and its effect on strain localization. Metallurgical and Materials Transactions A 1981122235241

17. Dudderar TF, Koch, Doerries DE.Measurement of the shapes of foil bulge-test samples. Experimental Mechanics 1977174133140

18. J. P. Boehler, S. Demmerle, S. A. Koss, Direct. New, Testing. Biaxial, for. Machine, Materials. Anisotropic, Experimental Mechanics 194434119

19. E. Mönch, D. Galster, ". A. Method, Producing. a. for, Uniform. Defined, Tensile. Biaxial, Field.". Stress, Journal. British, Applied. Of, 1. Physics, no, 1963

20. Z. Sacharuk, Critères de rupture et optimisation des éléments en matériaux composites. PhD Thesis. Université de Sherbrooke. Sherbrooke 1990

21. Y. Youssef, Résistance des composites stratifiés sous chargement biaxial: validation expérimentale des prédictions théoriques. PhD Thesis, Sherbrooke: Université de Sherbrooke, 1995

22. D. Arellano, G. Sepúlveda, H. Elizalde, R. ". Ramírez, Enhanced cruciform specimen for biaxial testing of fibre-reinforced composites." International Materials Research Congress. Cancun, 2007

23. Yong Yu, Min Wan, Xiang-Dong Wu, Xian-Bin Zhou.Design of a cruciform biaxial tensile specimen for limit strain analysis by FEM. Journal of Materials Processing Technology 2002

24. A. E. Antoniou, D. Van Hemelrijck, P. Philippidis, Failure prediction for a glass/epoxy cruciform specimen under static biaxial loading. Composite Science and Technology 201070812321241

25. ASTM Standard D6856-03, Standard Guide for Testing Fabric-Reinforced TextileComposite Materials, 2003.

26. H. J. Kwon, P. Y. B. Jar, Z. Xia, Characterization of bi-axial fatigue resistance of polymer plates. Journal of Materials Science 2005404965972

27. Gower. M. Gdoutos, R. Shaw, R. Mera, Development of a Cruciform Specimen Geometry for the Characterisation of Biaxial Material Performance for Fibre Reinforced Plastics. Experimental Analysis of Nano and Engineering Materials and Structures. Springer Netherlands, 2007937938

28. Chaudonneret. M. [28], P. Gilles, R. Labourdette, H. Policella, Machine d'essais de traction biaxiale pour essais statiques et dynamiques. La

Recherche Aérospatiale 19771977299305

29. David Alejandro Arellano Escárpita.Experimental investigation of textile composites strength subject to biaxial tensile loads. Ph.D. Thesis. Instituto Tecnológico y de Estudios Superiores de Monterrey. Monterrey 2011

30. Clay SB. Biaxial Testing Apparatus.United States of America Patent 5905205. 18 May 1999

31. G. Ferron, A. Makinde, Design and Development of a Biaxial Strength Testing Device. Journal of Testing and Evaluation (ASTM) 16, 3May 1988

32. Pascoe KJ, de Villiers JWR.Low Cycle Fatigue of Steels Under Biaxial Straining. Journal of Strain Analysis 1967

33. Welsh JS, Adams DF.Development of an Electromechanical Triaxial Test Facility for Composite Materials. Experimental Mechanics 2000403312320

34. H. Fessler, J. K. A. . Musson, Biaxial. Ton, Testing. Tensile, Machine, Professional Engineering Publishing 1969412226

35. A. Makinde, L. Thibodeau, K. Neale, Development of an apparatus for biaxial testing using cruciform specimens. Experimental Mechanics 1992322138144

36. D. R. A. Hayhurst, Creep. Biaxial-Tension-Rupture, Machine. Testing, Professional Engineering Publishing 197382119123

37. G. Ferron, Dispositif perfectionné d'essais de traction biaxiale. France. 26 Septembre 1986

38. Doyle JF, Phillips JWP.Manual on experimental stress analysis. Society for Experimental Mechanics (U.S.). Society for Experimental Mechanics, 1989

39. D. Post, B. Han, 2. Chap, ". Moire, Handbook. Interferometry,", Experimental. on, W. Mechanics, N. Sharpe, Jr., ed., Springer-Verlag, NY, 2008

40. M. Gre´diac, The use of full-field measurement methods in composite material characterization: interest and limitations. Composites: Part A 2004

41. D. Lecompte, A. Smits, S. Bossuyt, H. Sol, J. Vantomme, D. Van Hemelrijck, A. M. Habraken, Quality assessment of speckle patterns for digital image correlation. Optics and Lasers in Engineering 2006441111321145

42. P. Bing, Q. Kemao, X. Huimin, A. Anand, Two-dimensional digital image correlation for in-plane displacement and strain measurement: a review. Measurement Science and Technology 2009206117
43. ASTM Standard D 3039/D 3039M, Standard Test Method for Tensile Properties of Polymer Matrix Composite Materials, 2000.

# Chapter 3

MOLECULAR SIMULATIONS ON INTERFACIAL SLIDING OF CARBON NANOTUBE REINFORCED ALUMINA COMPOSITES

Yuan Li[1], Sen Liu[2], Ning Hu[2], Weifeng Yuan[3] and Bin Gu[3]

[1]Department of Nanomechanics, Tohoku University, Aramaki-Aza-Aoba, Aoba-ku, Sendai,, Japan

[2]Department of Mechanical Engineering, Chiba University, Yayoi-cho, Inage-ku, Chiba,, Japan

[3]School of Manufacturing Science and Engineering, Southwest University of Science and Technology, Mianyang,, P.R.China

INTRODUCTION

With remarkable physical and mechanical properties [1, 2], carbon nanotube (CNT), either single-walled carbon nanotube (SWCNT) or multi-walled carbon nanotube (MWCNT), has prompted great interest in its usage as one of the most promising reinforcements in various matrices (e.g., polymers, metals and ceramics) [3-11]. However, the dramatic improvement in mechanical properties has not been achieved so far. The reason can be attributed to several critical issues: (1) insufficient length and quality of CNT, (2) poor CNT dispersion and alignment, and (3) weak interface between CNT and matrix. Although great progress has been made to improve the first two issues by developing newly cost-effective CNT synthesis methods and exploring specific fabrication methods of composites (e.g., spark plasma sintering [12], sol-gel process [13]), the proper control of interfacial properties is still a challenge as the inherent characteristics is unclear.

Up to date, large amounts of investigations have been focused on the interfacial properties of polymer-based composites by using direct pull-out experiments with the assistance of advanced instruments (e.g., transmission electronic microscopy (TEM) [14,15], atom force microscope (AFM) [16,17], Raman spectroscopy [18], scanning probe microscope (SPM) [19]), or theoretical analysis based on continuum mechanics (e.g., cohesive zone model [20], Cox's model [21], shear lag model [22,23] and pull-out model [24,25]), or atomic simulations [26-33]. However, in contrast, much less work has been focused on the interfacial properties of alumina-based composites [34-38]. For example, it has been reported that there are three hallmarks of toughening behavior demonstrated in CNT-reinforced alumina composites (CNT/Alumina) as below [34]: crack deflection at the CNT/Alumina interface; crack bridging by CNT, and CNT pull-out on the crack plane, which is consistent with that in conventional micron-scale fiber reinforced composites. Therefore, a fundamental understanding on the interfacial sliding between CNT and alumina matrix (i.e., CNT pull-out from alumina matrix) is important for clarifying the interfacial properties, and therefore the mechanical properties of bulk CNT/Alumina composites.

Current experimental works have reported two common sliding behaviors in CNT/Alumina composites: the pull-out of SWCNT [35] and sword-in-sheath mode [36, 37] of MWCNT (i.e. "pull-out of the broken outer walls of CNT with matrix", or "pull-out of inner walls of CNT with matrix after the breakage of the outer walls" in relativity). Therefore, clarifying the above two distinguished pull-out behaviors is of critical importance for understanding the interfacial properties of CNT-reinforced composites.

In this Chapter, a series of pull-out simulations of either SWCNT or MWCNT from alumina matrix are carried out based on molecular mechanics (MM) to investigate the corresponding interfacial sliding behaviors in CNT/Alumina composites. By systematically evaluating the variation of potential energy increment during the pull-out process, the effects of grain boundary (GB) structures of alumina matrix, nanotube length, nanotube diameter, wall number and capped structure of CNTs are explored for the first time.

COMPUTATIONAL MODEL

As experimentally identified, CNTs are generally located in the GB of alumina [34-37, 39], which can be schematically illustrated in Fig. 1. Note that the GB structure is generally characterized by a multiplicity index Σ based on the geometrical concept of three-dimensional (3D) coincidence between two crystals named the coincidence site lattice (CSL) model [40], which is defined as the ratio of the crystal lattice sites density to the density of the two grain superimposed lattices. The corresponding computational model by using the commercial software of Materials Studio (Accelrys) can be constructed as follows:

1. Building a hexagonal primitive cell of neutral alumina;
2. Cleaving the required GB planes and joining them together;
3. Inserting a CNT into the GB;
4. Relaxing the constructed model to obtain the equilibrated configuration.

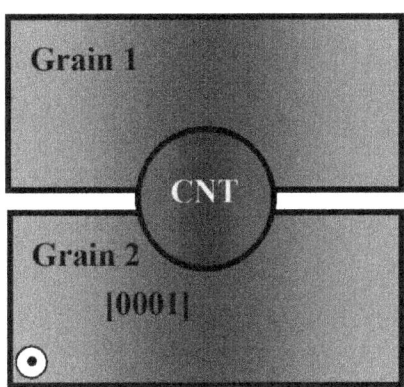

Figure 1: Schematic GB with CNT.

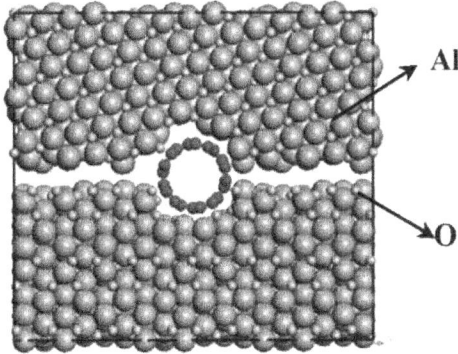

Figure 2: Simulation cell of GB with CNT.

As an example, the equilibrated model of $\Sigma 7$ ($14\bar{5}\bar{5}0$)//($4\bar{1}\bar{5}\bar{5}0$) [41] GB is shown in Fig. 2, in which the inserted open-ended SWCNT (5,5) has the length of l=5.17nm and diameter of D=0.68nm.

The pull-out process of CNT is schematically given in Fig.3, which is mainly divided into the following two steps:

1. Applying the fixed boundary conditions to the left end of alumina matrix;

2. Pulling out the CNT gradually along its axial (x-axis) direction with a constant displacement increment Δx of 0.2nm.

After each pull-out step, the structure should be relaxed in order to obtain the minimum systematic potential energy E.

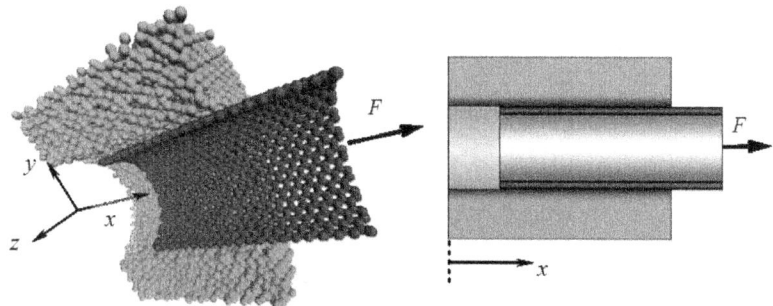

Figure 3: Schematic pull-out process of CNT from alumina matrix(green balls: atoms in alumina matrix, purple balls: atoms of CNT).

PULL-OUT SIMULATIONS OF OPEN-ENDED SWCNTS FROM ALUMINA MATRIX

Effect of GB Structure

To a large extent, GBs play a significant role on the microstructure formation and properties control of polycrystalline materials. To explore the influence of GB structure on the interfacial siding behavior between CNT and alumina matrix, three representative GB structures with a common rotation axis of [0001] ($\Sigma3(1120)//(1120)$ [42], $\Sigma7(1450)//(4150)$ [41,43,44], and $\Sigma31(47110)$ // (74110) [44]) are modeled. Note that the same fragment of SWCNT(5,5) with the length of l=5.17nm and diameter of D=0.68nm is employed.

The obtained variations of energy increment ΔE between two consecutive pull-out steps are plotted inFig. 4, where three distinct stages can be clearly seen for each case. In the initial ascent stage I, ΔE increases sharply until the pull-out displacement x reaches up to about 1.0nm. After that ΔE undergoes a long platform stage II followed by the quick descent stage III until the complete pull-out. It is noticeable that both stages I and III have the same range corresponding to the pull-out displacement of approximately a=1.0nm, which is very close to the cut-off distance of vdW interaction (i.e., 0.95nm). This feature of ΔE is similar to that for the pull-out process of CNT from polymer matrix [33] and that for the sliding among nested walls in a MWCNT [45].

Moreover, ΔE in these three curves are almost identical in stages I and III, and have the same average value at stage II although Σ7 GB results in a slightly higher ΔE. This suggests that the GB structure of alumina matrix has only a limited effect on the energy increment ΔE between two adjacent pull-out steps.

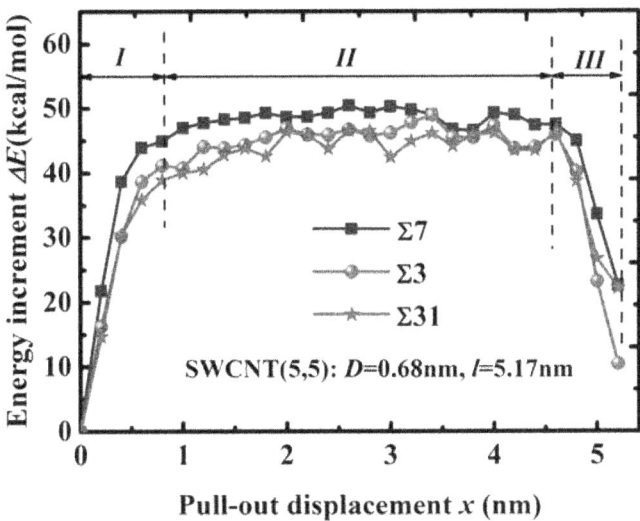

Figure 4: Effect of GB structure on the variation of energy increment during the pull-out process.

Here, as discussed in Refs. [33,45], the stable pull-out stage II is focused on, in which the average energy increment in stage II is referred to as ΔEII hereinafter. Obviously, ΔEII is independent of GB structure of alumina matrix. Therefore, in the following simulations, the Σ31 GB structure is employed to investigate the effect of nanotube length and diameter on the pull-out process.

Effect of Nanotube Length

To investigate the effect of nanotube length on the pull-out process, three SWCNTs (5,5) with different lengths are embedded in the same Σ31 GB of alumina matrix, respectively. The obtained variations of energy increment ΔE between two adjacent pull-out steps are given in Fig.5, in which the same trend is clearly observed for each case as that in Fig. 4. Moreover, the identical ΔEII of three cases indicates itsindependence of nanotube length. Therefore, in the following simulations, CNTs with the same length of 5.17nm are employed.

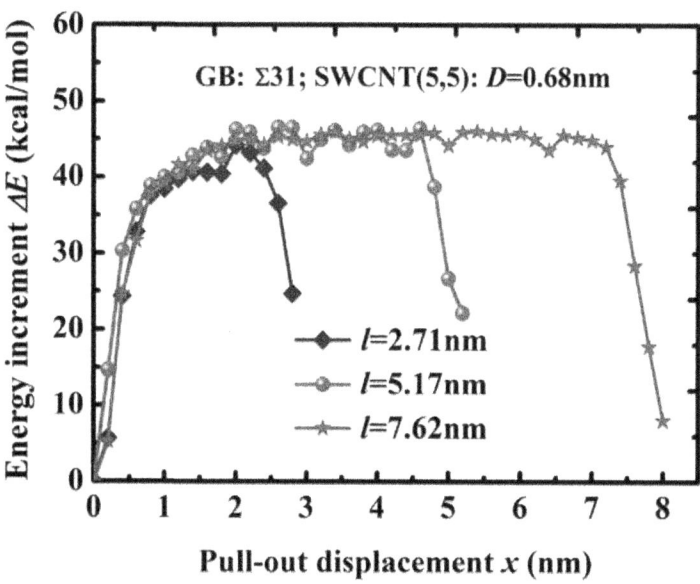

Figure 5: Effect of nanotube length on the variation of energy increment during pull-out process.

Effect of Nanotube Diameter

Based on the above length-independent behavior, four SWCNTs (i.e., (5,5), (10,10), (15,15), (20,20)) with the same length of 5.17nm but different diameters are embedded into alumina matrix with Σ31GB structure. The corresponding relationship between energy increment ΔE and pull-out displacement x is shown in Fig.6a. Unlike the length-independent behavior, ΔEII increases linearly with nanotube diameter as fitted in Fig.6b with the following formula

$$\Delta E_{II} = 52.04 \times D + 9.04 \tag{1}$$

where

ΔE_{II} is in kcal/mol, and D is in nm.

This phenomenon can be attributed to the number of atoms in circumferential direction, which increases linearly with nanotube diameter. For a CNT with larger diameter, there will be stronger vdW interactions needed to be overcome for the possible pull-out, which subsequently induces the higher energy increment in the same pull-out displacement.

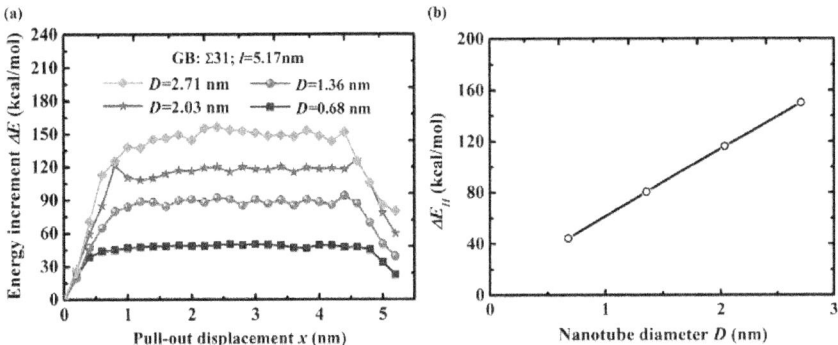

Figure 6: Effect of nanotube diameter on the variation of energy increment during the pull-out process(a) Relationship of energy increment and pull-out displacement; (b) Relationship of ΔEII and nanotube diameter.

Pull-Out Force and Surface Energy Density

As discussed above, for the pull-out of a SWCNT, the corresponding average energy increment in stageII, i.e., ΔE_{II}, is independent of GB structure and nanotube length, but is proportional to nanotube diameter. In view of that the work done by the pull-out force is equal to the energy increment in each pull-out step by neglecting some other minor energy dissipations, the pull-out force can be approximately calculated as

$$F_{II} = \frac{\Delta E_{II}}{\Delta x} \tag{2}$$

On this sense, we can conclude that the pull-out force of CNT from alumina matrix related to energy increment is also independent of GB orientation and nanotube length, but is proportional to nanotube diameter. From Eqs. (1) and (2), the corresponding empirical formula to predict the pull-out force is proposed as

$$F_{II} = 1.81 \times D + 0.31 \tag{3}$$

where

FIIFII

is in the unit of nN, and D of nm.

It should be noted that two new surface regions are generated at two ends of CNT after each pull-out step (i.e., the inner surface of the matrix at the left side of CNT, and the outer surface of CNT on the right side). Therefore, the corresponding surface energy should be equal to the energy increment. Therefore, the surface energy density can be calculated as

$$\gamma_{II} = \frac{\Delta E_{II}}{2\pi D \Delta x} = \frac{F_{II}}{2\pi D} \qquad (4)$$

Initially, this value is dependent on the diameter of SWCNT. However, as nanotube diameter increases, it will decrease gradually and then saturate to a constant. The converged value of surface energy density is approximately 0.3N/m. Note that this surface energy density is newly reported for the interface of SWCNT and alumina matrix, although there have been some reports about that for the interface of SWCNT and polymer matrix with the value of 0.09~0.12N/m [26, 30] or for sliding interface among nested walls in a MWCNT with the value of 0.14N/m [45]. It can be found that the surface interface density in CNT/Alumina composites is much higher than those of CNT/Polymer composites or CNT walls, implying its stronger interface.

Interfacial Shear Stress

Based on the above discussion, the corresponding interfacial shear stress is analyzed in the following.

The pull-out force is equilibrated with the axial component of vdW forces which induces the interfacial shear stress. Conventionally, if we employ the common assumption of constant interfacial shear stress with uniform distribution along the whole embedded region of CNT, the pull-out force FIIwill vary with the embedded length of CNT, which is obviously in contradiction to the above length-independent behavior of average energy increment ΔE_{II} in stage II. For the extreme case of a CNT with an infinite length, the interfacial shear stress tends to be zero, which is physically unreasonable. This indicates that the conventional assumption of interfacial shear stress is improper for the perfect interface of CNT/Alumina composites with only consideration of vdW interactions.

For this problem, the interfacial shear stress should be analyzed according to the different stages in the variation of energy increment ΔE. In stage I, the interfacial shear stress exists within a region of the length a=1.0nm at each end of CNT as described in Ref. [45] since the length of CNT in the model is equal to that of alumina matrix. In stage II, the situation may be different since the left end of CNT is deeply embedded into the alumina matrix with the pull-out displacement x much larger than a=1.0nm. To address the interfacial shear stress in this stage II, a simple simulation is performed here.

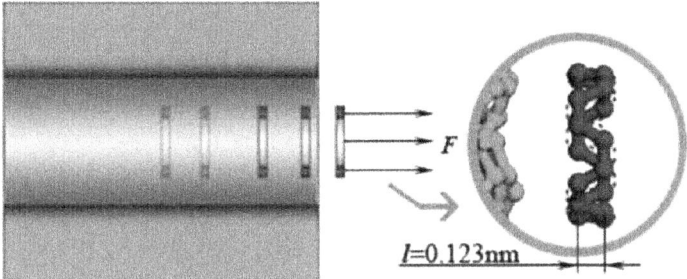

(a) Model for the pull-out of a simple CNT unit cell

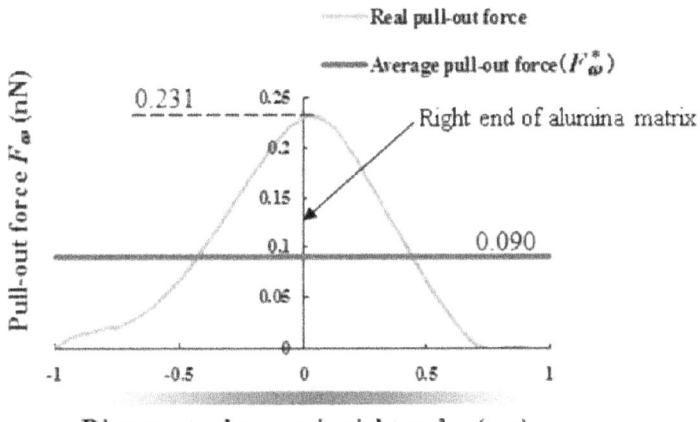

(b) Variation of pull-out force F_ω and average pull-out force F_ω^*

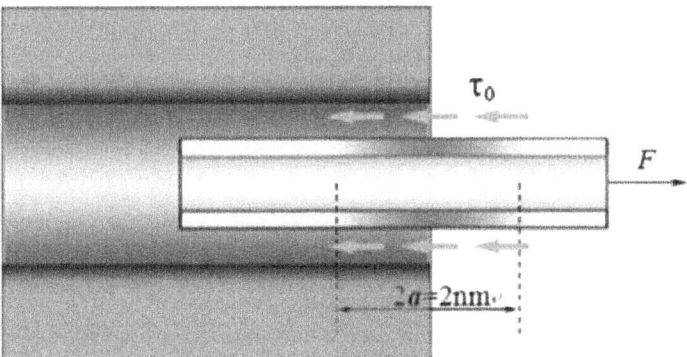

Figure 7: Analysis of interfacial shear stress.

As shown in Fig. 7a, a SWCNT(5,5) with only a half repeat unit is completely embedded in the middle position of alumina matrix. Then this SWCNT fragment is pulled out gradually with a constant increment of Δx=0.2nm to obtain the variation of systematic energy increment ΔEω, and the corresponding pull-out force Fω. As this SWCNT fragment is very short, the obtained pull-out force Fω, which is equilibrated by the shear force induced by the interfacial shear stress, can be used to characterize the distribution of interfacial shear stress. The obtained distribution of pull-out force Fω at various pull-out steps is shown in Fig.7b. At the initial stage of the pull-out, the pull-out force keeps value at zero. When the CNT unit cell moves into the range of a=1.0nm measured from the right end of alumina matrix, the pull-out force increases sharply. It reaches the maximum when the CNT unit cell is just located on the right end of the matrix. As the CNT unit cell is further pulled out, it decreases gradually to zero. In virtue of the above results, as shown in Fig. 7c, the interfacial shear stress is solely distributed within the region of 2a centered by the right end of matrix in stage II. The pull-out force during the pull-out process is further averaged within the range of 2a, i.e., F_ω^* =0.09nN in Fig. 7b.

By assuming that the interfacial shear stress is uniform within the above defined region for simplicity, the average of interfacial shear stress τ_0 in stage II can be defined from the pre-defined average pull-out force(i.e., F_ω^*

$$\tau_0 = \frac{F_\omega^*}{2\pi Da} \tag{5}$$

Obviously, τ_0 is dependent on the diameter D of SWCNT. However, it tends to be a constant as nanotube diameter increases gradually. The obtained converged interfacial shear stress τ_0 from various unit cells of SWCNT with different diameters is 303 MPa. Note that it only exists within the range of 2a centered by the right end of the matrix.

PULL-OUT SIMULATIONS OF OPEN-ENDED MWCNTS FROM ALUMINA MATRIX

Usually, there are two typical sliding behavior for MWCNT-reinforced composites: one is the complete pull-out of MWCNT, while the other is the so-called sword-in-sheath mode, e.g., in which the broken outer walls are pulled out (i.e., sheath) leaving the intact inner walls (i.e., sword) in the matrix. Therefore, based on the above information, two simple typical cases are firstly investigated: Case 1: pull-out of the whole MWCNT (e.g., Fig. 8a); Case 2: pull-out of only the outermost wall of MWCNT (e.g., Fig. 8b).

In view of the extremely high computational cost, several double-walled carbon nanotubes (DWCNTs) with wall number n=2 and triple-walled carbon nanotubes (TWCNTs) with wall number n=3 are discussed in the present simulation. The obtained average energy increment ΔE_{II} in stage II related to the pull-out force is also found to be proportional to the diameter of the outermost wall of MWCNT

D_o, which can be fitted as

Case 1

$$\begin{cases} \Delta E_{II} = 57.54 \times D_o + 4.36, \ F_{II} = 2.00 \times D_o + 0.15 \ (n = 2) & (a) \\ \Delta E_{II} = 58.26 \times D_o + 6.50, \ F_{II} = 2.03 \times D_o + 0.23 \ (n = 3) & (b) \end{cases} \quad (6)$$

Case 2

$$\begin{cases} \Delta E_{II} = 93.61 \times D_o + 10.17, \ F_{II} = 3.26 \times D_o + 0.35 \ (n = 2) & (a) \\ \Delta E_{II} = 96.60 \times D_o + 10.50, \ F_{II} = 3.33 \times D_o + 0.37 \ (n = 3) & (b) \end{cases} \quad (7)$$

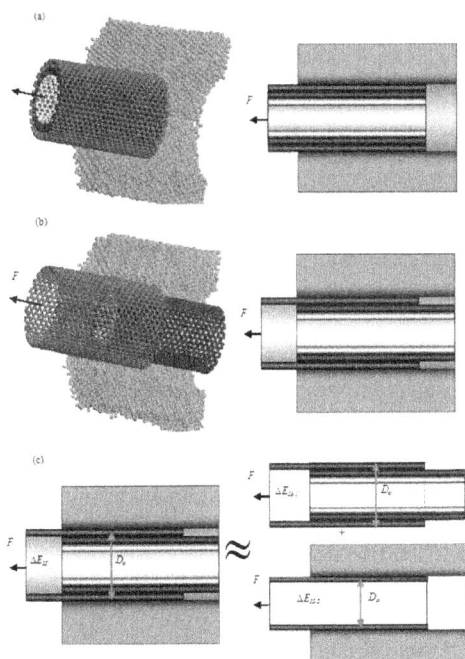

Figure 8: Two typical pull-out cases for an open-ended TWCNT(a) Case 1: pull-out of the whole MWCNT; (b) Case 2: pull-out of the outmost wall of MWCNT; c) Decomposition of Case 2 into two independent sub-problems.

For Case 1, the relationship of ΔEII and nanotube diameter for the complete pull-out of SWCNT (Eq. 3), DWCNT (Eq. 6a), and TWCNT (Eq.7a) are plotted in Fig. 9, which indicates the effect of wall number from some aspect. The slope for DWCNT is about 9.56% higher than that for SWCNT, which highlights the contribution of the first adjacent inner wall to ΔEII. However, the slope of TWCNT is only about 1.24% higher than that for DWCNT, which implies that the contribution of the second inner wall is gradually weakened as the distance from the sliding interface increases. Therefore, it can be concluded that the pull-out of MWCNT from alumina matrix is mostly affected by its two adjacent walls from the sliding interface, which indicates that for the whole pull-out of any MWCNT with more walls over 3, ΔEII can be approximately assumed to be equal to that of TWCNT (i.e., Eq. 6b).

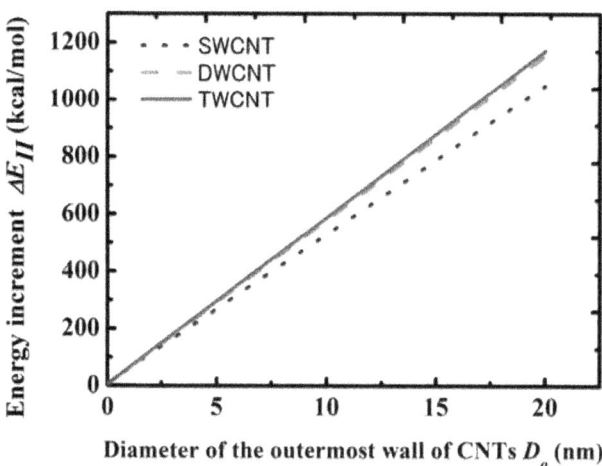

Figure 9: Effect of wall number on the energy increment for the pull-out of whole MWCNT.

For Case 2, taking TWCNT(5,5)/(10,10)/(15,15) as an example, it is surprising to find that the corresponding ΔEII is approximately equal to the sum of those of two sub-problems, i.e., the pull-out of a SWCNT(15,15) from alumina matrix, and the pull-out of the outermost wall in the TWCNT. Therefore, the corresponding ΔEII for the pull-out of any TWCNT from alumina matrix (i.e., Eq. 7b) can be approximately decomposed into the following two items as given in Fig. 8c: ΔEII-1 for the pull-out of the outermost wall of TWCNT against the other two inner walls (i.e., Eq. 5 in Ref. [45]), and ΔEII-2 for the pull-out of a SWCNT from alumina matrix (i.e., Eq. 3) whose diameter is equal to the outermost wall of the TWCNT.

It should be noted that for the real sword-in-sheath fracture mode, there are more than 3 walls pulled out. For example, as shown in Fig. 10a, several purple outer walls of a MWCNT are pulled out leaving the yellow inner walls within the matrix. Here, there are two sliding interfaces: one is between CNT and matrix, the other is between outer walls and inner wall. According to the above discussion, it can also be thought of as the superimposition of the following two sub-problems in Fig. 10b: one is the pull-out of the TWCNT which is composed of the outer three walls (i.e., Eq. 6b), and the other is the pull-out of outer three walls in a MWCNT with five walls (i.e., Eq. 6 in Ref. [45]). It indicates that the corresponding ΔEII and pull-out force FII can be calculated as

$$\Delta E_{II} = 58.26 \times D_o + 37.56 \times D_c - 4.00 \tag{8}$$

$$F_{II} = 2.04 \times D_o + 1.31 \times D_c - 0.14 \tag{9}$$

Here, Do is the diameter of the outermost wall of MWCNT, and Dc is the diameter of the green critical wall in Fig. 10 (i.e., the immediate outer wall at the sliding surface between outer walls and inner walls).

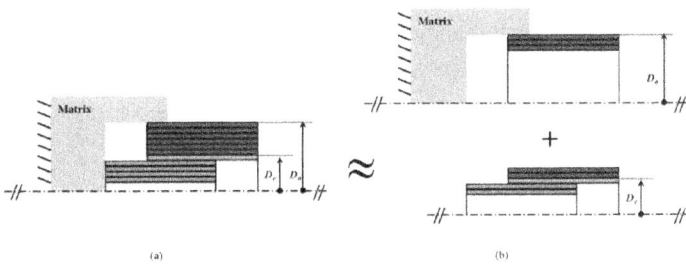

Figure 10: Real case of sword-in-sheath mode.

PULL-OUT SIMULATIONS OF A CAPPED MWCNT FROM ALUMINA MATRIX

It is noted that open-ended CNTs are employed in the above simulations. On the other hand, it has been reported that CNT cap makes great effect on its field emission properties [46], load transferring ability among nested walls of MWCNT [45, 47, 48]. However, to our best knowledge, there is no any detailed report on the effect of CNT caps on the interfacial properties of CNT-reinforced composites. Therefore, the pull-out of capped MWCNTs from alumina matrix in a sword-in-sheath mode is discussed here.

The schematic model is given in Fig. 11. By using the principal of superimposition, this pull-out process can be decomposed into the following three parts: pull-out of outer walls against matrix (i.e., part I); pull-out of inner walls against outer walls, which can be further decomposed into the open-ended part of inner/outer walls (i.e., part II) and the capped part of inner/outer walls (i.e., part III) as each wall in a MWCNT is composed of open-ended part and capped part.

Generally, the number of broken outer walls and intact inner walls are more than 3. Therefore the corresponding pull-out forces for the above three parts are analyzed as below.

Pull-out of outer walls against matrix (i.e., part I in Fig. 11): According to Eq. 2, the corresponding pull-out force F_1 can be predicted by using Eq. 6b, i.e.,

$$F_1 = 2.03 \times D_o + 0.23 \tag{10}$$

Pull-out of open-ended part of inner walls against outer walls (i.e., part II in Fig. 11): According to Eq. 6 in Ref. [45], the corresponding pull-out force F_2 can be predicted as

$$F_2 = 1.31 \times D_c - 0.37 \tag{11}$$

Pull-out of capped part between inner walls and outer walls (i.e., part III in Fig. 11): This part can be transferred as the interfacial sliding among nested walls in a capped MWCNT.

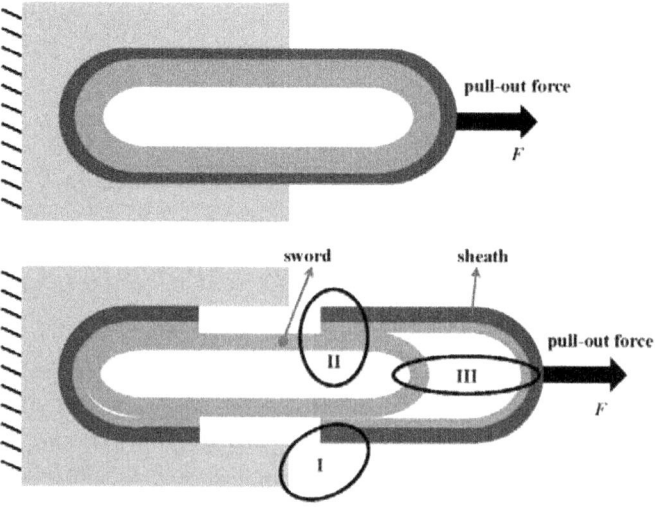

Figure 11: Schematic model for the pull-out of a capped MWCNT from alumina matrix.

As illustrated in Fig. 12a, after fixing the atoms of the outer cap, the inner wall is pulled out along its axial direction by applying a constant displacement increment of $\Delta x_2 = 0.01$ nm on the atoms of the right end of inner wall. Note that the present displacement increment Δx_2 is smaller than the above Δx of 0.2nm, which is used for making the effect of CNT caps on energy increment clearly. After each pull-out step, the structure is relaxed to obtain the minimum potential energy E. As discussed in Ref. [45], the pull-out force of an open-ended CNT is only proportional to nanotube diameter, and independent of nanotube length. For this reason, five DWCNTs with different diameters but same length are built up to investigate the effect of CNT cap. The calculated energy increments between two consecutive pull-out steps of three DWCNTs are shown in Fig. 13a, where Dc is the diameter of critical wall (i.e., the outer wall of DWCNT). It can be seen that for each DWCNT the energy increment ΔE increases rapidly up to a peak value at a specified displacement, and then decreases. The same feature is also observed in the simulations of two other DWCNTs with larger diameters, i.e., (54,54)/(59,59) with Dc=8.0nm and (83,83)/(88,88) with Dc=11.93nm. The maximum energy increment (i.e., ΔE_{max}) for the five DWCNTs is shown in Fig. 13b. The relationship between ΔE_{max} and Dc can be perfectly fitted into a quadratic function of

$$\Delta E_{max-DWCNT} = 2.09 \times D_c^2 - 2.15 \times D_c + 0.94 \qquad (12)$$

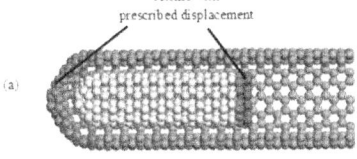

(a)

(b)

(c)

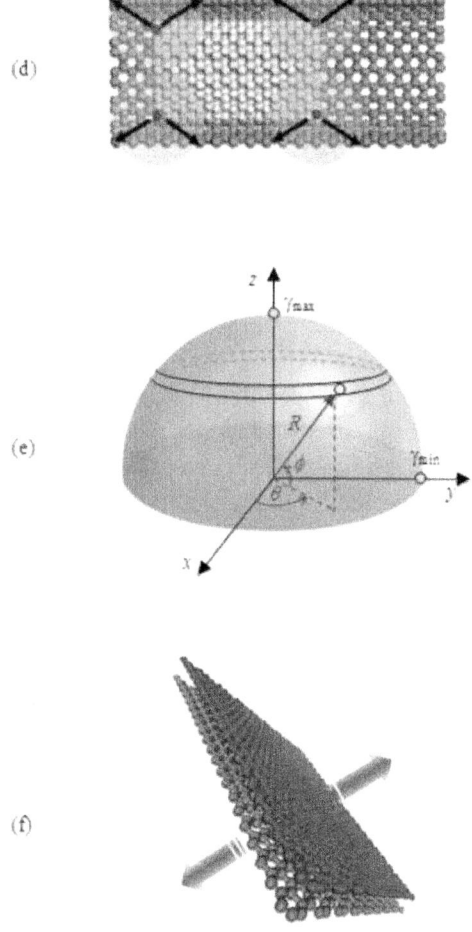

Figure 12: Interfacial sliding in a capped DWCNT(a) Schematic model of a capped DWCNT; (b) Pull-out of the capped part; (c) Pull-out of the open-ended part;(d) Force state of open-ended part of CNT; (e) Estimation of energy variation of a cap; (f) Pull-out of graphite sheets.

To understand this potential energy increment in detail, we further divided the inner wall (Fig. 12a) into two parts, i.e., the capped (Fig. 12b) and the open-ended part (Fig. 12c). The corresponding pull-out forces for these two parts are F_3^1 and F_3^2, which means $F_3 = F_3^1 + F_3^2$. The pull-out of open-ended part (Fig. 12c) does not cause any change of the potential energy, i.e., $F_3^2=0$. It means that the contribution of capped part, i.e., F_3^1 dominates the total pull-out force F_3. The reason can be explained using Fig. 12d-12e.

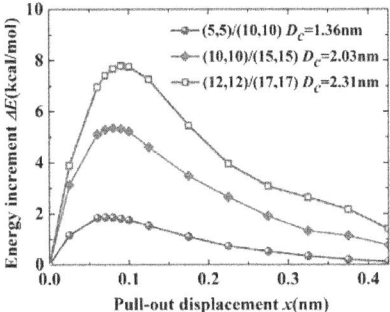

(a) Energy increment versus pull-out displacement for model of Fig.12a

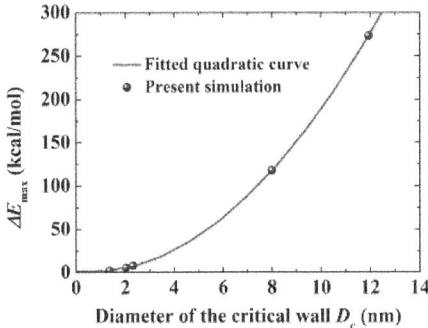

(b) Relationship between maximum energy increment and nanotube diameter

Figure 13: Variation of energy increment for the interfacial sliding in a capped DW-CNT.

First, in Fig. 12d, if the length of the outer wall of DWCNT is long enough, the carbon atoms of the inner wall are always in force equilibrium. For example, in Fig. 12d, the atoms in red are balanced by the symmetrical horizontal forces form the atoms of the outer wall, which are within the cut-off distance of Lennard-Jones potential [49,50]. During a pull-out process, the relative motion of the atoms between the inner wall and outer wall creates repetitive breaking and reforming of the vdW interactions and no resultant resistance force can be generated on the inner wall, i.e., $F_3^2=0$.

The quadratic form of the energy increment in Eq. 12 due to the capped effect is associated with the surface energy density. Considering a cap model shown in Fig. 12e, the bottom edge is just located on the boundary between the capped and open-ended part. If we use γ_{max} and γ_{min} to represent the maximum and minimum surface energy density (i.e., potential energy variation per unit area) under a specified separation displacement, γ_{max} is at the top of the cap while γ_{min} appears at the bottom of the cap. Then, the surface energy density is assumed to vary from the top to the bottom of the cap in the function of $\gamma(\phi) = \gamma_{max} \cos(90° - \phi) = \gamma_{max} \sin\phi$., which implies that $\gamma_{min}=0$., which implies that $\gamma_{min}=0$. This is reasonable as $F_3^2=0$. Then the total surface energy variation of the cap can be calculated as

$$U_{cap} = \int_0^{2\pi} (\int_0^{\frac{\pi}{2}} \gamma(\phi) R\cos\phi \times Rd\varphi)d\theta = \frac{\pi D_c^2}{4}\gamma_{max}$$

(13)

From Eq. 13, regardless of the function of $\gamma(\phi)$, the surface energy is always proportional to πDc^2. As a result, the energy increment induced by the pull-out of the cap can be described by a quadratic function of Dc, which is consistent with Fig. 13b and Eq. 12. Approximately, the γ_{max} at the small top flat area of the cap during the pull-out process can be predicted in the same way by simulating the separation of two flat graphite sheets in Fig. 12f. It is confirmed by the displacement-energy increments curves obtained from the simulation of two graphite sheets which is quite similar with those in Fig. 13a. The corresponding $\gamma_{max-cap}$ is around 0.03N/m under 0.01 nm separation displacement in the normal direction of two graphite sheets. Substituting this value into Eq. 13 leads to the total surface energy change as:

Ucap=πD2c4γmax−cap=2.76D2cUcap=πDc24γmax−cap=2.76Dc2

, which is approximately equivalent to Eq. 12. Therefore, it indicates the quadratic form of Eq. 12 is appropriate from the other aspect.

After validating the effectiveness of Eq. 12, the corresponding maximum pull-out force can be simply evaluated by equaling the work done by the pull-out force to the $\Delta E_{max-DWCNT}$ with the formula of

$$F_{3-DWCNT} = 1.45 \times D_c^2 - 1.49D_c + 0.65$$

(14)

It should be noted that the above analysis is for a capped DWCNT. For the case of MWCNT, a simplified model in Fig. 14 is developed, as only the immediate two outer and inner walls from the sliding interface can affect the corresponding pull-out interface [45]. The evaluated pull-out force is found to be approximately 29% higher than that for DWCNT due to the contribution of the immediate two outer and inner walls, which means

$$F_{3-MWCNT} = 1.87 \times D_c^2 - 1.92D_c + 0.84$$

(15)

Therefore, for the pull-out of a capped MWCNT from alumina matrix in a sword-in-sheath mode, the corresponding pull-out force can be assumed to be the sum of those for the above three parts (i.e., Eq. 10 for part I, Eq. 11 for part II, Eq. 15 for part III):

$$F = F_1 + F_2 + F_{3-MWCNT} = 1.87 \times D_c^2 + 2.03 D_o - 0.61 \times D_c + 0.7 \qquad (16)$$

For the pull-out of a MWCNT numbered as sample 14 in Ref. [37] which has the outermost wall with diameter of Do=94nm and the critical wall at the sliding interface with diameter about Dc=90nm, the calculated pull-out force using Eq. 16 is 15.28μN, which is in the same scale of experimental value of 19.7 μN [37].

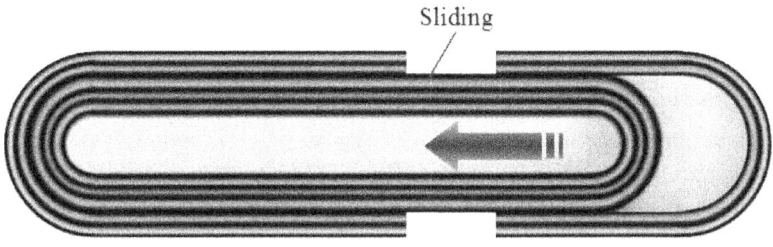

Figure 14: Schematic model for interfacial sliding in a capped MWCNT in sword-in-sheath mode

CONCLUSIONS

We systematically investigate the pull-out process of open-ended and capped CNTs from alumina matrix using MM simulations, aiming at clarifying the interfacial sliding behavior in CNT/Alumina composites. The effects of grain boundary structure of alumina matrix, nanotube length, nanotube diameter, wall number and capped structure of CNTs are explored systematically.

A set of universal formulae with the newly obtained surface energy density is proposed to approximately predict the pull-out force from nanotube diameter. The philosophy behind these simple empirical formulae is that the pull-out force is only proportional to nanotube diameter, and independent of nanotube length and GB structure of alumina matrix. The detailed interfacial shear stress is studied in this work, which indicates that the conventional definition of the interfacial shear strength is inappropriate in CNT/Alumina composites. Moreover, there are at most two adjacent walls at each side of the sliding interface which will affect this interfacial sliding in CNT/Alumina composites. Furthermore, it also indicates that CNT caps play a very important role in the pull-out process. These findings will be helpful for clarifying the toughening mechanism for mechanical properties of bulk CNT/Alumina composites and providing useful insight into the design of ideal materials.

ACKNOWLEDGEMENT

The authors are grateful to be partly supported by Tohoku Leading Women's Jump Up Project for 2013 (J1 10002158) and Grand-in-Aids for Scientific Research (No. 19360045 and No. 22360044) from the Ministry of Education, Culture, Sports, Science and Technology (MEXT) of Japan. The authors acknowledge Prof. C.B. Fan (Beijing Institute of Technology, China) for kindly providing the computational resources.

REFERENCES

1. Dresselhaus. G. Dresselhaus, P. Avouris, 2001Carbon Nanotubes: Synthesis, Structure, Properties, and Applications. Springer-Verlag Berlin Heidelberg.
2. Dresselhaus. G. Dresselhaus, J. C. Charlier, E. Hernandez, 2004Electronic, Thermal and Mechanical Properties of Carbon Nanotubes. Philosophical Transactions of the Royal Society A. 36220652098
3. Thostenson ET, Ren ZF, Chou TW2001Advances in the Science and Technology of Carbon Nanotubes and their Composites: A Review. Composites Science and Technology. 611318991912
4. O. Breuer, U. Sundararaj, 2004Big Returns from Small Fibers: A Review of polymer/carbon nanotube composites. Polymer Composites. 256630645
5. J. N. Coleman, U. Khan, W. J. Blau, Y. K. Gunko, 2006Small but Strong: A Review of the Mechanical Properties of Carbon Nanotube-Polymer Composites. Carbon. 4416241652
6. M. Moniruzzaman, K. I. Winey, 2006Polymer Nanocomposites Containing Carbon Nanotubes. Macromolecules. 391651945205
7. J. N. Coleman, U. Khan, Y. K. Gun'ko, 2006Mechanical Reinforcement of Polymers Using Carbon Nanotubes. Advanced Materials. 186689706
8. A. Peigney, C. Laurent, E. Flahaut, et al.2000Carbon Nanotubes in Novel Ceramic Matrix Nanocomposites. Ceramics International. 266677683
9. Curtin WA, Sheldon BW.2004CNT-reinforced Ceramics and Metals. Materials Today. 7114449
10. S. S. Samal, S. Bal, 2008Carbon Nanotube Reinforced Ceramic Matrix Composites- a Review. Journal of Minerals & Materials Characterization & Engineering. 74355370
11. J. Cho, A. R. Boccaccini, M. Shaffer, 2009Ceramic Matrix Composites Containing Carbon Nanotubes. Journal of Materials Science. 44819341951

12. G. D. Zhan, Wan. J. Kuntz, Ak. Mukherjee, 2003Single-wall Carbon Nanotube as Attractive Toughening Agents in Alumina-based Nanocomposites. Nature Materials. 23842
13. Chan BM, Cha SI, Kim KT, Lee KH & Hong SH.2005Fabrication of Carbon Nanotube Reinforced Alumina Matrix Nanocomposite by Sol-gel Process. Materials Science and Engineering A. 395124128
14. D. Qian, E. C. Dickey, R. Andrews, T. Rantell, 2000Load Transfer and Deformation Mechanisms in Carbon Nanotube-Polystyrene Composites. Applied Physics Letters. 762028682870
15. F. Deng, 2008Investigation of the Interfacial Bonding and Deformation Mechanism of the Nano Composites Containing Carbon Nanotubes. Tokyo University, PhD Dissertation.
16. A. H. Barber, S. R. Cohen, S. Kenig, H. D. Wagner, 2003Measurement of Carbon Nanotube-Polymer Interfacial Strength. Applied Physics Letters. 822341404142
17. A. H. Barber, S. R. Cohen, S. Kenig, H. D. Wagner, 2004Interfacial Fracture Energy Measurements for Multi-walled Carbon Nanotubes Pulled from a Polymer Matrix. Composite Science and Technology. 6422832289
18. Schadler LS, Giannaris SC, Ajayan PM.1998Load Transfer in Carbon Nanotube Epoxy Composites. Applied Physics Letters. 732638423844
19. Cooper CA, Cohen SR, Barber AH, Wagner HD.2002Detachment of Nanotubes from a Polymer Matrix. Applied Physics Letters. 812038733875
20. L. Y. Jiang, Y. Huang, H. Jiang, G. Ravichandran, H. Gao, K. C. Hwang, et al.2006A Cohesive Law for Carbon Nanotube/Polymer Interfaces based on the van der Waals Force. Journal of the Mechanics and Physics of Solids. 5424362452
21. Xiao KQ, Zhang LC.2004The Stress Transfer Efficiency of a Single-walled Carbon Nanotube in Epoxy Matrix. Journal of Materials Science. 3944814486
22. X. L. Gao, K. Li, 2005A Shear-lag Model for Carbon Nanotube-Reinforced Polymer Composites. International Journal of Solids and Structures. 4216491667
23. J. Tsai, T. Lu, 2009Investigating the Load Transfer Efficiency in Carbon Nanotubes Reinforced Nanocomposites. Composite Structures. 90172179
24. K. Lau, 2003Interfacial Bonding Characteristics of Nanotube/Polymer Composites. Chemical Physics Letters. 370399405

25. T. Natsuki, F. Wang, Q. Q. Ni, M. Endo, 2007Interfacial Stress Transfer of Fiber Pullout for Carbon Nanotubes with a Composite Coating. Journal of Materials Science. 4241914196
26. V. Lordi, N. Yao, 2000Molecular Mechanics of Binding in Carbon-Nanotube-Polymer Composites. Journal of Materials Research. 151227702779
27. K. Liao, S. Li, 2001Interfacial Characteristics of a Carbon Nanotube-Polystyrene Composite System. Applied Physics Letters. 792542254227
28. S. J. V. Frankland, A. Caglar, D. W. Brenner, M. Griebel, 2002Molecular Simulation of the Influence of Chemical Cross-links on the Shear Strength of Carbon Nanotube- Polymer Interfaces. Journal of Physical Chemistry B. 10630463048
29. J. Gou, B. Minaie, B. Wang, Z. Liang, C. Zhang, 2004Computational and Experimental Study of Interfacial Bonding of Single-walled Nanotube Reinforced Composites. Computational Materials Science. 31225236
30. Q. Zheng, D. Xia, Q. Xue, K. Yan, X. Gao, Q. Li, 2009Computational Analysis of Effect of Modification on the Interfacial Characteristics of a Carbon Nanotube- Polyethylene Composites System. Applied Surface Science. 25535243543
31. A. Al-Ostaz, G. Pal, P. R. Mantena, A. Cheng, 2008Molecular Dynamics Simulation of SWCNT-Polymer Nanocomposite and its Constituents. Journal of Materials Science. 43164173
32. S. C. Chowdhury, T. Okabe, 2007Computer Simulation of Carbon Nanotube Pull-out from Polymer by the Molecular Dynamics Method. Composites Part A. 38747754
33. Y. Li, Y. Liu, X. Peng, C. Yan, S. Liu, N. Hu, 2011Pull-out Simulations on Interfacial Properties of Carbon Nanotube-reinforced Polymer Nanocomposites. Computational Material Science. 5018541860
34. Z. Xia, L. Riester, W. A. Curtin, et al.2004Direct Observation of Toughening Mechanisms in Carbon Nanotube Ceramic Matrix Composites. Acta Materialia. 524931944
35. Fan, JP, Zhuang DM, Zhao DQ, et al.2006Toughening and Reinforcing Alumina Matrix Composite with Single-wall Carbon Nanotubes. Applied Physics Letters B. 89:121910(3).
36. G. Yamamoto, M. Omori, T. Hashida, H. Kimura, 2008A Novel Structure for Carbon Nanotube Reinforced Alumina Composites with Improved Mechanical Properties. Nanotechnology. 19:315708(7).

37. G. Yamamoto, K. Shirasu, T. Hashida, et al.2011Nanotube Fracture during the Failure of Carbon Nanotube/Alumina Composites. Carbon. 4937093716
38. L. Li, Z. Xia, W. Curtin, Y. Yang, 2009Molecular Dynamics Simulations of Interfacial Sliding in Carbon-Nanotube/Diamond Nanocomposites. Journal of the American Ceramic Society. 921023312336
39. A. L. Vasiliev, R. Poyato, N. P. Padture, 2007Single-wall Carbon Nanotubes at Ceramic Grain Boundaries", Scripta Materialia. 566461463
40. W. Bollmann, 1970Crystal Defects and Crystalline Interfaces. Springer, Berlin.
41. K. Matsunaga, H. Nishimura, S. Hanyu, et al.2005HRTEM Study on Grain Boundary Atomic Structures Related to the Sliding Behavior in Alumina Bicrystals. Applied Surface Science. 2417579
42. Y. Ikuhara, 2001Grain Boundary and Interface Structure in Ceramics. Journal of the Ceramic Society of Japan. 109(7): S110S120.
43. H. Nishimura, K. Matsunaga, T. Saito, et al.2003Atomic Structures and Energies of $\Sigma 7$ Symmetrical Tilt Grain Boundaries in Alumina Bicrystals. Journal of the American Ceramic Society. 864574580
44. H. Nishimura, K. Matsunaga, T. Saito, et al.2003Grain Boundary Structures and High Temperature Deformations in Alumina Bicrystals. Journal of the Ceramic Society of Japan. 1119688691
45. Y. Li, N. Hu, G. Yamamoto, Z. Wang, et al.2010Molecular Mechanics Simulation of the Sliding Behavior between Nested Walls in a Multi-walled Carbon Nanotube. Carbon. 4829342940
46. Wang MS, Wang JY, Peng LM.2006Engineering the Cap Structure of Individual Carbon Nanotubes and Corresponding Electron Field Emission Characteristics. Applied Physics Letters. 88(24): 243108(1-3).
47. G. A. Shen, S. Namilae, N. Chandra, 2006Load Transfer Issues in the Tensile and Compressive Behavior of Multi-walled Carbon Nanotubes. Materials Science and Engineering A. 429(1-2):66-73.
48. Z. Xia, W. A. Curtin, 2004Pullout Forces and Friction in Multiwall Carbon Nanotubes. Physical Review B. 69:2333408(1-4).
49. N. Hu, H. Fukunaga, C. Lu, M. Kameyama, B. Yan, 2005Prediction of Elastic Properties of Carbon Nanotube Reinforced Composites. Proceedings of the Royal Society A. 46116851710
50. N. Hu, K. Nunoya, D. Pan, T. Okabe, H. Fukunaga, 2007Prediction of Buckling Characteristics of Carbon Nanotubes. International Journal of Solids and Structures. 4465656550

Chapter 4

MECHANICAL COATING TECHNIQUE FOR COMPOSITE FILMS AND COMPOSITE PHOTOCATALYST FILMS

Yun Lu[1], Liang Hao[2] and Hiroyuki Yoshida[3]

[1]Graduate School & Faculty of Engineering, Chiba University,, Japan

[2]Graduate School, Chiba University,, Japan

[3]Chiba Industrial Technology Research Institute,, Japan

INTRODUCTION

Coating Techniques for Film Materials and Their Applications

In the field of materials science and engineering, the investigation on film materials is becoming increasingly important. By film materials, we can develop a variety of new material properties in the fields of electrics and electronics, optics, thermotics, magnetic, and mechanics, among others (S.Yoshida et al., 2008). In recent years, without the development of film materials we could not make any great progress in the renewable energy, environment improvement, exploitation of space, and so on. The coating techniques for film materials can fall into several categories as shown in Table 1.

In these techniques, physical vapor deposition (PVD) and chemical vapor deposition (CVD) are most widely applied. PVD are atomistic deposition processes in which material is vaporized from a solid or liquid source in the form of atoms or molecules and transported in the form of a vapor through a vacuum or low pressure gaseous (or plasma) environment to the substrate, where it condenses. PVD can be used to deposit films of elements and alloys as well as compounds using reactive deposition processes (Mattox, 2010). On the other hand, CVD may be defined as the deposition of a solid on a heated surface from a chemical reaction in the vapor phase. It belongs to the class of vapor-transfer processes which is atomistic in nature, which is the deposition species are atoms or molecules or a combined of these (Pierson, 1999). Microfabrication processes widely use CVD to deposit film materials in various forms including monocrystalline, polycrystalline, amorphous and epitaxial depending on the deposition materials and the reaction conditions. As listed in Table 1, there are other coating techniques for film materials such as liquid absorption coating, thermal spraying and mechanical coating. However, their applications are narrow comparing with PVD and CVD due to their features.

Table 1: Classification of the coating techniques for film materials

Physical vapor deposition (PVD)	Vacuum deposition	Resistance heating, Flash Evaporation, Vacuum Arc, Laser heating, Highfrequency heating, Electron beam heating
	MBE (Molecular Beam Epitaxy)	
	Laser deposition	
	Sputter deposition	Ion beam sputtering, DC sputtering, Highfrequency sputtering, Magnetron sputtering, Microwave ECR plasma deposition
	Ion beam plating	Highfrequency ion plating, Activated reactive evaporation, Arc ion plating
	Ion beam deposition	
	Ionized cluster beam deposition	
Chemical vapor deposition (CVD)	Thermal CVD	Atmospheric pressure CVD, Low pressure CVD
	Plasma CVD	DC plasma CVD, Highfrequency plasma CVD, Microwave plasma CVD, ERC plasma CVD
	Photo-excited CVD	
Liquid absorption coating	Plating	Electroplating, Electroless plating
	Anodic oxide coating, Painting, Sol-gel method	
	Spin coating, Dip coating, Roll coating, Spry coating	
Thermal spraying	Flame spraying, Electrical spraying (Arc, Plasma)	
Mechanical coating	Shot coating, Powder impact plating, Aerosol deposition, Gas deposition	

Advantages and Limitations Of These Coating Techniques

Any film coating technique has its advantages and limitations. The features of these coating techniques make their application fields different. Their advantages and limitations are summarized and shown in Table 2. Thickness control and coating of large-area films can be achieved in PVD which has become the major film coating technique in the fields of electronics, electrics and optics industries due to its high production efficiency, high film purity and low production cost. However, complicated and large scale equipments are necessary. In addition, films cannot be deposit on the substrates with complex profiles. In CVD processes, films can deposit on the substrates with complex profiles and the adhesion between films and substrates is generally strong. However, the processes are performed at temperature of 600 °C and above.

Large scale equipments are also needed just as PVD processes. The advantages and limitations of other coating techniques can also be found in Table 2. We will not explain them in detail any more.

Table 2: Advantages and limitations of the film coating techniques

Film coating technique	Advantages	Limitations
PVD	Thickness control Large area coating	Large and complex equipment Coating on flat substrate
CVD	Coating on complex substrate Strong adhesion	Large and complex equipment Elevated temperature process
Liquid absorption coating	Simple equipment Coating on complex substrate	Complicated post treatment Elevated temperature process
Thermal spraying	Rapid coating Large specific area	Elevated temperature process Grain growth
Mechanical coating	Ambient preparation Nanoscale coating	Large and complex equipment Coating on flat substrate

A NOVEL COATING TECHNIQUE FOR COMPOSITE FILMS

Proposal of A Novel Coating Technique

Contamination Phenomena During Mechanical Alloying (Ma)

Powder metallurgy has been widely used in the manufacturing of mechanical parts. A typical process of powder metallurgy is shown in Fig. 1. As shown in the schematic diagram, blending of powder particles is necessary before compacting. Ball milling is often used to mix powder particles and make them homogeneous. Mechanical alloying, well known as ball milling, is frequently used to improve various material properties and to prepare advanced materials that are different or impossible to be obtained by traditional techniques (Suryanarayana, 2001). Ceramic balls are often used as grinding mediums which are indispensable in MA. However, the contamination of powder from grinding mediums and the adhesion of powder particles to the grinding mediums always harass engineers. Especially, the adhesion of powder particles to the grinding mediums and the bowl is really difficult to be eliminated.

Concept Proposal of Mechanical Coating Technique (Mct)

From the contamination discussed above, we proposed a novel film coating technique in 2005 called mechanical coating technique (MCT) with the diagram schematic shown in Fig. 2 (Lu et al., 2005). In this technique, metal powder and ceramic grinding mediums (balls, buttons and columns) are used as the coating material and the substrates respectively. Firstly, they are charged into a bowl made of alumina. The mechanical coating is performed by a pot mill or a planetary ball mill. In the process, friction, wear and impact among metal powder particles, ceramic grinding mediums and the inner wall of the bowl occur. That results in the formation of metal films on ceramic grinding mediums. In fact, some other researchers prepared metal or alloy films on grinding mediums by this technique. Kobayashi prepared metallic films on ZrO_2 balls (Kobayashi, 1995). Romankov et al. (2006) deposited Al and Ti-Al coatings on Ti alloy substrates. Gupta et al. (2009) reported the formation of nanocrystalline Fe-Si coatings on mild steel substrates. Farahbakhsh et al. (2011) prepared Cu and Ni-Cu solid solution coatings on Ni balls.

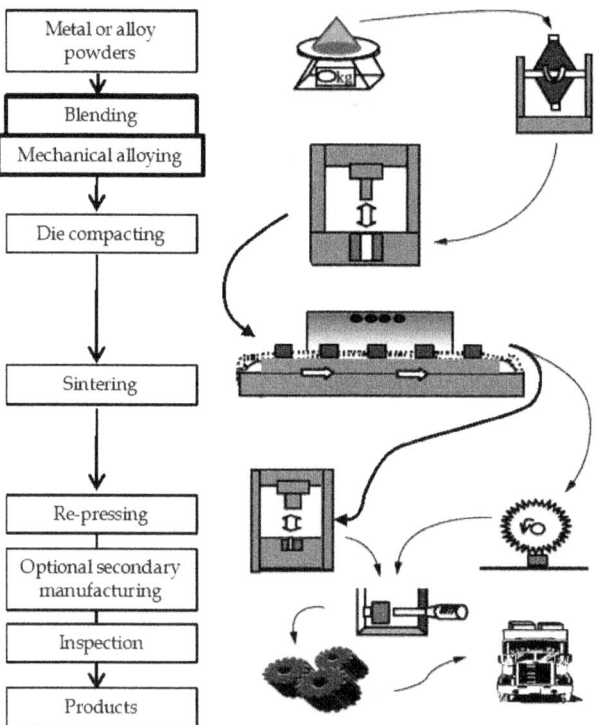

Figure 1: Simplified flowchart and schematic diagram of a typical powder metallurgy process.

Mechanical coating technique performed by ball milling has been established and known. After the proposal of MCT in 2005, we have made some important progress in advancing it. Titanium films on ceramic balls have been fabricated and their properties have been investigated (Yoshida et al., 2009 a). By MCT and its following high-temperature oxidation, TiO_2/Ti composite photocatalyst films have been successfully prepared (H. Yoshida et al., 2008). After that, we proposed 2-step MCT to fabricate TiO_2/Ti composite photocatalyst films without high-temperature oxidation (Lu et al., 2011). In addition, 2-step MCT was also used to fabricate TiO_2/Cu composite photocatalyst films (Lu et al., 2012). In this chapter, we will give a brief introduction to MCT, 2-step MCT and the relevant processes as a novel technique to fabricate composite films and TiO_2/metal composite photocatalyst films.

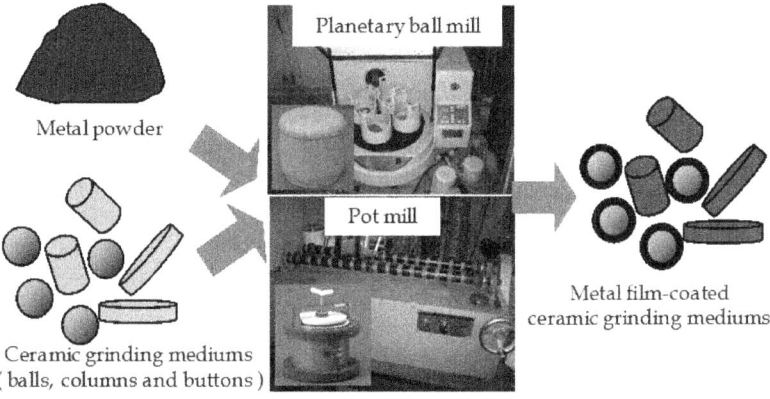

Figure 2: Schematic diagram of mechanical coating technique (MCT).

MCT and Its Influencing Parameters

In our early work, we fabricated Ti films on alumina (Al_2O_3) grinding mediums such as balls, buttons and columns. Ti powder and Al_2O_3 grinding mediums were used as the coating materials and the substrates respectively. They were charged into a bowl made of alumina with the dimension of Φ75 × 90 mm (400 *ml*). The mechanical coating was carried out by a pot mill with a rotation speed of 80 rpm for 1000 h. Fig. 3 (a) shows the appearance comparison of the Al_2O_3 grinding mediums before and after MCT. It can be clearly seen that the Al_2O_3 grinding mediums after MCT showed metallic luster which means metal films might be formed on these grinding mediums. Also, the appearances of metal-coated Al_2O_3 balls after high-temperature oxidation are given in Fig. 3 (b).

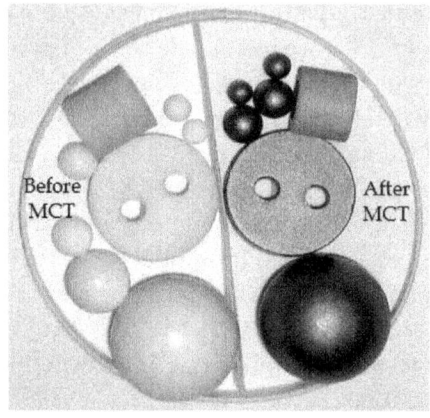
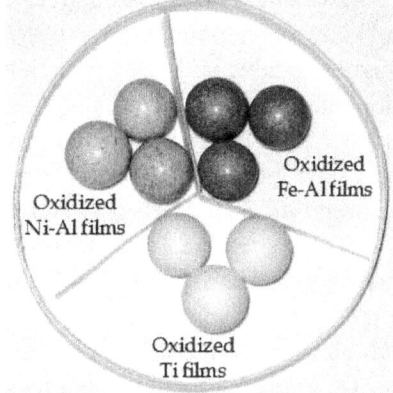

(a) Alumina objects before and after MCT

(b) Alumina balls after MCT and the following high-temperature oxidation

Figure 3: Appearances of Al_2O_3 grinding mediums after MCT and high-temperature oxidation.

The influencing parameters of MCT include:

1. Impact force or impact energy: type of mill, milling speed, milling container, milling time, grinding medium, extent of filling the bowl, ball-to-powder weight ratio
2. Physic, chemical and mechanical properties of powder and grinding mediums
3. Milling atmosphere and milling temperature

Characterization of Tl Films Fabricated By MCT

Appearance, Microstructure and Thickness of the Tl Films

Fig. 4 shows the appearances of the Al_2O_3 balls after MCT. It can be seen that the color of the Al_2O_3 balls changed from white to metallic gray as MCT time increased. It means that more Ti powder particles adhered to the surfaces of the Al_2O_3 balls. Impact force of only about 1 G can be obtained during MCT performed by pot mill. When it is carried out by planetary ball mill, impact force over 10 G or even 40 G can be realized. Therefore, the required time to form metal films can be decreased greatly in the case of planetary ball mill. The SEM images of the cross sections of the Ti-coated Al_2O_3 balls are also given in Fig. 5. With the increase of MCT time, the film thickness increased. The surface morphologies of the Ti film-coated Al_2O_3 balls are shown in Fig. 6. Discrete Ti particles adhered to the surfaces of Al_2O_3 balls and they connected with each other. The irregular surface can result in high specific area. The thickness evolution of Ti films during MCT was also monitored and is illustrated in Fig. 7. No matter in the case of pot mill or planetary ball mill, the film thickness increased with the increase of MCT time. They reached 10 and 12 μm respectively in pot mill and planetary ball mill after 1000 h and 26 h. Therefore, it is possible to control the film thickness by MCT.

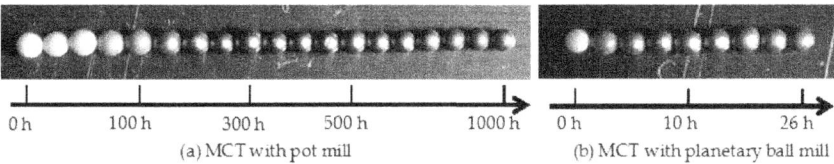

Figure 4: Appearances of the Ti-coated Al_2O_3 balls during MCT for different MCT time.

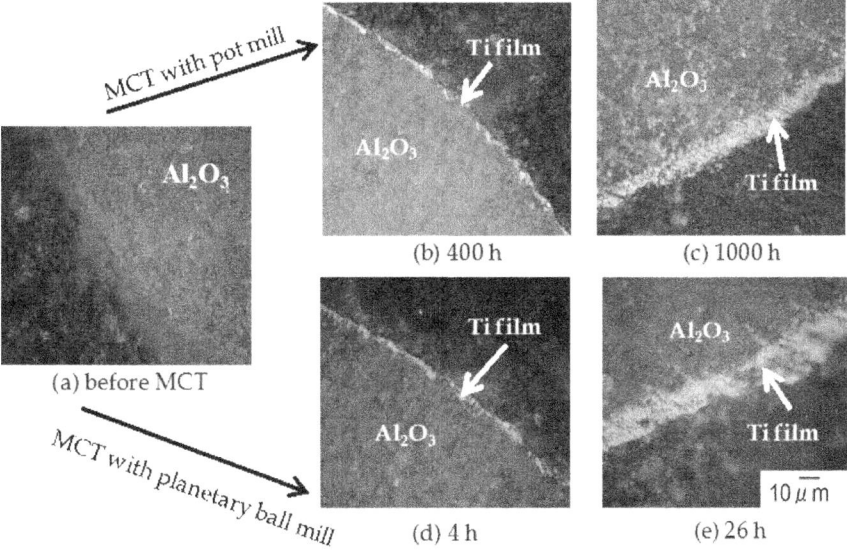

Figure 5: SEM images of the cross sections of the Ti film-coated Al_2O_3 balls during MCT.

Figure 6: Surface morphologies of Ti film-coated Al_2O_3 balls during MCT.

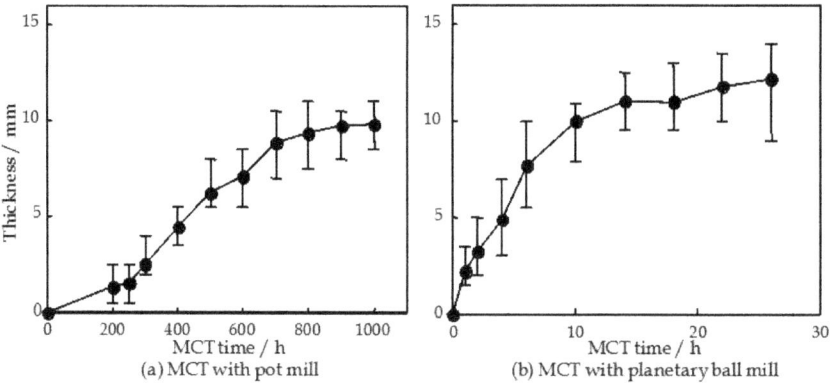

Figure 7: Thickness evolution of Ti films as a function of MCT time.

Electrical Resistance of The TI Films

Electrical resistance is one important property of film materials. Four-point probe method is frequently used to measure electrical resistance of films on a flat substrate (JIS K 7194, 1994). However, this method is only applicable to planar films. It cannot used to measure the electrical resistance of spherical films. Therefore, we proposed a new electrical resistance determination method of spherical films such as Ti films on Al_2O_3 balls. By this method, we established the relationship between electrical resistivity and film thickness.

The determination method is shown in Fig. 8. Two plate probes contacts the Ti film-coated Al_2O_3 ball (Φ 1 mm) along the direction of tangential line. To decrease contact resistance, a pressure force of 800 gf is loaded along the normal direction of the ball. The press force is determined in pre-experiments. Electrical resistance is measured for 10 times by changing the contact points between the two plate probes and the ball. In addition, the measurement on electrical resistance is carried out for three randomly chosen Ti film-coated Al_2O_3 balls. The average value of the measurements for 30 times is used as the electrical resistance of the Ti film. Fig. 9 shows the evolution of electrical resistance of the Ti film-coated Al_2O_3 balls during MCT by pot mill and planetary ball mill. For the both cases, the electrical resistance decreased with the increase of MCT time.

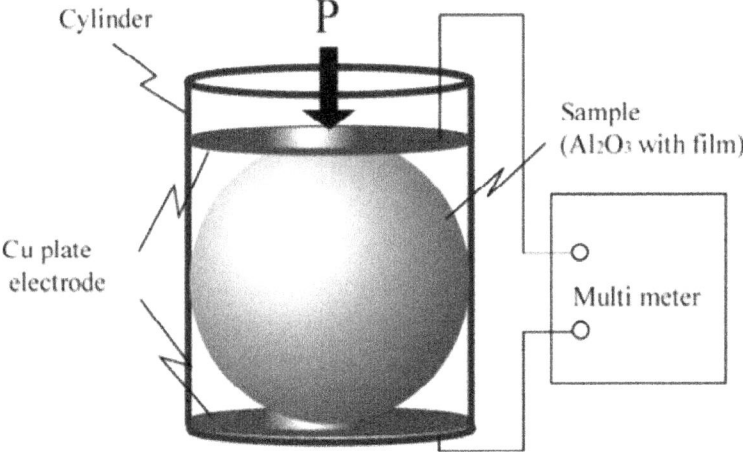

Figure 8: Measurement of electrical resistance of the Ti film-coated Al_2O_3 balls.

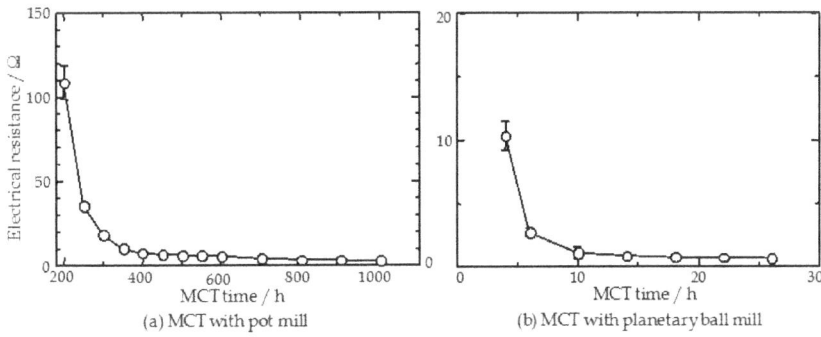

Figure 9: Electrical resistance of the Ti film-coated Al_2O_3 balls.

To establish the relationship between electrical resistance and film thickness, we proposed a spherical shell model for Ti film-coated Al_2O_3 ball as shown in Fig. 10. Here r is the radius of Al_2O_3 ball, h is the film thickness.

Therefore, the electrical resistance of the spherical Ti films on Al_2O_3 ball $z = +(r+h)$ to $z = -(r+h)$ can be given by

$$R = \int_{-(r+h)}^{+(r+h)} \rho \frac{dz}{A} = 2\int_0^{+(r+h)} (\rho \frac{dz}{A_1} + \rho \frac{dz}{A_2})$$

(1)

Where ρ is the electrical resistivity of the films, A_1 is the ring area of the films in the range of $0 \leq z \leq r$, and A_2 is the area of the circle crossed with vertical axis z in the range of $r < z \leq r+h$. Electrical resistance in the range of $0 \leq z \leq r$ and $r < z \leq r+h$ can be defined as R_1 and R_2 respectively and can be given by

$$R_1 = \int_0^r \rho \frac{dz}{A_1} = \int_0^r \frac{\rho}{\pi} \frac{dz}{(r+h)^2 - r^2}$$

(2)

$$R_2 = \int_r^{r+h} \rho \frac{dz}{A_2} = \int_r^{r+h} \frac{\rho}{\pi} \frac{dz}{(r+h)^2 - z^2}$$

(3)

During the measurement of electrical resistance shown in Fig. 8, the contact of the plate probes and Al_2O_3 ball should not be a point but a plane which has a certain area. It is proper to give the integral calculus from r to C ($r < C \leq r+h$).

$$R_2 = \int_r^C \frac{\rho}{\pi} \frac{dz}{(r+h)^2 - C^2} = \frac{\rho}{2\pi(r+h)} \ln \left| \frac{(r+h)^2 - C^2}{(r+h)^2 - r^2} \right|$$

(4)

The electrical resistance of the spherical Ti films on an Al_2O_3 ball can be given by

$$R = 2(R_1 + R_2) = 2 \left\{ \frac{\rho}{\pi} \frac{r}{(r+h)^2 - r^2} + \frac{\rho}{2\pi(r+h)} \ln \left| \frac{(r+h)^2 - C^2}{(r+h)^2 - r^2} \right| \right\}$$

(5)

Therefore, we can calculate electrical resistivity, ρ of the film by Eq. 5 using the measured electrical resistance of the films, R and the film thickness, h. Fig. 11 shows the relationship between the electrical resistivity of the Ti films and their thickness. It can be found that the electrical resistivity went down and then kept a constant. The evolution of the electrical resistivity should result from the density evolution of the films. In the case of planetary ball mill, the stable electrical resistivity was smaller than that in the case of pot mill. That should be due to the higher film density obtained in the circumstance of larger impact force in the case of planetary ball mill.

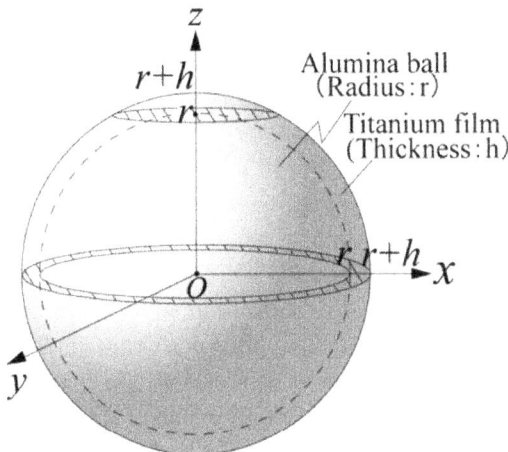

Figure 10: Spherical shell model for the electrical resistance of the Ti film-coated Al_2O_3 balls.

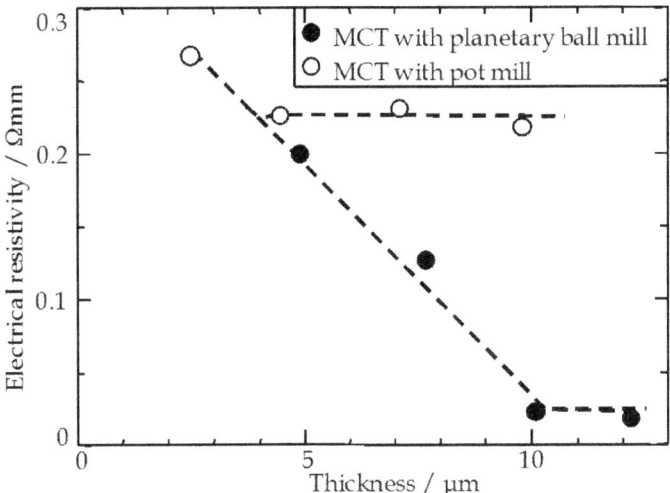

Figure 11: Relationship between electrical resistivity of Ti films and their thicknesses.

TIO$_2$/TI Composite Films Fabricated By Mct and The Following High-Temperature Oxidation

We successfully fabricated TiO$_2$/Ti composite films by MCT and the following high-temperature oxidation. Firstly, Ti films were prepared by MCT shown in Fig. 2. Subsequently, the Ti film-coated Al_2O_3 balls were oxidized at high temperatures. In this section, we will introduce the fabricate processes and the characterization of the TiO$_2$/Ti composite films.

Fabrication Processes

Ti powder with a purity of 99.9% and an average diameter of 30 μm was used as the coating material. Al_2O_3 balls with an average diameter of 1 mm were used as the substrates. A planetary ball mill (P5/4, Fritsch) was used to perform MCT (Yoshida, 2009 b). 40 g Ti powder and 60 g Al_2O_3 balls were charged into a bowl made of alumina with a dimension of $\Phi 75 \times 70$ mm (250 ml). MCT was carried out with a rotation speed of 300 rpm for 10 h. The obtained Ti film-coated Al_2O_3 balls were denoted as M10-Ti. To form TiO_2 films, the M10-Ti samples were oxidized in air at 573, 623, 673, 723, 773, 873 and 973 K for 20 h. Here the samples were denoted as M10-T-20. T means the oxidation temperature. The samples after MCT and the following high-temperature oxidation were examined by SEM (JEOL, JSM-6100) and XRD (JEOL, JDX-3530). Cu-K radiation in the condition of 30 kV and 30 mA was used for XRD. Before the characterization, all the samples were cleaned in acetone by ultrasonic (frequency: 28 kHz) to remove Ti and TiO_2 that did not adhere strongly.

Characterization of The Tio_2/TI Composite Films

Fig. 12 shows the appearances of the M10-Ti and M10-T-20 samples. The color of the M10-T-20 samples changed with increase of oxidation temperature and lost metallic luster comparing with M10-Ti. The color change indicates that the degree of oxidation was different at different oxidation temperature. The M10-573-20 samples showed brown color which was similar to that of TiO. That means the oxidation of Ti films at 573 K was insufficient. When oxidation temperature was increased to 673, 723, 773 and 873 K, the samples color changed from blue to gray. It was probably related to the growth of TiO_2 crystalline and the film thickness increase. Besides, the color of M10-973-20 samples was light yellow. It indicates that titanium was completely oxidized to titanium dioxide.

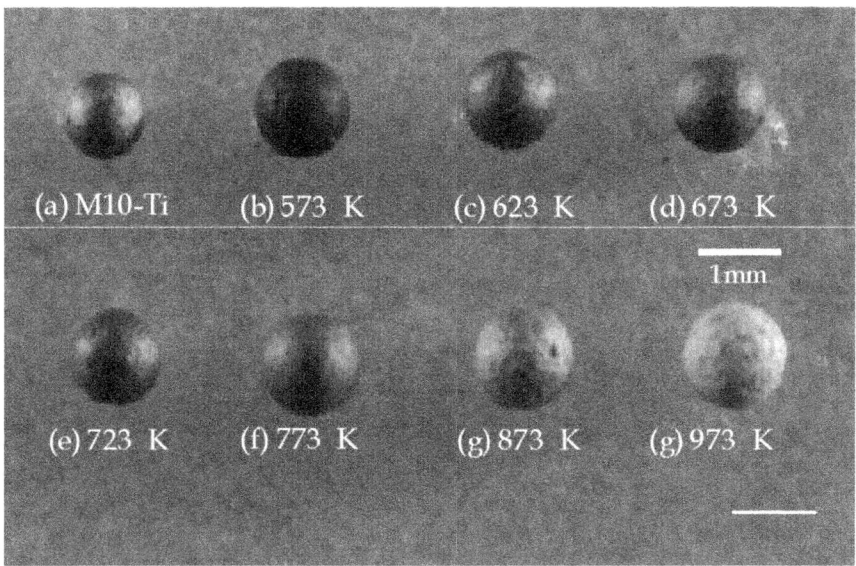

Figure 12: Appearances of M10-Ti and M10-T-20 samples.

The surface SEM images of the samples are shown in Fig. 13. From Fig. 13(a), the Ti films had uneven surfaces comparing with those prepared by PVD or CVD. The surface evolution with the increase of oxidation temperature can be seen from Fig. 13(b) to (f). It seems that the surface crystals grew up with the increase of oxidation temperature. However, column nanocrystals were formed at 973 K. Fig. 14 shows the XRD patterns of the samples after MCT and the following high-temperature oxidation. The diffraction intensity of Ti peaks decreased with the increase of oxidation temperature and the Ti peak at about 41° (2θ) disappeared when oxidation temperature was increased to 973 K. Conversely, the peaks of rutile TiO_2 appeared when oxidation temperature was above 673 K and the diffraction intensity became stronger as oxidation temperature increased. From the above results, it can be concluded that the films had a composite microstructure of Ti and rutile TiO_2 when oxidation temperature was between 673 and 873 K. TiO_2/Ti composite films on Al_2O_3 balls were fabricated by MCT and the following high-temperature oxidation.

Figure 13: Surface SEM images of M10-Ti and M10-*T*-20 samples.

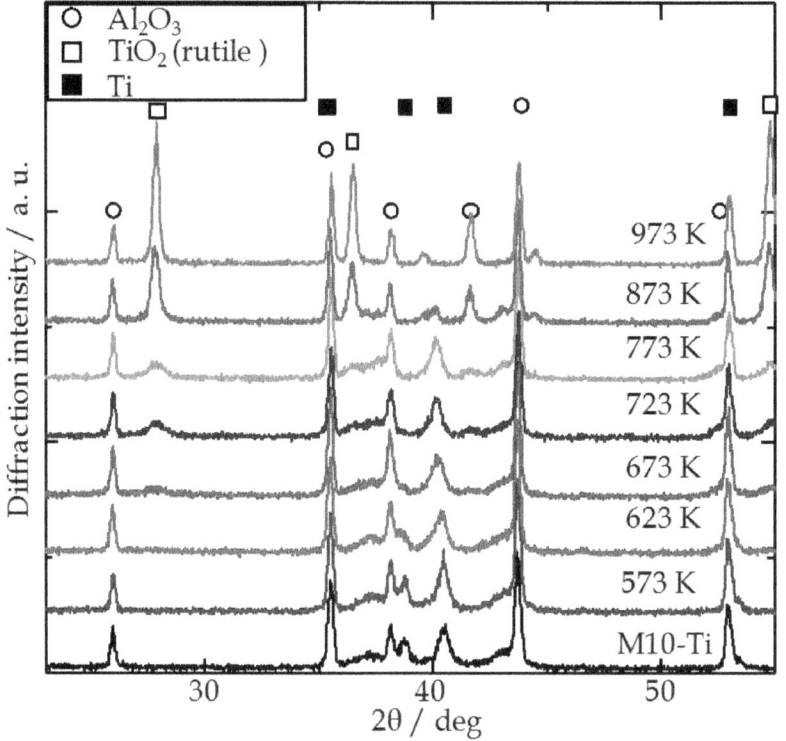

Figure 14: XRD patterns of M10-Ti and M10-T-20 samples.

Step Mechanical Coating Technique (2-Step Mct)

An advanced mechanical coating technique called 2-step mechanical coating technique (2-step MCT) was developed to fabricate TiO_2/Ti composite films. As anatase TiO_2 cannot be easily obtained by oxidation, we aim to deposit anatase TiO_2 on Ti films directly by MCT. In this section, we will introduce the processes of 2-step MCT and characterize the composite films. The influences of 2^{nd}step MCT time and the introduction of ceramic impact balls on the composite films and their photocatalytic activity were also discussed.

Processes of 2-Step Mct

The schematic diagram of 2-step MCT is shown in Fig. 15. In the first step, Ti films are prepared on the surfaces of Al_2O_3 balls as our previous work (Lu et al., 2005 & Yoshida et al., 2009 a). The source materials and their relevant parameters are listed in Table 3. 40 g Ti powder and 60 g Al_2O_3 balls are used as the coating material and the substrates respectively. A planetary ball mill (P5/4, Fritsch) is used to perform MCT. The experimental condition has been discussed in Section 2.4.1. In the second step, TiO_2/Ti composite films are fabricated. 15 g Ti film-coated Al_2O_3 balls and 13 g TiO_2 powder are used as the substrates and the coating material respectively.

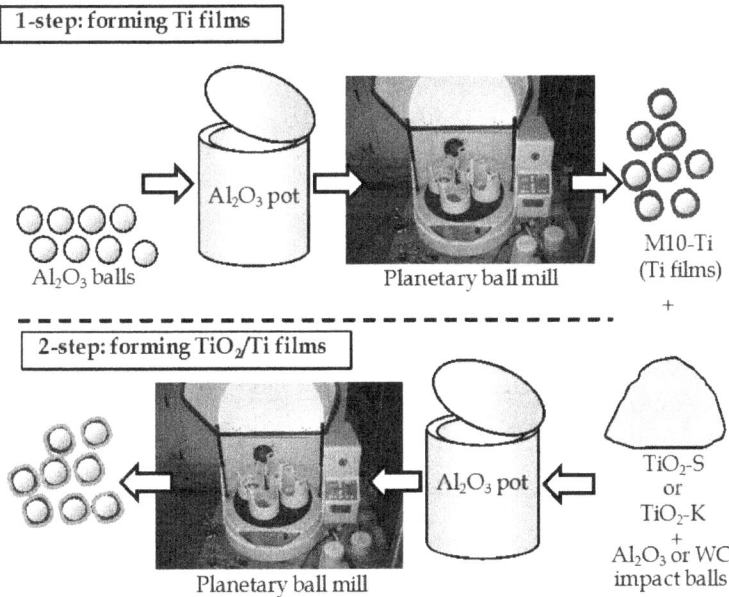

Figure 15: Schematic diagram of 2-step MCT for the fabrication of TiO_2/Ti composite films.

The relevant parameters can be found in Table 3. They are charged into a bowl made of alumina. Then the coating of TiO_2 is performed by the same planetary ball mill with a rotation speed of 300 rpm for 1, 3, 6 and 10 h. To investigate the influence of average diameter of TiO_2 powder on the photocatalytic activity of the composite films, two kinds of anatase TiO_2 powder with different average diameter are used as the coating materials. To understand the influence of impact force on the formation and the photocatalytic activity of TiO_2/Ti composite films, Al_2O_3 or WC impact balls with the diameter of 10 mm are also introduced into the second step of 2-step MCT. The relevant denotations are listed in Table 4.

Table 3: Source materials for the fabrication of TiO_2/Ti composite films by 2-step MCT

Ti powder	Purity: 99.1% Average diameter: 30 μm
Substrate	φ1 Al_2O_3 balls Purity: 93.0%
TiO_2 powder (anatase)	Average diameter: 7 nm (TiO_2-S)
	Average diameter: 0.45 μm (TiO_2-K)

Table 4: Sample denotations for the fabrication of TiO_2/Ti composite films by 2-step MCT

TiO_2 powder	Impact ball	Sample denotation
TiO_2-S		CMxS
	Al_2O_3	φ10A-CMxS
	WC	φ10W-CMxS
TiO_2-K		CMxK
	Al_2O_3	φ10A-CMxK
	WC	φ10W-CMxK

Note: x is the 2nd step MCT time.

TiO_2/Ti Composite Films Fabricated By 2-Step MCT Without Impact Balls

Fig. 16 shows the appearances of the Al_2O_3 balls after 2-step MCT without ceramic impact balls. The color of the samples changed and lost metallic luster with the increase of the 2nd step MCT time. The CM3S samples showed different colors from place to place on the surface. However, the CM6S and CM10S samples showed uniform color respectively. It hints that uniform composite films might form at that time. Fig. 17 shows the surface SEM images of the samples fabricated by 2-step MCT without ceramic impact balls. From Fig. 17(a), the gray areas correspond to Ti. It can be seen that uniform Ti films have been formed on the surface of Al_2O_3 ball. From Fig. 17(b) and (c), the white and gray areas correspond to Ti and TiO_2 respectively. Continuous TiO_2 films were not form while TiO_2 deposited on Ti films in the form of discrete island. Meanwhile, the SEM images of the cross sections of the samples fabricated by 2-step MCT without ceramic impact balls are shown in Fig. 18. It can be clearly seen that Al_2O_3 balls were coated with Ti films and discrete islands of TiO_2 adhered to the Ti films. A composite microstructure of Ti and TiO_2 was formed. During the impact between Al_2O_3 balls or Al_2O_3 ball and the inner wall of the bowl, TiO_2 powder particles were trapped between them. Under the great impact force, TiO_2 particles were inlaid into the Ti films. It results in the formation of the TiO_2/Ti composite microstructure. Fig. 19 shows the XRD patterns of the samples after 2-step MCT without impact balls. When the 2nd step MCT time was 1 h, the peaks of anatase TiO_2 appeared which means that TiO_2 particles had adhered to Ti films. As it came to 3 h, the intensity of anatase TiO_2 peaks reached their highest values. It indicates that the loading amounts of TiO_2 in the TiO_2/Ti composite films reached the maximum values. After that, the TiO_2 peaks became lower which should be due to the exfoliation of TiO_2 that coated Ti films. From the above results, the films had a composite microstructure of Ti and TiO_2. The loading amounts of TiO_2 in the composite films changed with the increase of the 2nd step MCT time. 2-step MCT is a simple and applicable technique to fabricate TiO_2/Ti composite films.

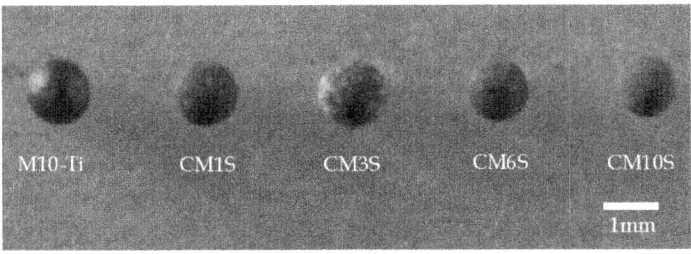

Figure 16: Appearances of the samples after 2-step MCT.

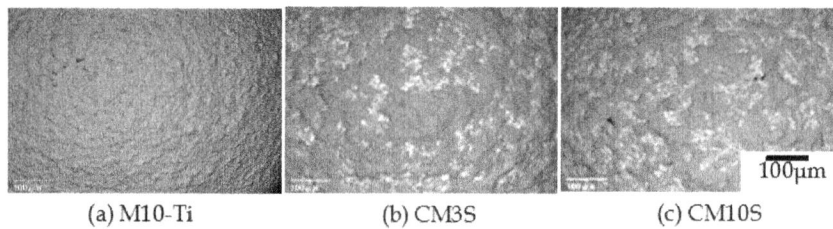

Figure 17: SEM images of the surfaces of the samples fabricated by 2-step MCT.

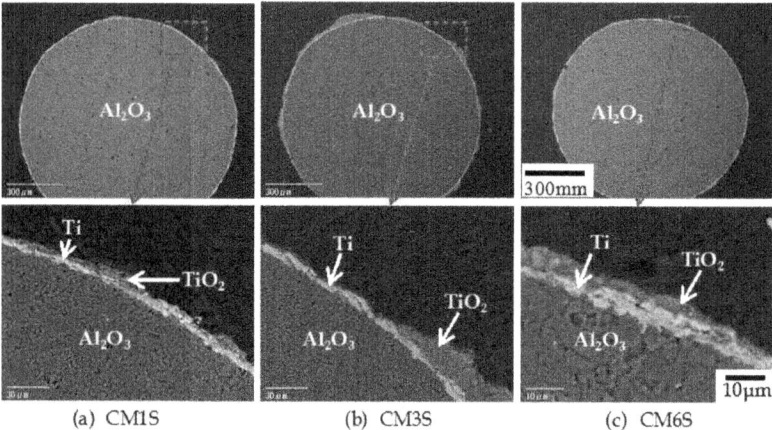

Figure 18: SEM images of the cross sections of the samples fabricated by 2-step MCT.

Tio$_2$/Ti Composite Films Fabricated By 2-Step Mct With Impact Balls

Fig. 20 shows the appearances of the Al$_2$O$_3$ balls after 2-step MCT with ceramic impact balls. The samples lost metallic luster and their colors changed. That hints TiO$_2$/Ti composite films might be formed. Fig. 21 shows the SEM image of the cross section of the TiO$_2$/Ti composite films fabricated by 2-step MCT with WC impact balls. It can be seen that the films had a composite microstructure of Ti film and discrete islands of TiO$_2$. The surface condition of the composite films fabricated with TiO$_2$ powder with different average diameters are compared in Fig. 22. The distribution of TiO$_2$ powder particles with the average diameter of 0.45 μm were more uneven under the impact of WC balls compared with those with the average diameter of 7 nm. Φ10W-CM6K sample had a high hardness of 472 (dynamic hardness) on the cross section. It was close to that of alumina. It hints that the composite films were very hard.

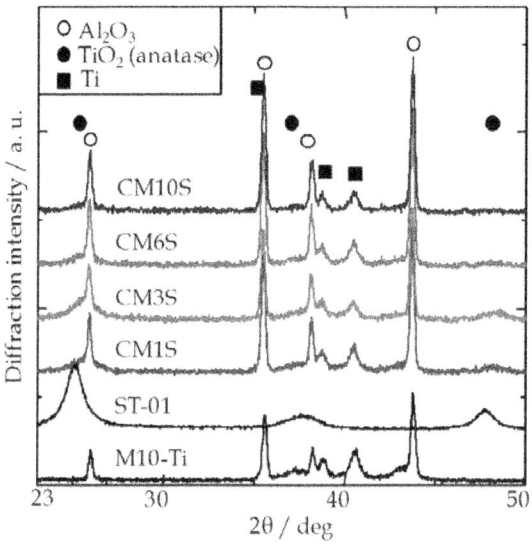

Figure 19: XRD patterns of the samples fabricated by 2-step MCT.

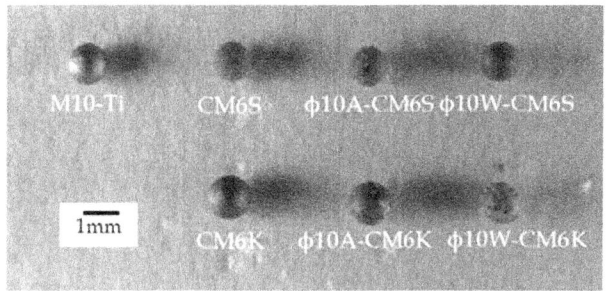

Figure 20: Appearances of the samples fabricated by 2-step MCT.

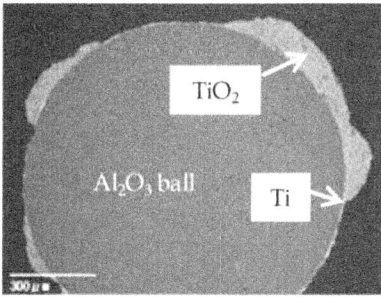

Figure 21: SEM image of the cross section of the Φ10W-CM6K sample fabricated by 2-step MCT.

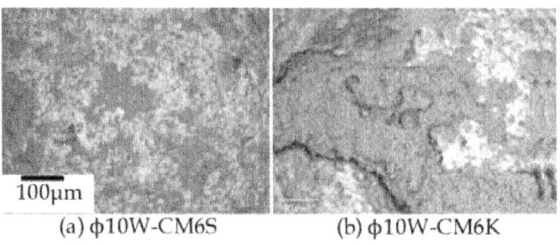

(a) φ10W-CM6S   (b) φ10W-CM6K

Figure 22: Surface SEM images of the samples fabricated by 2-step MCT.

The XRD patterns of the samples fabricated with TiO_2 powder of different average diameters by 2-step MCT with WC impact balls are given in Fig. 23. In the case of nano-sized TiO_2 powder (Fig.23 (a)), the peaks of Ti and TiO_2 can be found. On the other hand, the diffraction peaks of TiO_2 cannot be detected for the micron-sized TiO_2 powder (Fig.23 (b)). It means the loading amounts of TiO_2 in the composite films were rather small.

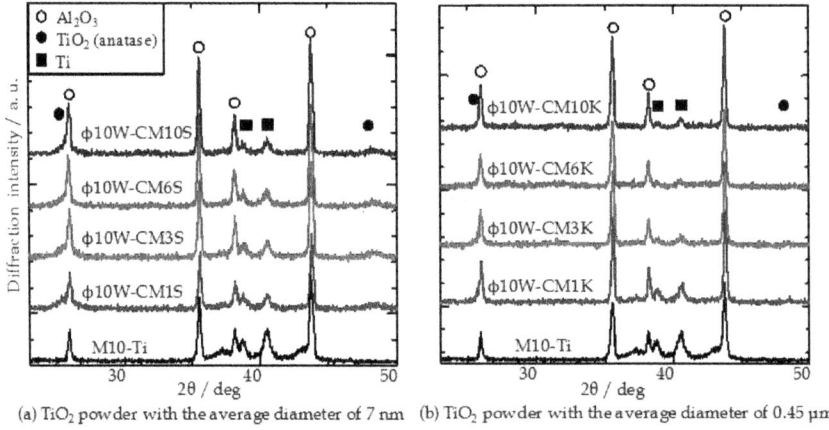

(a) TiO_2 powder with the average diameter of 7 nm   (b) TiO_2 powder with the average diameter of 0.45 μm

Figure 23: XRD patterns of the samples fabricated by 2-step MCT with WC impact balls.

Photocatalytic Activity of TiO_2/Metal Composite Films

Although the photocatalytic activity of TiO_2 under ultraviolet and visible light irradiation has been investigated around the world, only the photocatalytic activity of TiO_2/metal composite films under ultraviolet irradiation is involved and discussed in this section. Here, we developed evaluation method of photocatalytic activity of TiO_2/metal composite films by which we evaluated the photocatalytic activity of TiO_2/Ti and TiO_2/Cu composite films fabricated by MCT.

Evaluation Method of Photocatalytic Activity

We developed evaluation method of photocatalytic activity of the TiO_2/metal composite films by referring to Japan Industrial Standard (JIS R 1703-2, 2007). The evaluation procedure is as follows. Before the evaluation of photocatalytic activity, pre-adsorption of methylene blue (MB) is carried out to obtain the same initial evaluation condition for all the samples. Firstly, the cleaned samples are dispersed uniformly to form a layer of samples on the bottom of a cylinder-shaped cell with a dimension of $\Phi 18 \times 50$ mm. Subsequently, 3 *ml* MB solution with a concentration of 20 *µmol/l* is poured into the cell. The cell with the samples and MB solution is kept in a totally dark place for 12 h. Then, the evaluation of the photocatalytic activity will be carried out. The samples after pre-adsorption are laid uniformly on the bottom of a same cell to form a layer of samples and 7 *ml* MB solution with a concentration of 10 *µmol/l* is poured into the cell. The schematic diagram of the evaluation of photocatalytic activity is shown in Fig. 24. A colorimeter (Sanshin Industrial Co., Ltd) of 660 nm in UV radiation wavelength, which is near the peak of absorption spectrum of MB solution (664 nm), is used to measure the absorbance of MB solution. The UV irradiation time is 24 h. Besides, both of the pre-adsorption and the evaluation of photocatalytic activity are carried out at room temperature. The gradient, k of MB solution concentration-irradiation time curve is calculated by the least-squares method with the data from 1 to 12 h and k is used as the degradation rate constants. The higher the degradation rate constants k, the higher the photocatalytic activity.

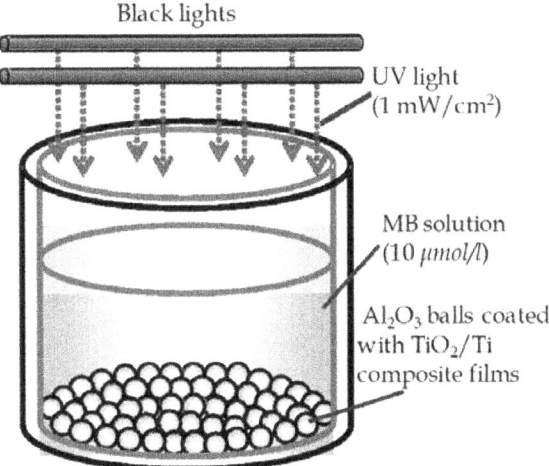

Figure 24: Photocatalytic activity evaluation of TiO_2/metal composite films fabricated by MCT.

Photocatalytic Activity Of TiO_2/Ti Composite Films

Fig. 25 shows the evolution of MB solution concentration as UV irradiation time under the action of TiO_2/Ti composite films fabricated by MCT and its following high-temperature oxidation as described in Section 2.4. MB solution concentration slight increased in the case of M10-Ti and M10-573-20. Meanwhile, MB solution concentration decreased in varying degrees in the case of the other samples. That means MB was degraded under the action of UV light and TiO_2 samples. In other word, the samples fabricated by MCT and its following high-temperature oxidation showed photocatalytic activity.

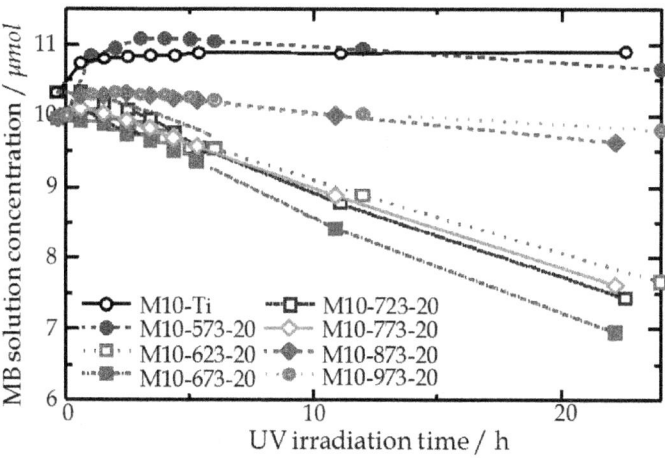

Figure 25: MB solution concentration as a function of UV irradiation time under the action of the samples fabricated by MCT and the following high-temperature oxidation.

The degradation rate constants, k are also given in Fig. 26. It can be seen that the degradation rate constants, k increased with the increase of oxidation temperature and reached the peak value at 723 K above which the degradation rate constants, k decreased. Combined with the XRD patterns in Fig. 14, the evolution of the photocatalytic activity can be discussed as follows. When oxidation temperature was below 623 K, the peaks of TiO_2 were not detected which means Ti films were not oxidized or were oxidized sufficiently. Therefore, the photocatalytic activity of the films was low as shown in Fig. 26. With the increase of oxidation temperature from 623 to 773 K, more Ti films were oxidized and TiO_2/Ti composite films were formed. The improvement of photocatalytic activity should relate to the composite microstructure of TiO_2 and Ti. According to charge separation effect (Rengaraj et al., 2007), electrons in TiO_2 may transfer to metals with higher work functions. The electron transfer can decrease the recombination velocity of electron-hole pairs in TiO_2. It can improve the photocatalytic activity of TiO_2. When oxidation temperature was 723 K, the composite films obtained the optimum ratio of TiO_2 to Ti. It resulted in the highest photocatalytic activity. When oxidation temperature was above 773 K, the oxidation degree of Ti films was further increased and Ti films were completely oxidized when oxidation temperature was above a certain value. Although the amounts of TiO_2 increased, the photocatalytic activity decreased due to the weakening of charge separation effect.

The degradation rate constants, k as a function of 2nd step MCT time are shown in Fig. 27. In the case of nano-sized TiO_2 powder (Fig. 27(a)), the samples fabricated without ceramic impact balls showed the highest photocatalytic activity and the degradation rate constants, k exceeded 350 nmol l^{-1} h^{-1}. After the introduction of ceramic impact balls into 2-step MCT, the photocatalytic activity was decreased. On the other hand, the samples fabricated with Al_2O_3 impact balls showed the greatest photocatalytic activity in the case of micron-sized TiO_2 powder (Fig. 27(b)). Compared with the TiO_2/Ti composite films fabricated with micron-sized TiO_2 powder, the composite films fabricated with nano-sized TiO_2 powder showed much higher photocatalytic activity. It should relate to the higher specific area of nano-sized TiO_2 powder particles. From Fig. 26 and 27, anatase TiO_2/Ti composite films showed much higher photocatalytic activity than that of rutile TiO_2/Ti composite films. It is well known that anatase TiO_2 generally shows higher photocatalytic activity than rutile TiO_2.

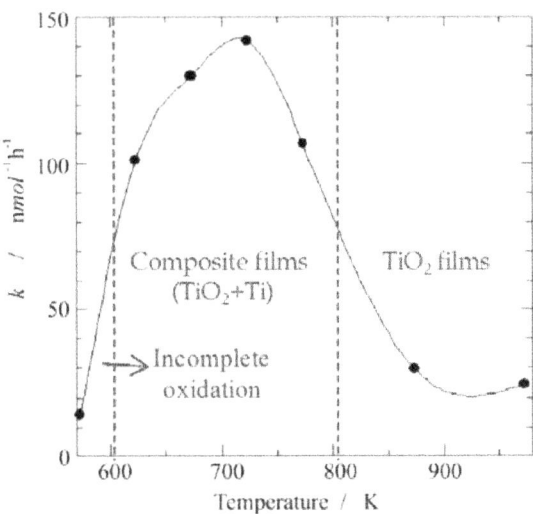

Figure 26: Degradation rate constants, k as a function of oxidation temperature.

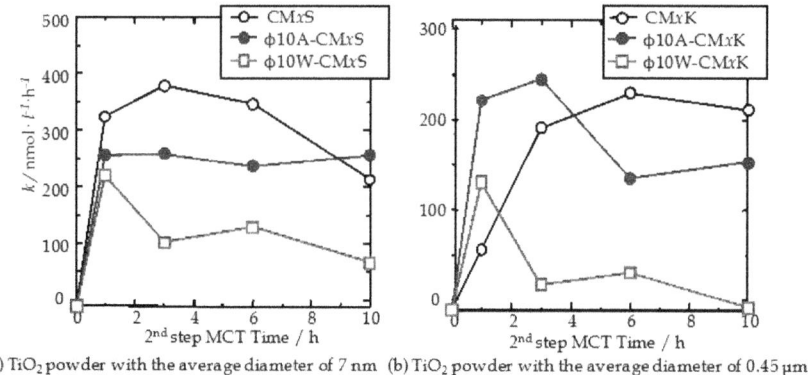

(a) TiO₂ powder with the average diameter of 7 nm   (b) TiO₂ powder with the average diameter of 0.45 μm

Figure 27: Degradation rate constants, k as a function of 2nd step MCT time

Improvement of Photocatalytic Activity By High-Temperature Oxidation

High-temperature oxidation was carried out to increase the crystallinity and the volumes of TiO_2 in TiO_2/Ti composite films fabricated by 2-step MCT and therefore improve the photocatalytic activity of the composite films (Lu et al., 2011 b). The composite films after the high-temperature oxidation were characterized and their photocatalytic activity was also evaluated. In addition, the effect on high-temperature oxidation on the microstructure and the photocatalytic activity of the composite films was also discussed.

Improved Fabrication Processes

Firstly, TiO_2/Ti composite films were prepared by 2-step MCT as described in Section 2.5.1. Ti powder with a purity of 99.1% and an average diameter of 30 μm was used as the coating material. Al_2O_3 balls with an average diameter of 1 mm were used as the substrates. After the formation of Ti films on Al_2O_3 balls, anatase TiO_2 powder with an average diameter of 0.45 m (Kishida Chemical Co. Ltd., Japan) was used to form TiO_2/Ti composite films. To make the composite films strong enough, Al_2O_3 or WC balls with the diameter of 10 mm were introduced into the fabrication of the composite films. The schematic diagram can be seen in Fig. 15. Subsequently, high-temperature oxidation was carried out for the TiO_2/Ti composite films fabricated by 2-step MCT. The oxidation temperature was set at 673, 773 and 873 K and the oxidation time was 10 h. The denotations of the samples fabricated by 2-step MCT and the following high-temperature oxidation are listed in Table 5.

Table 5: Denotations of the samples fabricated by 2-step MCT and the following high-temperature oxidation

TiO$_2$ powder	Impact ball	Sample maker
	---	CMxK-y
TiO$_2$-K	Al$_2$O$_3$	φ10A-CMxK-y
	WC	φ10W-CMxK-y

Note: x means the 2nd step MCT time, y is oxidation temperature.

Characterization of The Tio$_2$/Ti Composite Films

Fig. 28 shows the appearances of the samples fabricated by 2-step MCT and the following high-temperature oxidation. The samples lost metallic luster and became white compared with the Ti film-coated Al$_2$O$_3$ balls. Also, dark and uneven areas can also be seen. The surface SEM images of the samples are shown in Fig. 29. The surface color of the samples seems to be uniform except for some point areas. It indicates that the uniform TiO$_2$ films were formed. However, for the TiO$_2$/Ti composite films fabricated by 2-step MCT there are light and dark areas corresponding to Ti and TiO$_2$ respectively shown in Fig. 17.

The SEM images of the cross sections of the samples are given in Fig. 30. It can be seen that a composite microstructure of TiO$_2$ films and Ti films formed. In other words, TiO$_2$/Ti composite films were fabricated. The TiO$_2$ films consisted of the deposited TiO$_2$ in the second step of 2-step MCT and the TiO$_2$ formed in high-temperature oxidation. From the above results, it can be concluded that the loading amounts of TiO$_2$ in the composite films were increased by high-temperature oxidation.

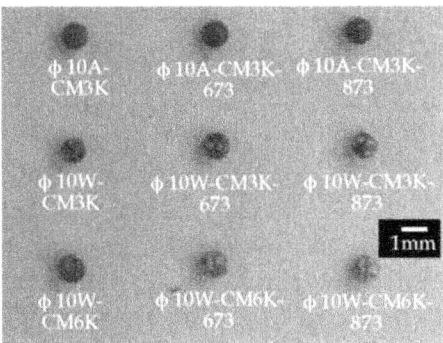

Figure 28: Appearances of the samples fabricated by 2-step MCT and the following high-temperature oxidation.

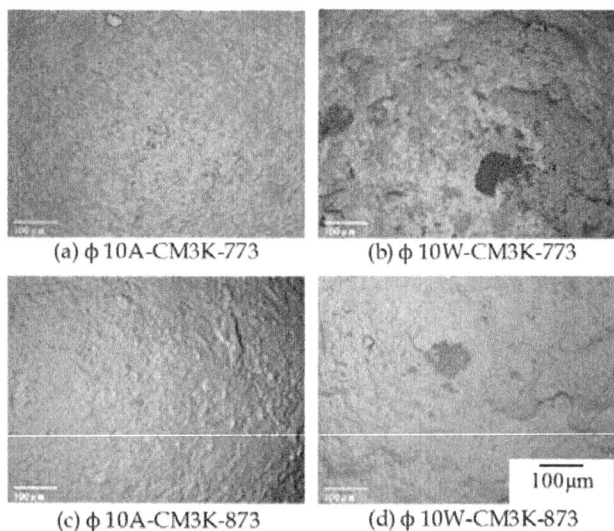

Figure 29: Surface SEM images of the samples fabricated by 2-step MCT and the following high-temperature oxidation.

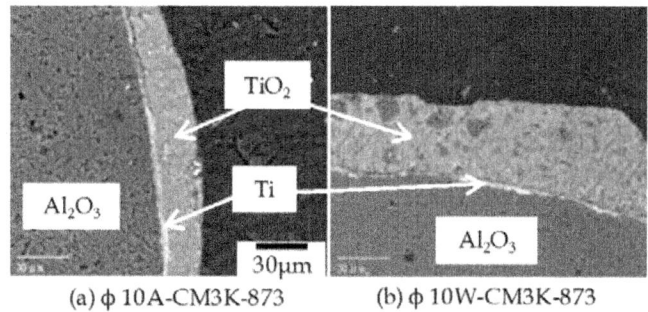

Figure 30: SEM images of the cross sections of the samples fabricated by 2-step MCT and the following high-temperature oxidation.

Fig. 31 shows the XRD patterns of the samples fabricated by 2-step MCT and the following high-temperature oxidation at 673 and 873 K. For all the samples, Ti peaks and anatase TiO_2 peaks were detected while the later was rather weak due to its small loading amounts during the second step in 2-step MCT. When oxidation temperature was 873 K, the peaks of rutile TiO_2 were detected which means rutile TiO_2 was formed in the high-temperature oxidation. For the sample Φ 10W-CM6K-673 K, a weak peak of rutile TiO_2 at about 27.5° (2θ) was be found. It indicates that rutile TiO_2 formed when oxidation temperature was 673 K although the amount was rather small.

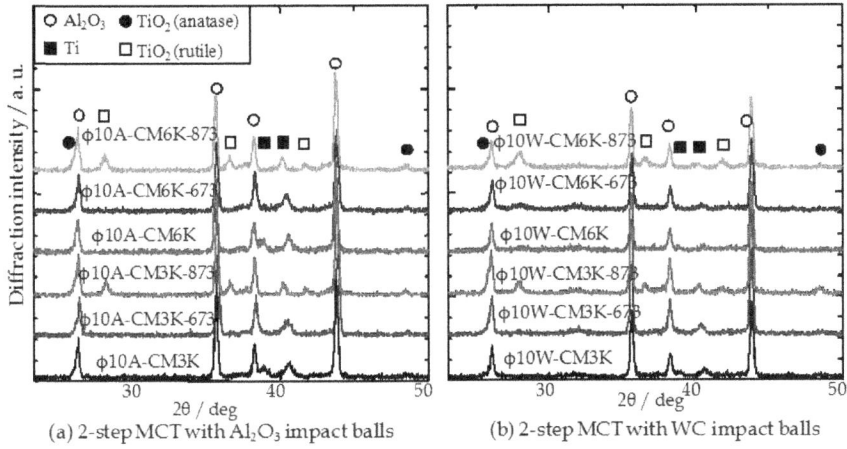

Figure 31: XRD patterns of the samples fabricated by 2-step MCT and the following high-temperature oxidation

Photocatalytic Activity of The TiO_2/Ti Composite Films

The degradation rate constants, k as a function of oxidation temperature are shown in Fig. 32. For the samples fabricated by 2-step MCT with Al_2O_3 impact balls (Fig. 32(a)), they showed their highest photocatalytic activity when oxidation temperature was 673 K. When the 2nd MCT was 3 h, the composite films showed the greatest photocatalytic activity. On the other hand, for those fabricated by 2-step MCT with WC impact balls (Fig. 32(b)), the samples showed the similar evolution of photocatalytic activity except the Φ 10W-CM3K samples. However, their photocatalytic activity was much lower than those fabricated by 2-step MCT with Al_2O_3 impact balls. It may relate to the smaller loading amounts of anatase TiO_2 as WC impact balls exerted greater impact force and resulted in the exfoliation of the adhered TiO_2.

The improvement of photocatalytic activity should result from the microstructure and phase evolution of the composite films. As discussed in Fig. 31, Ti films were oxidized partly and rutile TiO_2 was formed when oxidation temperature was 673 K and anatase TiO_2 was reserved at that temperature. Therefore, a composite microstructure of Ti, anatase TiO_2 and rutile TiO_2 was formed. Charge separation effect and mixed crystal effect might work which can improve photocatalytic activity (Cao et al., 2009). With the increase of oxidation temperature to 873 K, the volume ratio of rutile TiO_2 in the composite films increased. It might lead that the charge separation effect and mixed crystal effect were restrained and therefore decreased the photocatalytic activity of the composite films.

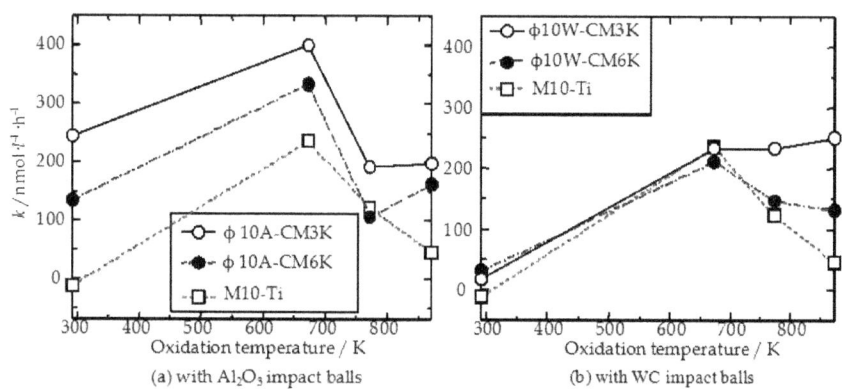

Figure 32: Degradation rate constants, k as a function of oxidation temperature

TIO$_2$/CU Composite Photocatalyst Films

In this section, we will describe and discuss the fabrication of TiO$_2$/Cu composite films by 2-step MCT. The composite films were characterized by XRD and SEM. The formation of Cu films and TiO$_2$/Cu composite films was also examined. The photocatalytic activity of the composite films was evaluated by measuring the degradation rate constants, k of MB under the UV irradiation.

Fabrication of Tio$_2$/Cu Composite Films

TiO$_2$/Cu composite films were fabricated by 2-step MCT as shown in Fig. 15. Firstly, Cu films were fabricated. The source materials and the experimental condition are listed in Table 6. To improve the production efficiency, Cu powders with different average diameters were used as the coating materials. Meanwhile, the loading amounts of Cu powder and the rotation speed was also changed as shown in Table 6. Secondly, TiO$_2$/Cu composite films were fabricated. 15 g Cu film-coated Al$_2$O$_3$ balls and 13 g anatase TiO$_2$ powder with the average diameter of 7 nm (ST-01, Ishihara Sangyo, Japan) were used as the substrates and the coating material respectively. To make the composite films stronger, 20 Al$_2$O$_3$ impact balls with the diameter of 10 mm were simultaneously put into the bowl. The rotation speed was set at 400 rpm and the 2nd MCT time was 1, 3, 6 and 10 h.

Table 6: Source materials and experimental conditions for fabrication of Cu films by MCT.

Experiment number	Cu powder mass [g]	Cu purity [%]	Average size of Cu [μm]	Al$_2$O$_3$ ball mass [g]	Al$_2$O$_3$ ball purity [%]	Rotation speed [rpm]
1	80	99.8	40	60	93.0	400
2	40	99.8	10	60	93.0	480

Characterization Of TiO$_2$/Cu Composite Films

Fig. 33 shows the appearances of the Cu-coated Al$_2$O$_3$ balls after MCT at 400 rpm with the relevant parameters in the experiment number 1 shown in Table 6. It can be seen that the color of the samples changed from white to red brown with the increase of 1st step MCT time. It means that more Cu powder adhered to the surfaces of Al$_2$O$_3$ balls. SEM results revealed that Cu films in this experimental condition were formed when it came to 54 h. Fig. 34 shows the appearances of the samples after 2-step MCT. The color of the samples also changed with the increase of 2nd step MCT time. It indicates that the loading amounts of TiO$_2$ in the composite films changed. The SEM images of the cross sections of the samples are given in Fig. 35. TiO$_2$ adhered to Cu films (Fig. 35 (a)) and some Cu particles inlaid into TiO$_2$ films (Fig. 35 (b)). Therefore, it can be said that a composite microstructure of TiO$_2$ and Cu formed.

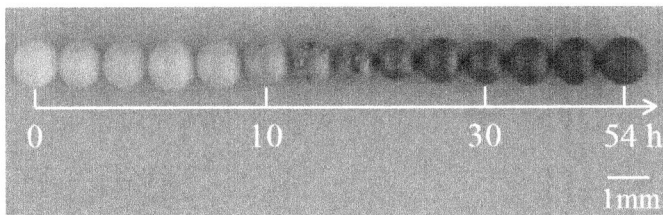

Figure 33: Appearances of the Cu-coated Al$_2$O$_3$ balls after MCT.

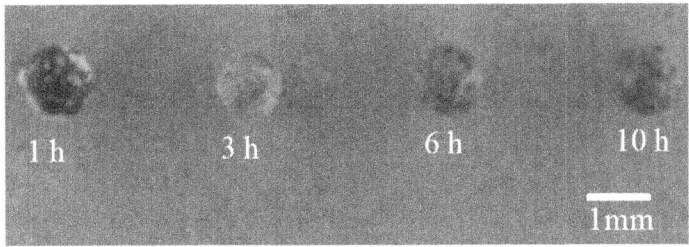

Figure 34: Appearances of the samples after 2-step MCT

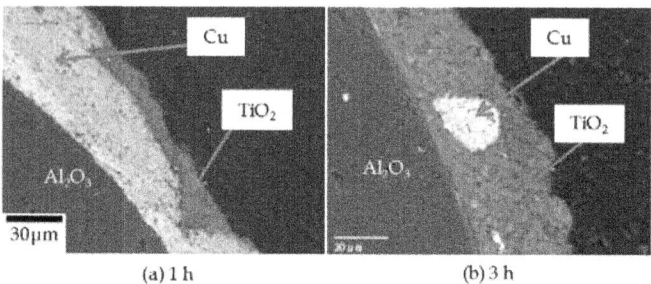

Figure 35: SEM images of the cross sections of TiO$_2$/Cu composite films fabricated by 2-step MCT with different the 2nd step MCT times

Photocatalytic Activity Of Tio$_2$/Cu Composite Films

MB solution concentration evolution in the evaluation of the photocatalytic activity of TiO$_2$/Cu composite films is shown in Fig. 36. The concentration of MB solution with the Cu film-coated Al$_2$O$_3$ balls was found to increase slightly with increase of UV irradiation time. It means Cu films did not have photocatalytic activity. On the other hand, under the action of TiO$_2$/Cu composite films and UV irradiation, the concentration of MB solution decreased in varying degrees. It suggests the composite films showed photocatalytic activity.

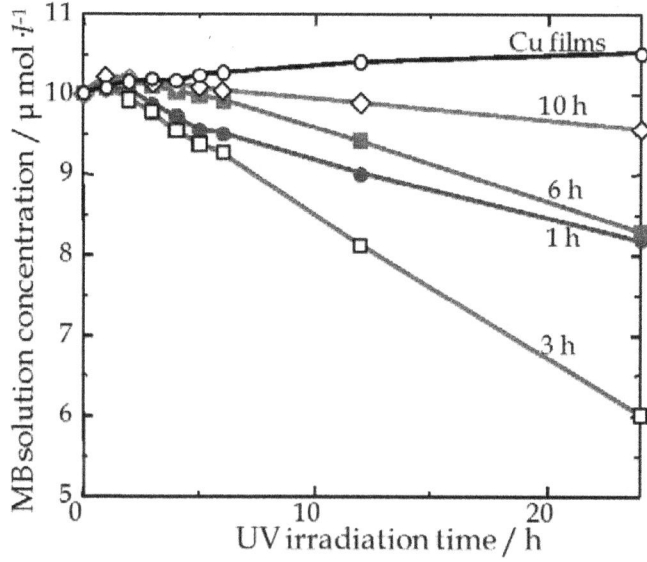

Figure 36: Evolution of MB solution concentration as a function of UV irradiation time in the evaluation of photocatalytic activity of TiO$_2$/Cu composite films.

For TiO$_2$/Cu composite films with 3 h of 2nd step MCT time, the MB solution concentration decreased to the minimum value after the same UV irradiation time for all the composite films. The degradation rate constants, k is illustrated in Fig. 37. The degradation rate constants, k increased with the increase of 2nd step MCT time and reached the peak value when it came to 3 h. After that, the degradation rate constants, k decreased with the increase of 2nd step MCT time. It means that the TiO$_2$/Cu composite films fabricated during MCT with 3 h of 2nd step MCT time showed the greatest photocatalytic activity for all the composite films.

The photocatalytic activity of TiO$_2$/Cu composite films should relate to the loading amounts of TiO$_2$ in the composite films. With increase in 2nd step MCT time, the loading amounts of TiO$_2$ increased. When it came to 3 h, the loading amounts of TiO$_2$ might reach the peak value. After the maximum value, TiO$_2$ that adhered to Cu films began to peel off. The more the loading amounts of TiO$_2$ that deposited on Cu films, the higher the photocatalytic activity. In other words, the photocatalytic activity of TiO$_2$/Cu composite films should be proportional to the loading amounts of TiO$_2$ in the composite films. The formation of TiO$_2$/Cu composite microstructure is considered to be another reason why the photocatalytic activity of the composite films was improved. After formation of the interface of TiO$_2$/Cu, electrons in the conduction of TiO$_2$ will migrate to Cu films through the interface, which can decrease the recombination rate of electron-hole pairs in TiO$_2$. It may result in the improvement of charge separation efficiency (Rengaraj et al., 2007). Then, more electrons are trapped in Cu films for reduction reaction and more holes are held in the valence band of TiO$_2$ for oxidation reaction.

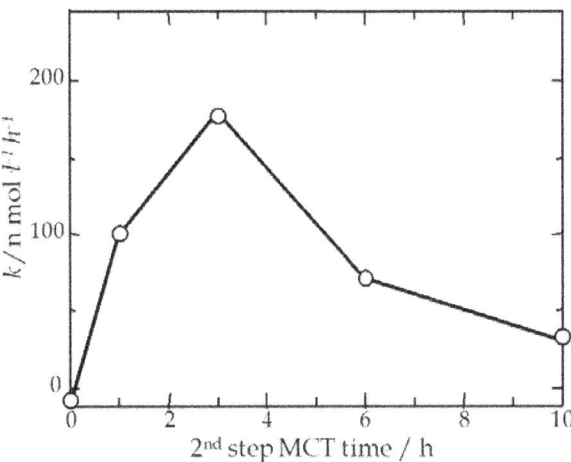

Figure 37: Degradation rate constants, k as a function of the 2nd step MCT time.

FORMATION PROCESS OF CU FILMS DURING MCT

Although Cu films and TiO_2/Cu composite films were fabricated by MCT, the formation process and its mechanism of Cu films are still unknown. Therefore, the formation process and its possible mechanism were examined and will be discussed in this section.

Because the formation process of Cu films happens in a closed and invisible bowl, it is difficult to determine the evolution of the films. However, it was considered to relate to the collision, friction and welding among the Cu powder particles, the inner wall of the bowl and the ceramic grinding mediums (Lü et al., 1995; Maurice and Courtney, 1990; Maurice and Courtney, 1994; Chattopadhyay et al., 2001). By now, we have tried to analyze the evolution of Cu films by observing the change of ceramic substrates. Fig. 38 shows the SEM images of Cu-coated Al_2O_3 balls after MCT. The areas of dark and light color correspond to alumina and copper respectively. More Cu particles adhered to the surfaces of Al_2O_3 balls with the increase of 1st step MCT time. When it came to 54 h, continuous Cu films formed and the thickness was about 10 μm. In other words, the surfaces of Al_2O_3 balls were totally coated by Cu films. The result is in good agreement with that in Fig. 33.

The coverage of Al_2O_3 ball surface with Cu is illustrated in Fig. 39. The evolution of Cu films during MCT can fall into five ranges. In the first range, the coverage hardly increased. However, Al_2O_3 balls became dark. Micron-sized Cu particles on the surfaces of Al_2O_3 balls were not found by SEM. It means that a small quantity of Cu atom clusters might transfer to the surfaces of Al_2O_3 balls. Under impact and friction force, the atom clusters adhered to the surfaces of Al_2O_3 balls and nucleated. In the second and third range, more Cu atom clusters adhered to the nuclei of Cu; these nuclei gradually grew up and could be observed by SEM (Fig. 38 (a)). Then discrete islands of Cu connected with each other (Fig. 38 (b)). The growth of Cu nuclei and the connection of discrete islands of Cu resulted in the coverage increase. After the first three ranges, the surfaces of Al_2O_3 balls were nearly coated with Cu and the coverage was close to 100%. In other words, continuous Cu films formed. Although the experiments were stopped when MCT time reached 54 h, the fourth and fifth ranges are considered to exit as the similar evolution of Fe films has been established in our published work (Hao et al., 2012). In the fourth range, the thickness of continuous Cu films may increase. As deformation of Cu particles, Cu particles become hard and adhesion between Cu particles may become difficult. Finally, exfoliation of Cu films would dominate. From the above results, the evolution of Cu films can fall into nucleation, growth of nuclei and connection, formation of continuous films and thickening, exfoliation of continuous films.

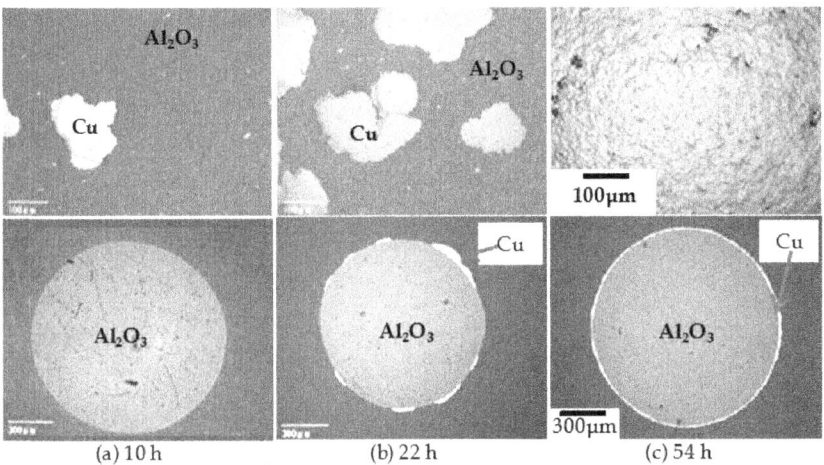

Figure 38: SEM images of the surfaces and the cross sections of Cu films fabricated by MCT.

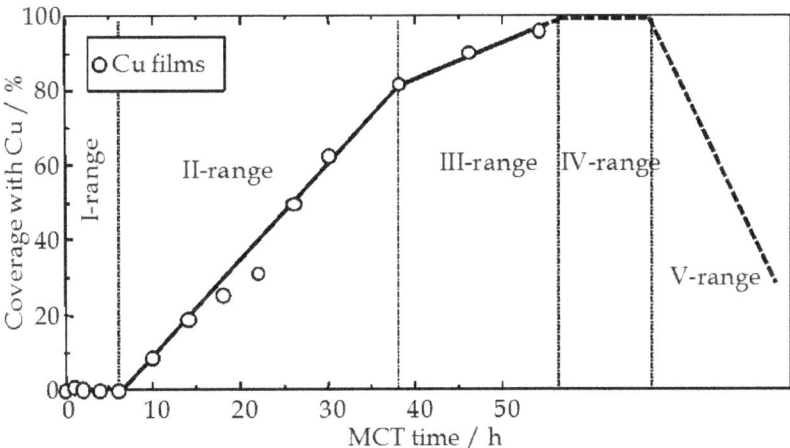

Figure 39: Coverage of Al_2O_3 ball' surface with Cu as a function of MCT time during MCT.

Fig. 40 shows the SEM images of the surfaces and the cross sections of the Cu-coated Al_2O_3 balls after MCT with 480 rpm. It can be observed that continuous Cu films formed when MCT time was 20 h and the average thickness of the films was about 80 μm. Compared with the fabrication of Cu films with Cu powder of 40 μm in average particle size by MCT at 400 rpm (Fig. 38(c)), the fabrication of Cu films with Cu powder of 10 μm in average particle size by MCT at 480 rpm was quicker. In other words, the condition of

Cu powder of 10 μm in average particle size and a rotation speed of 480 rpm accelerated the formation of Cu films.

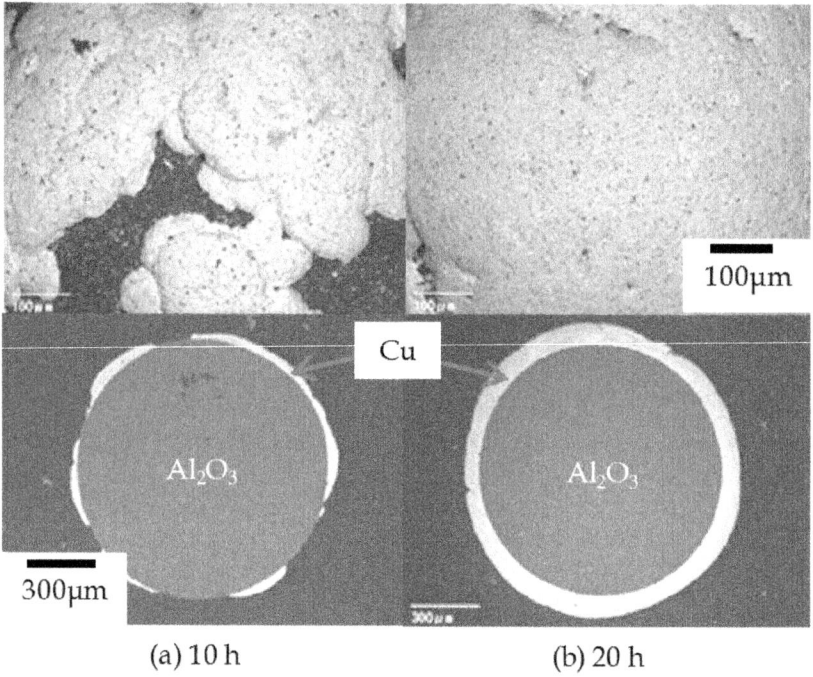

(a) 10 h   (b) 20 h

Figure 40: SEM images of the Cu-coated Al_2O_3 balls fabricated with Cu powder of 10μm in average particle size by MCT at 480 rpm

PROSPECT OF MCT

Compared with the traditional film coating techniques such as PVD and CVD, our proposed mechanical coating technique (MCT) shows many advantages including inexpensive equipments, simple process, low preparation cost and large specific area, among others. In addition, it can be performed in air atmosphere at ambient temperature. It can not only fabricate metal/alloy films but also nonmetal/metal composite films such as TiO_2/Ti composite photocatalyst films. It is expected to fabricate other functional film materials in the near future.

We will continue to advance the development of MCT in the fabrication of composite films and promote their applications. Our main research subjects within next few years are listed as follows.

1. Analysis on evolution and the relevant mechanism of metal films

2. Theory construction on film formation and numerical simulation
3. Fabrication of visible light-responsive TiO_2/metal composite films
4. Improvement on photocatalytic activity of TiO_2/metal composite films
5. Application investigation of TiO_2/metal composite films in sterilization, environment purification, and so on.

REFERENCES

1. Y. Q. Cao, H. J. Long, Y. M. Chen, Y. A. Cao, 2009Photocatalytic activity of TiO2 films with rutile/anatase mixed crystal structures. Acta Physico-Chimica Sinica, 2510881092
2. P. P. Chattopadhyay, I. Manna, S. Talapatra, S. K. Pabi, 2001A mathematical analysis of milling mechanics in a planetary ball mill. Materials Chemistry and Physics, 688594
3. I. Farahbakhsh, A. Zakeri, P. Manikandan, K. Hokamoto, 2011Evaluation of nanostructured coating layers formed on Ni balls during mechanical alloying of Cu powder. Applied Surface Science, 25728302837
4. G. Gupta, K. Mondal, R. Balasubramaniam, 2009In situ nanocrystalline Fe-Si coating by mechanical alloying. Journal of Alloys and Compounds, 482118122
5. L. Hao, Y. Lu, H. Asanuma, J. Guo, 2012The influence of the processing parameters on the formation of iron thin films on alumina balls by mechanical coating technique. Journal of Materials Processing Technology, 21211691176
6. Industrial. Japan, J. I. S. K. Standard, 1994
7. Industrial. Japan, J. I. S. R. Standard, 170322007
8. K. Kobayashi, 1995Formation of coating film on milling balls for mechanical alloying. Materials Transactions, 36134137
9. Y. Lu, M. Hirohashi, S. Zhang, 2005Fabrication of oxide film by mechanical coating technique, Proceedings of International Conference on Surfaces, Coatings and Nanostructured Materials, Aveiro, Portugal.
10. 2011 aY. Lu, H. Yoshida, H. Nakayama, L. Hao, M. Hirohashi, (20, a. , Formation of TiO2/Ti composite photocatalyst film by 2-step mechanical coating technique. Materials Science Forum, 675-67712291232
11. 2011 bY. Lu, H. Yoshida, K. Toh, L. Hao, M. Hirohashi, (20, b. , Performance improvement of TiO2/Ti composite photocatalyst film by heat oxidation treatment. Materials Science Forum, 675-67712331236
12. Y. Lu, L. Hao, K. Toh, H. Yoshida, 2012Fabrication of TiO2/Cu composite

photocatalyst thin film by 2-step mechanical coating technique and its photocatalytic activity. Advanced Materials Research, 415-41719421948

13. L. Lü, M. O. Lai, S. Zhang, 1995Modeling of the mechanical-alloying process. Journal of Materials Processing Technology, 52539546

14. D. M. Mattox, 2010Handbook of physical vapor deposition (PVD) processing, William Andrew Publication, 978-0-81552-037-5Burlington, USA

15. D. R. Maurice, T. H. Courtney, 1990Physics of mechanical alloying, a first report. Metallurgical and Materials Transactions, A21289303

16. D. R. Maurice, T. H. Courtney, 1994Modeling of mechanical alloying: part I. deformation, coalescence, and fragmentation mechanisms. Metallurgical and Materials Transactions, A25147158

17. H. O. Pierson, 1999Handbook of chemical vapor deposition, Noyes Publications and William Andrew Publication, 0-81551-432-8York, USA

18. S. Rengaraj, S. Venkataraj, J. W. Yeon, Y. Kim, X. Z. Li, G. K. H. Pang, 2007Preparation, characterization and application of Nd-TiO2 photocatalyst for the reduction of Cr (VI) under UV light illumination. Applied Catalysis, Vol. B 77, 157165

19. S. Romankov, W. Sha, S. D. Kaloshkin, K. Kaevitser, 2006Formation of Ti-Al coatings by mechanical alloying method. Surface Coatings Technology, 20132353245

20. C. Suryanarayana, 2001Mechanical alloying and milling. Progress of Materials Science, 461184

21. H. Yoshida, Y. Lu, H. Nakayama, H. Sano, M. Hirohashi, 2008Fabrication and evaluation of composite photocatalytic film by mechanical coating technique, Proceedings of the 6th International Forum on Advanced Material Science and Technology, Hong Kong, China

22. 2009 aH. Yoshida, Y. Lu, H. Nakayama, M. Hirohashi, (20, a. , Analysis of Ti films fabricated by mechanical coating technique (In Japanese). Journal of Materials Science Society of Japan, 46141146

23. 2009 bH. Yoshida, Y. Lu, H. Nakayama, M. Hirohashi, (20, b. , of. Fabrication, O. Ti, by. film, coating. mechanical, technique, photocatalytic. its, activity, Journal of Alloys and Compounds, 475383386

24. S. Yoshida, Y. Taga, A. Kinbara, et al.2008Handbook of thin films, Ohmsha Publication, 978-4-27420-519-4Tokyo, Japan

Chapter 5

GENERATION OF R-CURVE FROM 4ENF SPECIMENS: AN EXPERIMENTAL STUDY

V. Alfred Franklin[1] and T. Christopher[2]

[1]Faculty of Mechanical Engineering, Sardar Raja College of Engineering, Alangulam, Tirunelveli 627808, India

[2]Faculty of Mechanical Engineering, Government College of Engineering, Tirunelveli 627007, India

ABSTRACT

The experimental determination of the resistance to delamination is very important in aerospace applications as composite materials have superior properties only in the fiber direction. To measure the interlaminar fracture toughness of composite materials, different kinds of specimens and experimental methods are available. This article examines the fracture energy of four-point end-notched flexure (4ENF) composite specimens made of carbon/epoxy and glass/epoxy. Experiments were conducted on these laminates and the mode II fracture energy, G_{IIC}, was evaluated using compliance method and was compared with beam theory solution. The crack growth resistance curve (R-curve) for these specimens was generated and the found glass/epoxy shows higher toughness values than carbon/epoxy composite. From this study, it was observed that R-curve effect in 4ENF specimens is quite mild, which means that the measured delamination toughness, G_{IIC}, is more accurate.

INTRODUCTION

Owing to their high stiffness and strength combined with low weight, polymer matrix composites have become more appropriate structural materials for aerospace applications. However, the usual laminated nature and the relatively low matrix strength make them particularly susceptible to delamination. The highly anisotropic nature of laminated composite structures causes a mismatch in mechanical properties between individual lamina within the laminate, which in turn can produce interlaminar crack initiation and propagation. For example, low velocity impact can generate relatively large delamination, which is highly detrimental to compressive load because of localized buckling phenomena [1]. Testing of thin skin stiffened panels designed for aircraft fuselage applications has shown that bond failure at the tip of the frame flange is a very important failure mode. Debonding also occurs when a thin-gage composite fuselage panel is allowed to buckle in service [2]. The growing use of composite materials in aircraft and spacecraft applications has motivated researchers to understand their fracture behaviour and damage mechanisms. Hence, fracture characterization of composites structures attains special relevancy.

Most of the composites currently in service contain only two-dimensional (in-plane) reinforcement and delamination remains an important failure mode in such composites. The development of standardised test methods to characterize the resistance to interlaminar crack propagation is necessary for two main reasons: (i) such tests offer the possibility to compare existing and new materials on the same basis and (ii) such tests offer reliable input data for new damage tolerance models. Unlike mode I DCB testing, mode II testing is not fully standardized by ASTM. There are standard test methods which are available for ENF [3], ELS [4], and calibrated endloaded split (C-ELS) test [5] specimens. In fact, the ENF test requires $a/L > 0.7$ to obtain stable crack propagation [6], whereas in the ELS test $a/L > 0.55$ is sufficient [7]. Stabilized ENF was not popular in round-robin trials [5]. Synthesizing the previous findings, it is shown that accurate and repeatable toughness values are obtained provided that the ratio of crack [8]. length to half-span length (a/L) is 0.6. Accuracy of data reduction considerations indicated a common range of $21 \leq L/h \leq 29$ or, for $L = 50$ mm, 3.4 mm $\leq 2h \leq 4.7$ mm for ENF specimens [8].

Martin and Davidson [9] proposed another version of mode II configuration, shown in Figure 1, called the 4ENF test, and this geometry encourages stable crack propagation so that an experimental compliance calibration can be applied for data reduction. Moreover just from the data of one test, mode II crack growth resistance curve can be generated. The advantages of 4ENF test are (i) simple coupon geometry, (ii) simple closed-form solution, (iii) propagation toughness which can be evaluated, and (iv) pure bending at the crack tip. The main drawback of 4ENF configuration is longitudinal sliding. During the year 1998, a draft mode II 4ENF test protocol [10] for an international round robin exercise was prepared [11], but it was not accepted as a standard test method because of frictional effects [12].

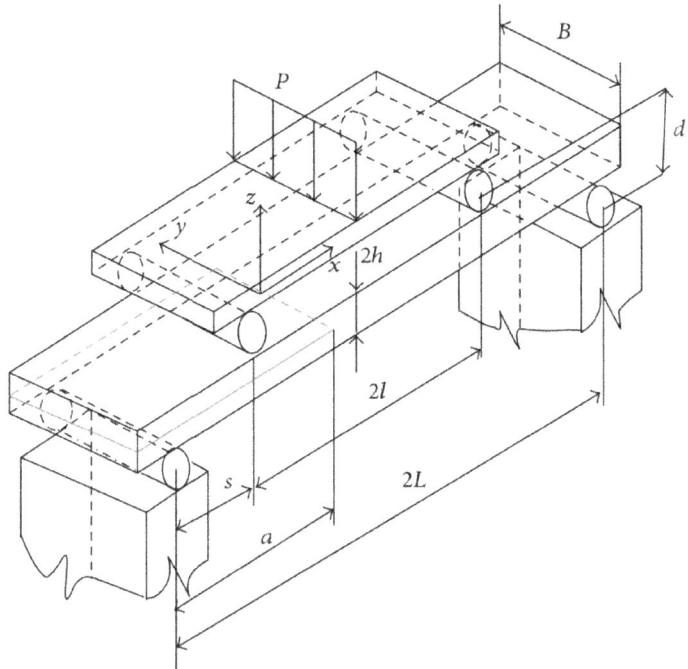

Figure 1: Four-point end-notched flexure (4ENF) specimen geometry.

Fan et al. [13] introduced a new test method, named internal-notched flexure (INF) test, which was designed to measure the critical energy release rate of fiber reinforced composites under shear mode. The test applies three-point bending to a beam specimen that has an insert film embedded in the midthickness, thus named internal-notched flexure (INF) test, similar to central-notched flexure (CNF) test [14]. The main difference between the two tests is the placement of the specimen in the three-point bend device. Both tests use symmetrical three point loading, but the CNF test applies the load in the middle of the insert film with the span length longer than the insert film. The INF test, on the other hand, applies the load asymmetrically with respect to the insert film, with one end of the insert film being much closer to the loading pin than the other end. The delamination crack in the INF specimen is subject to concentrated forces only, without any moment, which is the same condition as that for 4ENF specimens of the same fiber lay-up. Therefore, the fracture mode introduced in the INF test should be the same as that in the 4ENF test, and the two tests should be subject to a pure shear mode of fracture. The stability during crack growth of different mode II fracture specimens is shown in Figure 2. Among these mode II specimens, 4ENF specimen shows better propagation stability.

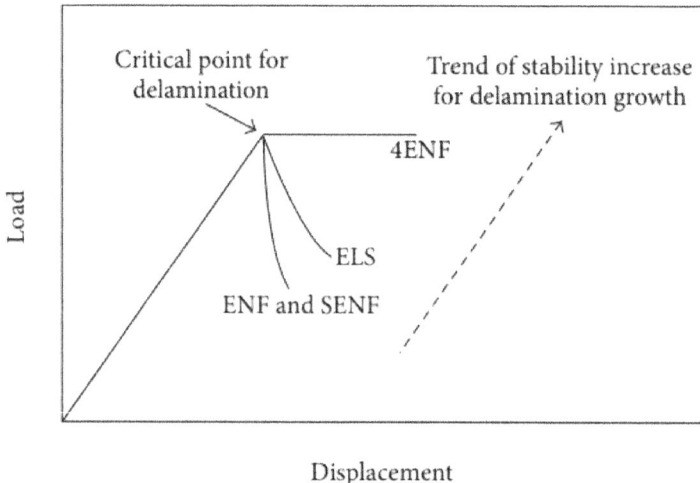

Figure 2: Schematic showing the trend of crack growth stability under mode II [13].

The R-curve is the relationship between the fracture toughness and crack length. Applying 4ENF test, crack growth is stable under displacement control and the complete R-curve can be determined from the data of just one test [15]. In many works the R-curve behavior, R-curve resistance, or J-resistance [16] was observed, which means that the fracture toughness increases with the crack size; this can be attributed to the fiber bridging and fiber pullout during the fracture [17, 18]. These fibers increase the resistance to delamination.

In the present study, tests were conducted on carbon/epoxy and glass/epoxy laminates to estimate mode II fracture energy and the R-curves have been generated from the test data. However, the R-curve effect noticed was mild and that for carbon/epoxy specimen was almost insignificant.

FACTORS AFFECTING MODE II FRACTURE TOUGHNESS

Influence of Fiber Volume Fraction

Hunston et al. [19] studied the influence of fibre content on mode I delamination toughness and showed much higher G_{IIC} values for a composite which was resin rich than those for the same material with lower resin content. In the same way, experiments using the edge cracked torsion (ECT) specimen show a physically powerful influence of fibre content on mode III fracture toughness [20]. Davies et al. [21] examined mode II fracture toughness of glass/epoxy composites over a large range of fibre volume fraction using

4ENF test. The initiation and propagation fracture toughness versus fibre content are shown in Figures 3 and 4. From these figures, it is noted that the toughness decreases quite substantially with increasing fibre content. At high fiber contents, the delamination toughness assume a level of stability which is apparently dominated by fiber/matrix interfacial adhesion rather than matrix plasticity. Obviously, Table 1 summarises that there is a strong effect of fibre volume fraction on the initiation and propagation toughness. There are a lot of potential explanations for the large influence of fibre volume fraction on mode II interlaminar fracture toughness. But a more believable reason for the high toughness of the low fibre content specimens is plasticity effects at crack tip [19]. Shear fracture tests on adhesively bonded metal specimens were conducted by Chai [22] and a strong increase in mode II toughness was found as the adhesive layer thickness was increased. Similarly, a strong effect of resin film thickness on interlaminar fracture toughness of interleaved carbon fibre composites was shown by Carlsson [23].

Table 1: Influence of fiber volume fraction on mode II toughness [21].

Range of fiber volume fraction, V_f	Initiation toughness (kJ/m^2) Mean (SD)	Propagation toughness (kJ/m^2) Mean (SD)
40–45%	2.13 (0.07)	3.04 (0.45)
>55%	0.98 (0.18)	1.64 (0.33)

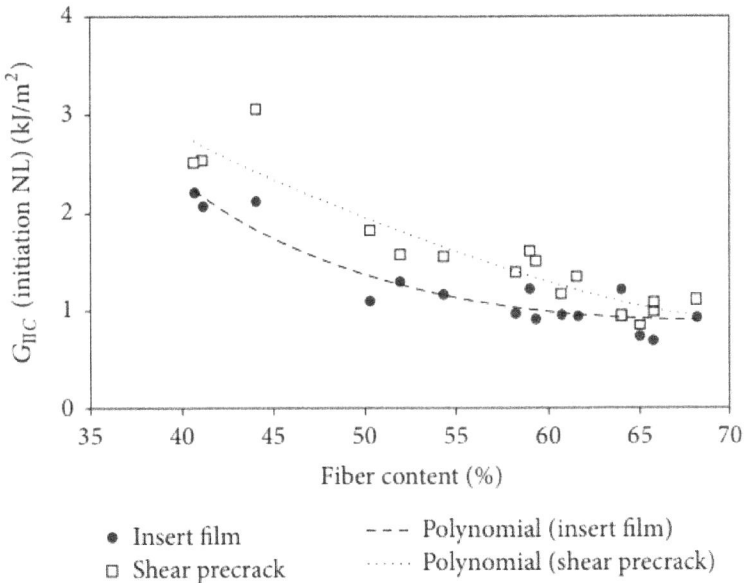

Figure 3: Influence of fiber volume fraction on initiation toughness [21].

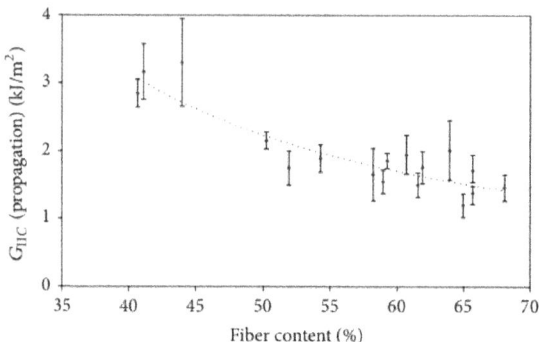

Figure 4: Influence of fiber volume fraction on propagation toughness [21].

Influence of Nose Distance

Kageyama et al. [12] studied the effect of nose distance (d) on load-displacement characteristics with two fixtures A (d = 56 mm) and B (d = 12 mm). Figure 5 shows the load versus displacement curves obtained by using fixtures A and B. In both specimens, PTFE film was inserted between delamination surfaces. Stick-slip was observed when fixture A was used. While using fixture B, the effect of geometrical nonlinearity was minimized and linear relations were noticed at initial stage of loading and no stickslips occurred. Effect of geometrical nonlinearity was clearly observed when fixture A was used and that was not observed in fixture B. However, unsymmetrical loading can reduce the effect of geometrical nonlinearity, but data reduction method becomes very complicated. Hence, the nose distance should be as small as possible in order to avoid the effect of large deformation.

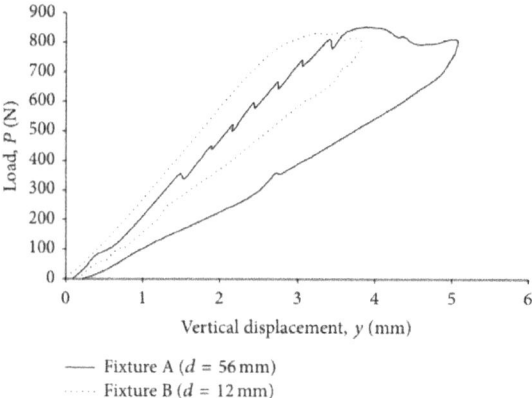

Figure 5: Influence of nose distance on load-displacement history [12].

Influence of Inner Span Length

It was found [15] that when $l/L = 0.4$, the propagation toughness is an average of 30% lower than the initiation toughness, and when $l/L = 0.5$, the propagation toughness is 21% lower than the initiation toughness. This shows the influence of inner span length on fracture toughness. Moreover, Table 2 summarizes mode II fracture toughness, $GIIC$, results for two inner span lengths, in which the averaged results of the 4ENF specimens with PTFE films at loading and supporting noses and between delamination surfaces are compared with those obtained by SENF test under crack shear displacement control [24, 25] according to JIS K7086 standards. From Table 2, it is noticed that if inner span length ($2l$) increases, $GIIC$ also increases and vice versa.

Table 2: Influence of inner span on critical mode II interlaminar fracture toughness of carbon/epoxy 4ENF specimens.

Method	Inner span 2l (mm)	G_{IIC} (kJ/m²)		
		NL	5% or max	Propagation
4ENF [39]	50	0.682	0.902	1.002
	70	0.854	1.011	1.047
4ENF [12]	50	0.525	0.692	0.767
	70	0.563	0.666	0.690
SENF [24, 25]		1.041	1.261	1.225

Influence of Friction

In a finite element study, Schueker and Davidson [26] investigated that the differences in delamination toughness obtained by the ENF and 4ENF tests could be attributed to friction between the crack faces. Moreover the effect of friction increases with span ratio [27, 28]. Kageyama et al. [12] have shown that the experimental compliance does not agree well with the theoretical one. Assuming that the difference is due to friction between delamination surfaces, they estimated the coefficient of friction as 2.1 for an inner span of 50 mm, and this value is much higher than the expected value. However, as elementary beam theory gives smaller value of theoretical compliance (λ_{11}) than the exact elastic solution, the effect of friction might be larger than that estimated. The authors [12] presented a method to take into account the effect of friction of 4ENF specimen on delamination fracture toughness which is given by Accuracy of (1) depends on the accuracy of theoretical value of compliance, $\lambda 11$. They concluded that the effect of friction has larger effect on G_{IIC} in 4ENF specimen. But Schuecker and Davidson [15] analyzed the effect of friction by using finite element analysis (VCCT), in view of the work reported by Kageyama et al. [12]. Table 3 presents the results obtained for friction coefficients $\mu = 0.5$ and 1.0. The error caused by friction is obviously larger in 4ENF specimens, but it remains acceptably small. They argued that the introduction of friction combined with the damage model was the source of error in [12].

Table 3: Effect of friction on G_{IIC} in 4ENF specimen in terms of percentage error [31].

Coefficient of friction	Relative to nonfriction case VCCT	CBT	Relative to VCCT CBT
0.5	−1.36	1.18	2.04
1.0	−2.75	2.36	4.72

Cartie et al. [′ 29] studied the influence of hydrostatic pressure on delamination fracture toughness of carbon/epoxy (IM7/977-2) composite pressure vessels and noted that the effect of pressure on mode I fracture toughness is insignificant. But they observed that mode II delamination resistance was increased by up to 25% for an increase in pressure from 4 to 900 bars. They observed that the crack initiation was unstable and higher G_{IIC} values were measured at initiation than during subsequent propagation. Table 4 shows the results obtained for both initiation and propagation and it is noted that there is a significant increase in initiation and propagation toughness values with increasing pressure. Figure 6 shows that, under mode I loading, there may be a small decrease in the fracture toughness and this drop cannot be considered to be significant. A significant influence. of pressure on fracture toughness was noted under mode II loading. One possible reason for this is the influence of pressure on the friction between the two sliding surfaces. Friction forces will be directly proportional to the applied lateral pressure on the specimen faces so the work required to overcome these friction would be expected to increase the measured value of G_{IIC}.

Table 4: Influence of hydrostatic pressure on mode I and mode II delamination toughness [29]

	Initiation (5% max)			Propagation ($a = 55$ mm)		
Pressure (bar)	4	300	900	4	300	900
Average mode I delamination toughness, G_{IC} (J/m^2)	406	397	363	766	719	682
Average mode II delamination toughness, G_{IIC} (J/m^2)	1548	1842	1812	944	1165	1285

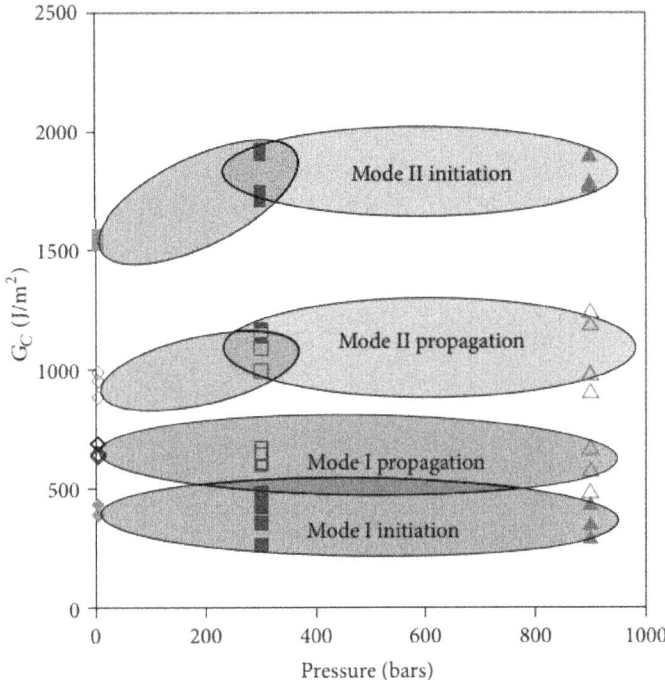

Figure 6: Influence of applied hydrostatic pressure on mode I and mode II fracture energy: a summary [29].

Influence of Measuring Techniques and Reduction Schemes.

An experimental study to investigate the accuracy of the 4ENF test for the determination of mode II delamination toughness was conducted by Schuecker and Davidson [15] and it was observed that the 4ENF test gave higher values than the ENF test; the percentage difference increases with increasing the ratio of the inner span versus the outer one. They assumed that the error might be due to the inexact measuring techniques for determining load, deflection, and crack length. With different span ratio, 4ENF and ENF tests were conducted on unidirectional carbon/epoxy specimens and results were compared. The initiation and propagation tests were conducted and the fracture toughness was evaluated by compliance calibration method. The crack length was measured visually, using c-scan system and LVDT. The specimen deflection was determined from the testing machine's actuator displacement and from the average of two LVDT readings. ENF initiation tests were also performed in the same way. In the 4ENF test the compliance versus crack length curve was linear but in ENF test it was fit with a third-order polynomial. Initiation results show that the 4ENF test gives 2–12.7% higher values than that of the ENF test. By 4ENF propagation test, it was observed that the propagation toughness is 20–30% higher than the initiation one. They concluded that if compliance and crack length are measured accurately then both 4ENF and ENF tests will produce the same toughness values.

In the same way, Fan et al. [13] conducted INF and ENF tests on glass/polyester material to estimate mode II fracture energy. The average value at the onset of delamination was found as J/m^2 and J/m^2, respectively. This also shows that the initiation toughness values are closer if the parameters are measured accurately.

Influence of Fracture Criteria

Even though there is a progress in interlaminar fracture testing, uncertainty remains on determination of the onset of delamination growth from the starter crack, commonly designated as "crack initiation." This is a predominantly significant issue, because initiation fracture toughness is believed to be the most relevant for design purposes. In fact, due to the nesting inherent to unidirectional specimens, fibers above and below the mid-plane tend to bridge the delamination as it grows from the insert, giving rise to an R-curve effect. However, fiber bridging is considered an artefact of the DCB specimen that does not occur in structural composites [30]. Ideally, crack initiation could be defined by observation at one of the specimen edges and this definitely introduces some degree of operator dependency. According to the nonlinearity (NL) criterion, initiation is taken at the point where the -δ curve deviates from linearity. The NL criterion yields the most conservative toughness values. On the other hand, the 5% offset or maximum load (5%-max) criterion defines initiation at the lowest displacement point among the 5% offset and the maximum load. The previous one is obtained by intersecting the load-displacement curve with a line corresponding to compliance 5% higher than the initial one (Figure 7). The maximum load criterion is unambiguous and seems to lead to lower scatter [30]. However, it yields higher toughness values than the NL criterion and the 5% C, which is also used in fracture testing of metals. The finite element model [31] predicted early deviations from linearity as a result of large process zones in ENF and ELS specimens. On the other hand, very accurate results were obtained with the 4ENF specimens, in spite of the relatively small starter crack. This is due to the crack length independent critical load. As described in [32], there is no clear physical meaning of the nonlinear load point in the ENF test. Thus, critical load will be taken as the maximum load point [8] since it coincides with actual initiation. Moreover from an international collaborative test program [11], it is noticed that ENF,

ELS, and 4ENF tests gave similar fracture toughness results from the insert when maximum load criteria were used.

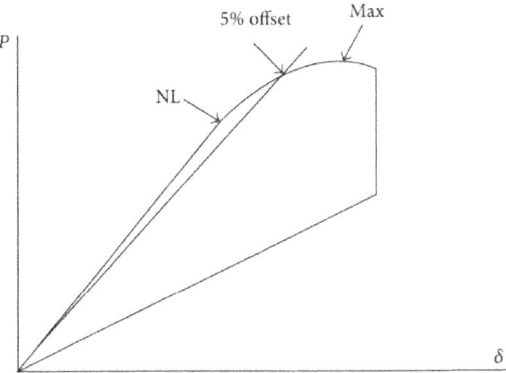

Figure 7: Schematic representing different initiation criteria [31].

EXPERIMENTAL WORK

Materials and Specimen Preparation

The materials used in the present study were unidirectional glass/epoxy and carbon/epoxy composites. The glass fibres of 300 gsm were impregnated by hand lay-up with an epoxy resin (LY 556), with hardener HY951 in the ratio of approximately 10 : 1 to control the curing action and the degree of hardness of the cured film. The initial crack was made by introducing a thin Teflon film of thickness 13 μm during stacking procedure. The laminate was prepared by hand lay-up and was postcured to stabilize the matrix property. Another material used in the present study was carbon/epoxy prepreg with a fiber volume fraction of 60%. As before, the starter crack was obtained by placing the PTFE film at the mid-plane during stacking procedure and the laminate was cured at 180° C under a pressure of about 6 bars.

Experimental Procedures

The specimens were tested under a displacement rate of 1 mm/min, on a test frame with 3-ton capacity load cell. Typewriter correction fluid was applied on the thickness side of specimens and markings were made at a regular interval of 1 mm. The machine crosshead displacement transducer was used to record upper loading point displacements, while crack length measurements were made visually with the help of travelling microscope. Zile and Tamuzs [33] experimentally found that the compliance data and their derivatives obtained from the continuous loading tests were very similar to load-unload-reload test. This means that both the loading techniques can be used to obtain critical energy release rate. Hence, the continuous loading test was adopted in the present study

EXPERIMENTAL DATA REDUCTION

Linear elastic fracture mechanics (LEFM) became a common practice to characterize the resistance to delamination. There are competing terminologies in literatures, such as fracture toughness, average fracture energy, J-integral, work of fracture, and critical strain energy release rate. In the present study, the critical strain energy release rate is used to characterize the R-curve.

Compliance Calibration

The compliance calibration method is superior to the compliance fitting method in case of non-self-similar crack advance [34]. It was shown [35] that, under displacement control, $\partial C/\partial a < 0$ and this means that the crack (delamination) growth is stable. According to beam theory and finite element results [26, 35], a linear relationship between the compliance and crack length was observed. One of the most accurate data reduction methods to determine delamination toughness from 4ENF test data is compliance calibration technique (as compared to beam theory and finite element analysis if the crack length is measured accurately). Because of that, it was shown that beam theory and finite element based data reduction techniques can produce errors in toughness values [27, 28] due to the fact that it is difficult to accurately determine the bending stiffness of each test specimen. But compliance calibration data reduction is a direct method that assumes linear elastic behaviour and self-similar crack advance. Hence, any errors due to uncertainties in the geometric and/or material properties of each test specimen can be avoided. Mode II interlaminar fracture toughness was reduced from 4ENF test data usingan experimental compliance calibration method with the compliance expression [10] as

$$C = C_o + ma, \qquad (3)$$

and the fracture toughness is evaluated by

$$G_{IIC} = \frac{P_c^2}{2B} \cdot \frac{\partial C}{\partial a} = \frac{P_c^2}{2B} \cdot m. \qquad (4)$$

Beam Analysis

Using beam theory, Zile and Tamuzs [33] derived an expression for mode II fracture toughness which is given byBy unit load theorem, an expression for mode II fracture toughness is given by [31]

$$G_{IIC} = \frac{3P_c^2 L^2}{32BE_{11}I}. \tag{5}$$

By unit load theorem, an expression for mode II fracture toughness is given by [31]

$$G_{IIC} = \frac{3P_c^2 s^2}{64BE_{11}I}. \tag{6}$$

RESULTS AND DISCUSSIONS

In Section 2 of this paper, various factors affecting mode II fracture toughness (namely, fiber volume fraction, loading nose distance, inner span length, coefficient of friction, measuring techniques, and initiation criteria on the fracture) of 4ENF specimens were revived. This will give an overview about 4ENF specimen and testing to the beginners. In the second part of the paper (Section 3), fracture tests were carried out on 4ENF specimens made of carbon/epoxy and glass/epoxy and the load-displacement plot was shown in Figure 8. These plots indicate essentially linear behaviour up to crack initiation. Figure 9 shows the compliance fitting plot established from fracture tests. The relation between compliance and crack length is observed as linear and the slope of these trend lines gives the value of $(\partial C/\partial a)$. Using these slope values, the G_{IIC} was evaluated and tabulated in Table 6. The propagation values are found little below the initiation values and the minimum initiation values are measured directly from the insert film during propagation. Thus, there was always some unstable growth during the initial crack increment. Afterwards, crack growth was always stable and the load at which the crack advanced remained nearly constant. A complete R-curve has been generated from this test data; however, as the critical load was almost equal during propagation, there would be little variation in energy release rate with crack length. Considering (4), the fracture toughness is directly proportional to the slope of the compliance versus crack length curve. Therefore, the error between the different ways of determining G_{IIC} depends on $\partial C/\partial a$ only.

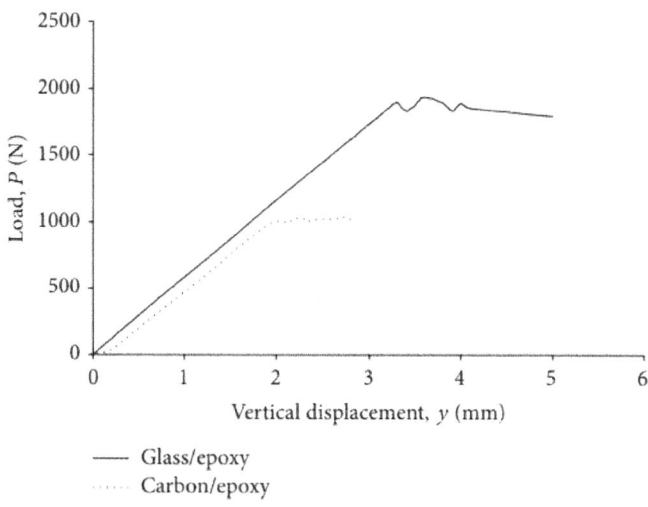

Figure 8: Load-displacement plot for carbon/epoxy and glass/epoxy specimens.

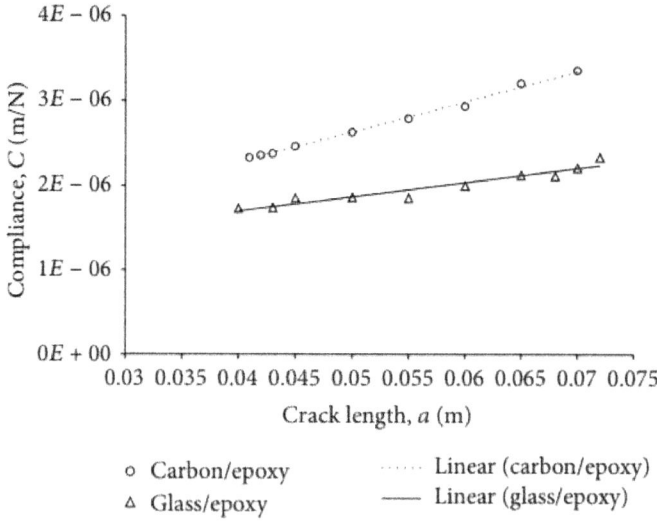

Figure 9: Linear curve fitting of compliance by compliance fitting method.

The present test produces very stable crack growth. Figure 10 shows mode II interlaminar toughness versus crack length curves (resistance curves) at fibre volume fractions of about 60%. For each composite specimen, the toughness increases to some extent during crack propagation. Also, it is noticed that glass/epoxy specimen shows a higher resistance to delamination than carbon/epoxy.

For glass/epoxy specimens, the R-curves show a mild upward trend with the delamination growth, whereas for carbon/epoxy it is almost flat. However, no extensive fiber bridging was noticed as mentioned in [21, 33].

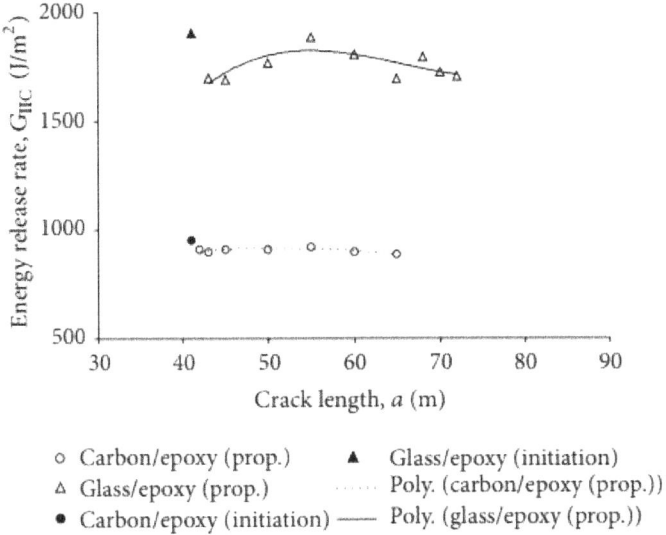

Figure 10: Crack growth resistance curve (R-curve) of 4ENF specimens.

The influence of specimen configuration and material on G_{IIC} is presented in Table 5 from previous literature. From Tables 1, 5, and 6, it is noticed that the toughness values obtained in the present study are comparable with published test data. From Table 6, it is observed that beam theory estimates a higher initiation and propagation toughness as compared to compliance calibration method. This difference may be due to the measurement inaccuracy in flexural rigidity or in crack length. However, a good agreement was noticed between these two methods. Captivatingly, CBT (see (6)) gave reasonably accurate results for the 4ENF specimen, in spite of the relatively small starter crack. But higher % of error in delamination toughness was noticed with (5). In fact, (6) predicts that the critical load does not depend on the crack length and thus remains constant throughout propagation. From this study, it is concluded that the R-curve effect noticed in 4ENF specimens was very mild as compared to mode I fracture test [36–38].

Table 5: Experimental delamination fracture energy from literatures.

Material	Initiation (J/m^2)				Propagation (J/m^2)			
	ENF	ELS	ONF	4ENF	ENF	ELS	ONF	4ENF
Carbon/epoxy [33]	—	—	—	610	—	—	—	1160
Carbon/epoxy [40] IM7/8552	—	—	—	1334	—	—	—	—
Glass/epoxy [40] S2/8552	—	—	—	1759	—	—	—	—
Glass/epoxy [41]	2115	—	—	—	—	—	—	—
Glass/polyester [42]	—	713	730	—	—	—	2787	—
Glass/polyester [43]	496	—	—	—	—	—	—	—
Carbon/PEEK [44]	—	1730	—	—	—	2890	—	—
Carbon/PES [45]	—	1250	—	—	—	1850	—	—
C/HG9106 [46]	335	565	—	—	—	800	—	—
Graphite/epoxy [47]	875	—	—	—	—	—	—	—
T300/976 [48]	450	—	—	—	—	—	—	—
IM7/977-2 [48]	994	—	—	—	—	—	—	—
IM7/977-2 [49]	910	—	—	—	—	—	—	—
Glass/polyprop. [18]	3800	—	—	—	—	—	—	—
CYCOM 982 [50]	1850	—	—	—	—	—	—	—
APC-2 [50]	2730	—	—	—	—	—	—	—

Table 6: Mode II delamination toughness from present experiments.

Material	E_{11} (GPa)	$2h$ (mm)	Initiation toughness (J/m^2)		Propagation toughness (J/m^2)	
			CC (4)	CBT (6)	CC (4)	CBT (6)
Carbon/epoxy	131	3.24	952	1105	910	1038
Glass/epoxy	38	5.96	1905	2109	1807	2022

Width, $B = 20$ mm; $2L = 100$ mm; $s = 20$ mm; $V_f = 60\%$.

CONCLUDING REMARKS

An experimental study was carried out on 4ENF specimens made of glass/epoxy and carbon/epoxy. From experiments, it is noticed that the 4ENF specimen is well suited for measuring mode II delamination toughness and provides full resistance curves. From Table 6 and Figure 10, it is observed that the 4ENF specimen made of glass/epoxy shows higher resistance to delamination than carbon/epoxy composite. Moreover the experimental resistance curves showed that there was no significant toughening and hence no considerable fibre bridging was observed. Thus the R-curve effect was quite mild, providing accurate value of G_{IIC}. Since the variation in load (after initiation) with crack length was almost constant, all initiation criteria, namely, NL, 5% C, and maximum load, estimate almost equal G_{IIC} values.

NOMENCLATURE

4ENF: Four-point bend end-notched flexure
a: Crack length
B: Specimen width
C: Compliance of the specimen
d: Nose distance
E: Longitudinal tensile modulus
G_{II}: Mode II strain energy release rate
G_{13}: Shear modulus
G_{IC}: Mode I fracture toughness or critical strain energy release rate
G_{IIC}: Mode II fracture/delamination toughness
$2h$: Specimen thickness
I: Area moment of inertia
$2L$: Outer span
$2l$: Inner span
m: Slope of C-a plot
P_c: Critical load
s: Distance from one of the specimen supports to the nearest loading point
V_f: Fiber volume fraction
y: Displacement at mid span
λ_{11}: Theoretical compliance of 4ENF specimen without friction
μ: Coefficient of friction.

Theoretical compliance of 4ENF specimen without friction

Coefficient of friction.Conflict of Interests

The authors declare that there is no conflict of interests regarding the publication of this paper.

REFERENCES

1. A. B. de Morais, C. C. Rebelo, P. M. S. T. de Castro, A. T. Marques, and P. Davies, "Interlaminar fracture studies in Portugal: past, present and future," Fatigue & Fracture of Engineering Materials & Structures, vol. 27, no. 9, pp. 767–773, 2004.

2. T. K. O'Brien, "Fracture mechanics of composite delamination," in ASM Handbook, Composites, vol. 21, pp. 241–245, ASM International, 2001.

3. "Testing methods for interlaminar fracture toughness of carbon fiber reinforced plastics," Japanese Industrial Standard Group, JIS K-7086-1993 ,1993.

4. B. R. K. Blackman, A. J. Brunner, and J. G. Williams, "Mode II fracture testing of composites: a new look at an old problem," Engineering Fracture Mechanics, vol. 73, no. 16, pp. 2443–2455, 2006.

5. ISO, "Fibre-reinforced plastic composites—determination of the mode II fracture resistance for unidirectionally reinforced materials using the calibrated end-loaded split (C-ELS) test and an effective crack length approach," ISO 15114:2014, 2014.

6. L. A. Carlsson, J. W. Gillespie Jr., and R. B. Pipes, "On the analysis and design of the end notched flexure (ENF) specimen for mode II testing," Journal of Composite Materials, vol. 20, no. 6, pp. 594–604, 1986.

7. H. Wang and T. Vu-Khanh, "Use of end-loaded-split (ELS) test to study stable fracture behaviour of composites under mode II loading," Composite Structures, vol. 36, no. 1-2, pp. 71–79, 1996.

8. B. D. Davidson and S. S. Teller, "Recommendations for an ASTM standardized test for determining G_{IIc} of unidirectional laminated polymeric matrix composites," Journal of ASTM International, vol. 7, no. 2, Article ID JAI102619, 2010.

9. R. H. Martin and B. Davidson, "Mode II fracture toughness evaluation using a four point bend end notched flexure test," in Proceedings of the 4th International Deformation and Fracture of Composites Conference, pp. 243–252, London, UK, 1997.

10. R. H. Martin, "Protocol for the determination of the mode II delamination resistance of unidirectional fiber reinforced polymer matrix composites using the four point bend end notched flexure (4ENF) specimen," MERL, reference 8-1-98. Protocol used for VAMAS Round robin tests, 1998.

11. P. Davies, G. D. Sims, B. R. K. Blackman, et al., "Comparison of test configurations for determination of mode II interlaminar fracture toughness results from international collaborative test programme,"Plastics, Rubber and Composites, vol. 28, no. 9, pp. 432–437, 1999.

12. K. Kageyama, I. Kimpara, T. Suzuki, H. Ohsawa, M. Kanai, and H. Tsuno, "Effects of test conditions on mode II interlaminar fracture toughness of four-point ENF specimens," in Proceedings of the 12th International Conference on Composite Materials, pp. 362–369, Paris, France, 1999.

13. C. Fan, P. Y. B. Jar, and J. J. R. Cheng, "Internal-notched flexure test for measurement of mode II delamination resistance of fibre-reinforced polymers," Journal of Composites, vol. 2013, Article ID 695862, 7 pages, 2013.

14. H. Maikuma, J. W. Gillespie, and J. M. Whitney, "Analysis and experimental characterization of the center notch flexural test specimen for mode II interlaminar fracture," Journal of Composite Materials, vol. 23, no. 8, pp. 756–786, 1989.

15. C. Schuecker and B. D. Davidson, "Evaluation of the accuracy of the four-point bend end-notched flexure test for mode II delamination toughness determination," Composites Science and Technology, vol. 60, no. 11, pp. 2137–2146, 2000.

16. K. Tohgo, D. Fukuhara, and A. Hadano, "The influence of debonding damage on fracture toughess and crack-tip field in glass-particle-reinforced Nylon 66 composites," Composites Science and Technology, vol. 61, no. 8, pp. 1005–1016, 2001.

17. C.-A. Wang, Y. Huang, and Z. Xie, "Improved Resistance to damage of silicon carbide-whisker-reinforced silicon nitride-matrix composites by whisker-oriented alignment," Journal of the American Ceramic Society, vol. 84, no. 1, pp. 161–164, 2001.

18. G. Reyes V. and W. J. Cantwell, "The mechanical properties of fibre-metal laminates based on glass fibre reinforced polypropylene," Composites Science and Technology, vol. 60, no. 7, pp. 1085–1094, 2000.

19. D. L. Hunston, R. J. Moulton, J. J. Johnston, and W. D. Bascom, "Matrix resin effects in composite delamination: mode I fracture aspects," in ASTM STP 937, N. J. Johnston, Ed., pp. 74–94, American Society for Testing and Materials, Philadelphia, Pa, USA, 1987.

20. X. Li, L. A. Carlsson, and P. Davies, "Influence of fiber volume fraction on mode III interlaminar fracture toughness of glass/epoxy composites," Composites Science and Technology, vol. 64, no. 9, pp. 1279–1286, 2004.

21. P. Davies, P. Casari, and L. A. Carlsson, "Influence of fibre volume fraction on mode II interlaminar fracture toughness of glass/epoxy using the 4ENF specimen," Composites Science and Technology, vol. 65, no. 2, pp. 295–300, 2005.

22. H. Chai, "Shear fracture," International Journal of Fracture, vol. 37, no. 2, pp. 137–159, 1988.

23. L. A. Carlsson, "Fracture of laminated composites with interleaves," in Fracture of Composites, E. A. Armanios, Ed., vol. 120-121 of Key Engineering Material, pp. 489–520, Trans Tech Publications, 1996.

24. K. Kageyama, M. Kikuchi, and N. Yanagisawa, "Stabilized end notched flexure test. Characterization of Mode II interlaminar crack growth," in Proceedings of the 3rd Symposium on Composite Materials: Fatigue and Fracture, T. K. O'Brien, Ed., ASTM STP 1110, pp. 210–225, American Society for Testing and Materials, Philadelphia, Pa, USA, November 1989.

25. K. Kageyama, I. Kimpara, I. Ohsawa, M. Hojo, and S. Kabashima, "Mode I and mode II delamination growth of interlayer toughened carbon/epoxy (T800H/3900-2) composite system," in Composite Materials: Fatigue and Fracture—Fifth Volume, R. H. Martin, Ed., vol. 5 of ASTM STP 1230, pp. 19–37, American Society for Testing and Materials, 1995.

26. C. Schueker and B. D. Davidson, "Effect of friction on the perceived mode II delamination toughness from three- and four-point bend end notched flexure tests," in Composite Structures: Theory and Practice, P. E. Grant and C. Q. Rousseau, Eds., vol. 1383, pp. 334–344, ASTM STP, American Society of Testing of Materials, 2000.

27. T. K. O'Brien, G. B. Murri, and S. A. Salpekar, "Interlaminar shear fracture toughness and fatigue thresholds for composite materials," in Composite Materials: Fatigue and Fracture, P. A. Lagace, Ed., vol. 2 of ASTM STP 1012, pp. 222–250, American Society of Testing of Materials, 1989.

28. B. D. Davidson, C. S. Altonen, and J. J. Polaha, "Effect of stacking sequence on delamination toughness and delamination growth behavior in composite end-notched flexure specimens," in Composite Materials: Testing and Design (12th Volume), ASTM STP 1274, R. B. Deo and C. R. Staff, Eds., pp. 393–413, American Society of Testing of Materials, Philadelphia, Pa, USA, 1996.

29. D. Cartié, P. Davies, M. Peleau, and I. K. Partridge, "The influence of hydrostatic pressure on the interlaminar fracture toughness of carbon/epoxy composites," Composites B: Engineering, vol. 37, no. 4-5, pp. 292–300, 2006.

30. P. Davies, B. R. K. Blackman, and A. J. Brunner, "Standard test methods for delamination resistance of composite materials: current status," Applied Composite Materials, vol. 5, no. 6, pp. 345–364, 1998. ·

31. A. B. de Morais and M. F. S. F. de Moura, "Evaluation of initiation criteria used in interlaminar fracture tests," Engineering Fracture Mechanics, vol. 73, no. 16, pp. 2264–2276, 2006.

32. B. D. Davidson, "Towards an ASTM standardized test for determining GIIc of unidirectional laminated polymeric matrix composites," in Proceedings of the 21st Annual American Society for Composites Technical Conference, P. K. Mallick, Ed., DEStech Publications, September 2006.

33. E. Zile and V. Tamuzs, "Mode II delamination of a unidirectional carbon fiber/epoxy composite in four-point bend end-notched flexure tests," Mechanics of Composite Materials, vol. 41, no. 5, pp. 383–390, 2005.

34. A. J. Vinciquerra and B. D. Davidson, "Effect of crack length measurement technique and data reduction procedures on the perceived toughness from four-point bend end-notched flexure tests,"Journal of Reinforced Plastics and Composites, vol. 23, no. 10, pp. 1051–1062, 2004. ·
35. R. H. Martin and B. D. Davidson, "Mode II fracture toughness evaluation using four point bend, end notched flexure test," Plastics, Rubber and Composites, vol. 28, no. 8, pp. 401–406, 1999.
36. V. A. Franklin and T. Christopher, "Fracture energy estimation of DCB specimens made of glass/epoxy: an experimental study," Advances in Materials Science and Engineering, vol. 2013, Article ID 412601, 7 pages, 2013.
37. V. A. Franklin and T. Christopher, "Generation and validation of crack growth resistance curve from DCB specimens: an experimental study," Strength of Materials, vol. 45, no. 6, pp. 674–683, 2013.
38. V. Alfred Franklin, T. Christopher, and B. Nageswara Rao, "Influence of root rotation on delamination fracture toughness of composites," International Journal of Aerospace Engineering, vol. 2014, Article ID 829698, 12 pages, 2014.
39. R. H. Martin, T. Elms, and S. Bowron, "Characterization of mode II delamination using the 4ENF," inProceedings of the 4th European Conference on Composites: Testing & Standardisation, pp. 161–170, 1998.
40. P. Hansen and R. H. Martin, "DCB, 4ENF and MMB delamination characterisation of S2/8552 and IM7/8552," in Proceedings of the 15th Annual Technical Conference on Composite Materials, 2000.
41. A. Korjakin, R. Rikards, F.-G. Buchholz, H. Wang, A. K. Bledzki, and A. Kessler, "Comparative study of interlaminar fracture toughness of GFRP with different fiber surface treatments," Polymer Composites, vol. 19, no. 6, pp. 793–806, 1998.
42. A. Szekrényes, Delamination of composite specimens [Ph.D. dissertation], Budapest University of Technology and Economics, Budapest, Hungary, 2005.
43. F. Ozdil and L. A. Carlsson, "Beam analysis of angle-ply laminate mixed-mode bending specimens,"Composites Science and Technology, vol. 59, no. 6, pp. 937–945, 1999.
44. S. Hashemi, J. Kinloch, and J. G. Williams, "The effects of geometry, rate and temperature on mode I, mode II and mixed-mode I/II interlaminar fracture toughness of carbon-fibre/poly(ether-ether ketone)

composites," Journal of Composite Materials, vol. 24, no. 9, pp. 918–956, 1990.

45. S. Hashemi, A. J. Kinloch, and J. G. Williams, "Mechanics and mechanisms of delamination in a poly(ether sulphone)-Fibre composite," Composites Science and Technology, vol. 37, no. 4, pp. 429–462, 1990.

46. H. Albertsen, J. Ivens, P. Peters, M. Wevers, and I. Verpoest, "Interlaminar fracture toughness of CFRP influenced by fibre surface treatment: part 1. Experimental results," Composites Science and Technology, vol. 54, no. 2, pp. 133–145, 1995.

47. J. J. Polaha, B. D. Davidson, R. C. Hudson, and A. Pieracci, "Effects of mode ratio, ply orientation and precracking on the delamination toughness of a laminated composite," Journal of Reinforced Plastics and Composites, vol. 15, no. 2, pp. 141–173, 1996.

48. C. Dahlen and G. S. Springer, "Delamination growth in composites under cyclic loads," Journal of Composite Materials, vol. 28, no. 8, pp. 732–781, 1994.

49. B. D. Davidson and K. L. Koudela, "Influence of the mode mix of precracking on the delamination toughness of laminated composites," Journal of Reinforced Plastics and Composites, vol. 18, no. 15, pp. 1408–1414, 1999.

50. J. W. Gillespie, L. A. Carlsson, R. B. J. Pipes, R. Rothschilds, B. Trethewey, and A. Smiley, "Delamination growth in composite materials," NASA-CR 178066, 1986.

Chapter 6

ON THE IMPACT OF MANUFACTURING UNCERTAINTY IN STRUCTURAL HEALTH MONITORING OF COMPOSITE STRUCTURES: A SIGNAL TO NOISE WEIGHTED NEURAL NETWORK PROCESS

Hessamodin Teimouri[1], Abbas S. Milani[1], Rudolf Seethaler[1], Amir Heidarzadeh[2]

[1]School of Engineering, University of British Columbia, Kelowna, Canada

[2]Department of Aerospace Engineering, Sharif University of Technology, Tehran, Iran

ABSTRACT

This article investigates the potential impact of manufacturing uncertainty in composite structures here in the form of thickness variation in laminate plies, on the robustness of commonly used Artificial Neural Networks (ANN) in Structural Health Monitoring (SHM). Namely, the robustness of an ANN SHM system is assessed through an airfoil case study based on the sensitivity of delamination location and size predictions, when the ANN is imposed to noisy input. In light of the observed poor performance of the original network, even when its architecture was carefully optimized, it had been proposed to weigh the input layer of the ANN by a set of signal-to-noise (SN) ratios and then trained the network. Both damage location and size predictions of the latter SHM approach were increased to above 90%. Practical aspects of the proposed robust SN-ANN SHM have also been discussed.

INTRODUCTION

The cornerstone of Structural Health Monitoring (SHM) in engineering design is the comparison of data measured over a pre-defined damaged structure to the same type of information obtained from the healthy (un-dam- aged) structure, when subjected to identical loading/testing conditions [1] - [4] . A main goal in SHM is to seek for abnormalities in the structure's behavior and try to classify or correlate them to the location and extent of damage during the actual service of the same or a similar structure. For this purpose, the machine learning techniques have been developed and widely used by researchers and industry experts [1], by means of simulating the learning ability of humans via computer algorithms to analyze the measured input data and gain the corresponding (output) knowledge and skills. More specifically, the ultimate purpose of machine learning algorithms is to design computer tools that can effectively find the inherent relations between the inputs and outputs of a given complex system, and subsequently predicting the desired unknown data (e.g., the presence or absence of a critical crack in the current state of a structure) or judging its characteristics (e.g., the crack length). Generalization and robustness of the learning algorithms are vital to SHM system designers and require the ability of a chosen algorithm to predict the structure's response when confronted with input data outside the nominal training set (i.e., the problem of uncertainty) [5].

To elaborate on the latter concept of uncertainty, let us consider a sample SHM framework shown inFigure 1. The uncertainty in this system may come from the sensing systems by means of inaccurate data transmitted from sensors or imprecise database developed during the damage signature development process, manufacturing errors, environmental noises, loading perturbations, or the feature extraction/classification toolboxes; all of which can be potentially misleading for the SHM alerts or result in imprecise predictions [2]. Thus, performance and robustness of the SHM system in high-risk applications, such as those in aerospace, should be examined in the presence of noise and uncertainty of input parameters. Such type of SHM uncertainty has been exemplified earlier via a numerical case study [5] on a composite T-joint [6], which suggested that "the variation caused in the response of the structure due to uncertainty sources could be as large as those by the damage it self", hence a clear need for developing more robust SHM systems.

As a step forward to address the above need, the main aim of this article is to conduct an investigation into the development of a robust SHM via a weighted Artificial Neural Network (ANN), which can be immune against potential manufacturing errors in the structure. The selected case study is on predicting the location and extent of delamination in a composite airfoil [5] with the NACA-0012 profile under tensile loading [6]. Section 2 describes the experimental setup of the airfoil, the finite element model of the structure, and the developed damage signature database (DSD) to train the SHM. Section 3 provides background information on defining the proposed robust ANN SHM of the airfoil, based on a concept of signal-to-noise (SN) weighting. Section 4 provides the analysis results from the DSD along with an ANOVA analysis for correlating strain responses to the pre-def- ined damage scenarios, and thereby to quantify the significance of SHM uncertainty parameters. The same section follows with results of different weighted and unweighted ANNs, designed to predict damage in the structure from strain signatures in the presence of manufacturing errors (here in the form of composite ply thickness variations). Finally, Section 5 summarizes the main findings of the study and outlines some practical notions regarding the implementation of the proposed SN-ANN SHM.

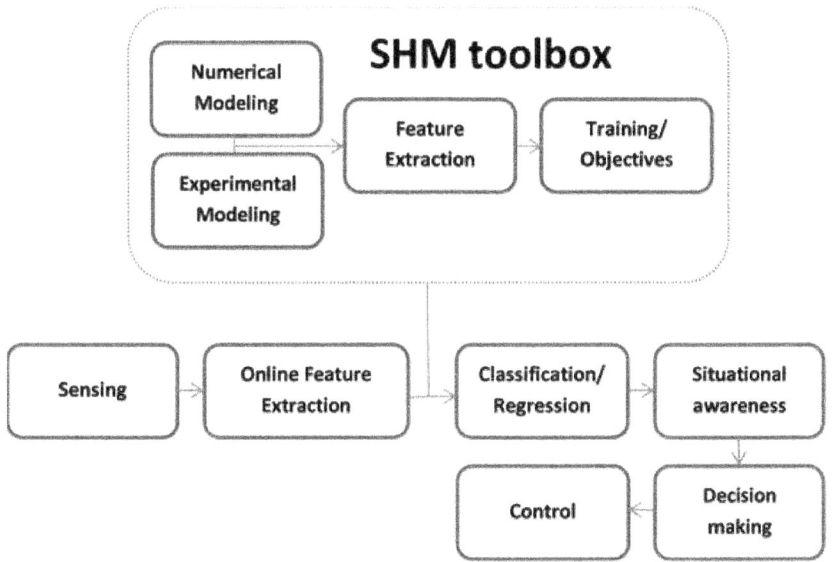

Figure 1: Example of a SHM system with different features [2].

CASE STUDY DESCRIPTION

The selected composite NACA-0012 airfoil is a sandwich structure consists of a 3 mm thick PVC foam, reinforced with E-glass and carbon woven fabrics (Figure 2). Table 1 lists the stacking sequence of the laminate schedule. A prototype structure was manufactured using the hand layup process and elastic material properties of the laminate components were estimated based on earlier studies [7]-[9] (Table 2).

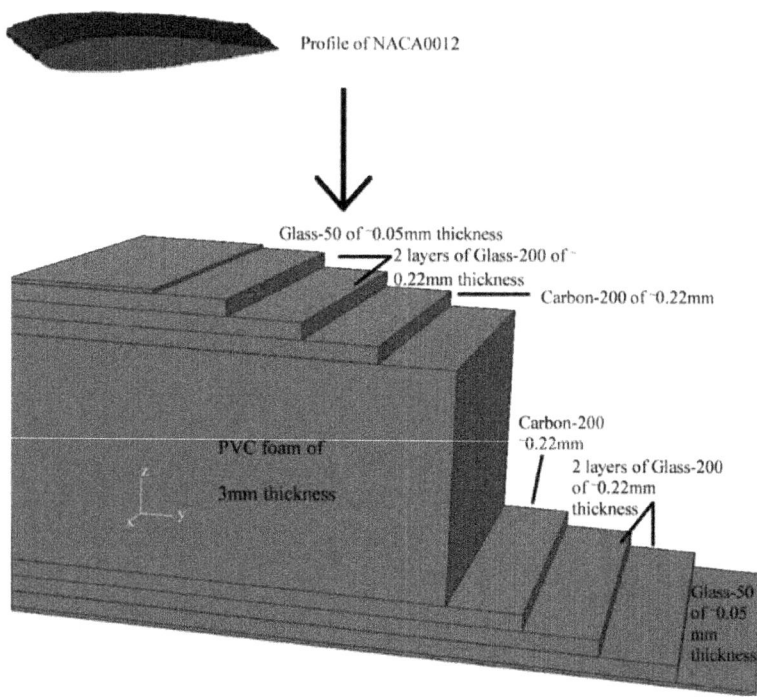

Figure 2: The NACA0012 composite airfoil lay-up used in the case study.

Table 1: Stacking sequence of the airfoil and nominal ply thicknesses

Layer no. (from top)	Type	Density (gr/m^2)	~Thickness (mm)
1	E-glass (woven)	50	0.06
2	E-glass (woven)	200	0.22
3	E-glass (woven)	200	0.22
4	Carbon (woven)	200	0.22
5	PVC foam	80	3.0
6	Carbon (woven)	200	0.22
7	E-glass (woven)	200	0.22
8	E-glass (woven)	200	0.22
9	E-glass (woven)	50	0.06

Table 2: The material properties used for modeling the airfoil plies (x-index refers to the fiber direction; woven fabrics in FE simulations were modeled as a cross-ply laminate)

Ply	Young's Modulus (MPa)	Poisson's Ratio	Shear Modulus (MPa)
CFRP	$E_1 = 62000$	$v_{xy} = 0.22$	$G_{xy} = 3270$
	$E_2 = 4800$	$v_{xz} = 0.22$	$G_{xz} = 3270$
	$E_3 = 4800$	$v_{yz} = 0.30$	$G_{yz} = 1860$
GFRP	$E_1 = 21000$	$v_{xy} = 0.26$	$G_{xy} = 1520$
	$E_2 = 7000$	$v_{xz} = 0.26$	$G_{xz} = 1520$
	$E_3 = 7000$	$v_{yz} = 0.30$	$G_{yz} = 2650$

The airfoil profile was initially tested under a pre-defined set of delamination scenarios under static tensile loading (Figure 3). Subsequently, a finite element (FE) model of the set-up with the material data in Table 2 was established in Abaqus/Standard and validated against experimental data. More details of the tensile experiments and the Finite Element verification can be found in [10]. The numerical model was employed as a virtual experimental tool to create more damage scenarios with varying ply thicknesses (mimicking a typical type of manufacturing error). Overall 166 damage scenarios were developed considering delaminations with different lengths, ranging from 1.5 cm to 4.5 cm, and at different locations along the chord line of the airfoil between the lower carbon ply and the middle PVC foam of the NACA0012 airfoil, with the internal chord of 31 cm and the external chord of 33.5 cm. Figure 4(a), Figure 4(b) illustrate a general scheme of the delaminations of different lengths and locations as used in establishing the damage signature database (DSD). It is noted that in each individual damaged airfoil scenario in the DSD, one single delamination (i.e., one size, one location) was implemented. It is also noted that 17 positions along the lower surface of the airfoil were considered as sensory points to estimate an accurate (semi-continuous) strain distribution in each DSD simulation.

To account for manufacturing uncertainty, ply thickness variations due to the hand layup production process were assumed to change from one sample to another. The thickness variation range was initially estimated based on tensile experiments. Namely, Table 3 shows the global displacement variations observed at different loading values for the tested airfoils with no delamination. It was assumed that these variations in the structure's global response have been equivalently caused by variations in thickness of different plies (carbon 200 gr/m^3 and glass 200 gr/m^3; both below and above the PVC foam). Subsequently, the FE model was used along with an inverse method to determine a reasonable thickness variation range for each of the above mentioned plies to cover at least 60% of experimental data scatter. The obtained lower and upper thickness limits (Table 4), which were also common during the hand laid-up trials, were next employed to generate random values in the subsequent stochastic simulations for each ply thicknesses, assuming a uniform probability density function. In summary, each damage scenario was simulated with five randomly varying ply thicknesses, in addition to the nominal thickness case listed in Table 1. Next, the goal was to employ the simulated DSD and develop a robust SHM to predict both damage size and location in the airfoil.

SIGNAL-TO-NOISE WEIGHTED ARTIFICIAL NEURAL NETWORK DEVELOPMENT

The Artificial Neural Networks (ANNs) are known as crude electronic models that have been inspired by neural structure of the brain. According to Gurney [11], ANNs "are interconnected assemblies of simple processing elements, units or nodes whose functionality is loosely based on the animal neuron. The processing ability of the network is stored in the inter-unit connection strengths, or weights, obtained by a process of adaption to, or learning from, a set of training patterns". The classical machine learning theory is classified into three main categories: classification, regression, and density estimation [12]. The ANNs under the classification category have been widely used in SHM features such as structural load monitoring, usage prediction, and damage diagnostics [13]-[19]. Given an SHM application, different ANN architectures can be defined and optimized based on feed-forward and recurrent networks.

Figure 3: A composite airfoil sample under tension [10].

The majority of the published work on the development of ANN SHMs has implemented feed-forward networks, and in particular the Multilayer Perceptron (MLP) networks [20] . There are limited reports, however, describing the application of recurrent networks, especially for NDT testing, SHM and material property characterization in general.

The input layer of a MLP network can receive, for instance, the strain measurements at different sensory point

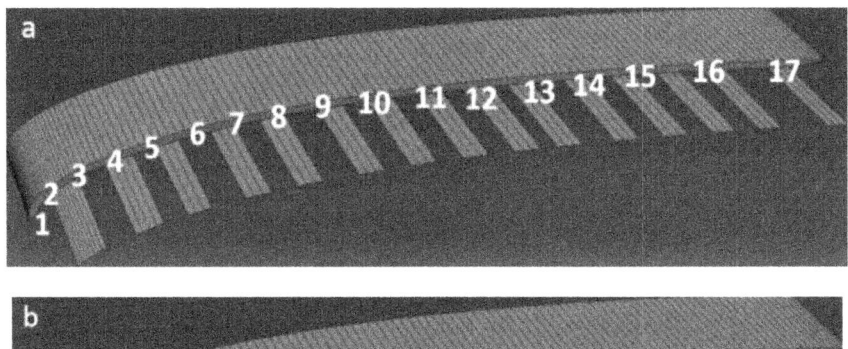

Figure 4: Sample delaminations with different lengths plotted adjacent to the PVC foam in the finite element model of the airfoil: (a) ~1 cm delamination, (b) ~2 cm delamination.

Table 3: Variation of displacement observed at different load magnitudes for different samples through repeats of the tensile tests [10]

Load [N]	Min. Displacement [mm]	Max. Displacement [mm]	Difference %	Mean [mm]
500	0.222	0.349	36.3	0.285
1000	0.465	0.692	32.8	0.579
1500	0.757	1.058	28.4	0.907
2000	1.072	1.479	27.5	1.275
2500	1.403	1.668	15.9	1.535

Table 4: Half of the calculated thickness ranges (in mm) of different composite layers to cover about 60% of the variation observed in the experiments [10]

		Glass 200 below PVC foam		
Load (N)	Minimum thickness	Displacement at max loading (corresponding to minimum thickness)	Maximum thickness	Displacement at max loading (corresponding to maximum thickness)
1500	0.16	1.033	0.39	0.8437
		Carbon 200 below PVC foam		
Load (N)	Minimum thickness	Displacement at max loading (corresponding to minimum thickness)	Maximum thickness	Displacement at max loading (corresponding to maximum thickness)
1500	0.18	1.031	0.36	0.8421
		Glass 200 above PVC foam		
Load (N)	Minimum thickness	Displacement at max loading (corresponding to minimum thickness)	Maximum thickness	Displacement at max loading (corresponding to maximum thickness)
1500	0.17	1.038	0.28	0.8424
		Carbon 200 above PVC foam		
Load (N)	Minimum thickness	Displacement at max loading (corresponding to minimum thickness)	Maximum thickness	Displacement at max loading (corresponding to maximum thickness)
1500	0.15	1.040	0.3	0.8408

of the structure, and the output layer would predict the location and extent of the existing damage corresponding to the measured input data. As will be shown in Section 4, a conventional MLP ANN can be potentially unable to provide accurate results when optimally trained with un-noisy damage scenarios (i.e., here based on nominal computer simulations), and asked to predict the damage status in practice in the presence of uncertainties. For this reason, in the present work an application of the Signal-to-Noise (SN) ratio coefficients is tested to weigh the input layer of the MLP network (Figure 5) and possibly improve the accuracy of damage predictions.

Signal-to-noise ratio analysis is a means for comparing the level (amplitude) of a desired signal (target value) to the corresponding level of background noise (fluctuations/variation) in measurements. This concept has been widely used in electrical and electromagnetics engineering where the log function of the ratio of signal to noise is defined as the SN factor. In general, there are four types of SN's [21] :

- Lower-the-better static SN: $-10\log_{10}\log_{10}\left(\frac{1}{m}\Sigma\frac{1}{y_{ij}^2}\right)$ also called SN-L;

- Higher-the-better static SN: $-10\log_{10}\log_{10}\left(\frac{1}{m}\Sigma y_{ij}^2\right)$ also called SN-H;

- Nominal-the-best static SN: $10\log_{10}\log_{10}\left(\frac{\overline{y_i}^2}{s_i^2}\right)$ where $\ln\ln s_i^2 = \ln\ln\frac{\Sigma(y_{ij}-\overline{y_i})^2}{m-1}$ also called SN-M.

Here for a given sensor on the structure, y_{ij} is defined as the sensor reading in the ith damage scenario and the jth thickness variation (test repeat). In another type of so-called 'dynamic' signal-to-noise ratio analysis, for each sensor point all the measured data are plotted in the ordered x-y plane (here damage scenario versus measurements) and the slope of a regression line passing through them is calculated (see [21] for theoretical details). If MSE indicates the mean square error of sample measurements, the dynamic SN ratio is then defined as:

- Zero-proportional Dynamic SN: $10\log_{10}\log_{10}\frac{slope^2}{MSE}$ also called SN-D.

For the present study, only the SN-M and SN-D methods are physically meaningful as the sensor values under each damage condition are always best if they are nominal (not the lower the better, not the higher the better). It is to note that in comparison with the SN-M mode, in the dynamic mode SN-D, the measurement values are not averaged (assumed constant) over the entire damage scenarios; instead it is assumed that the nominal value of measurements at each sensor point can vary with the damage scenario. As a result, it is expected that the SN-D method outperforms the SN-M method in such SHM applications (more on this topic to follow in the next section). Table 5 lists these SN ratios for the DSD of the composite airfoil. In fact, the obtained SN coefficients are

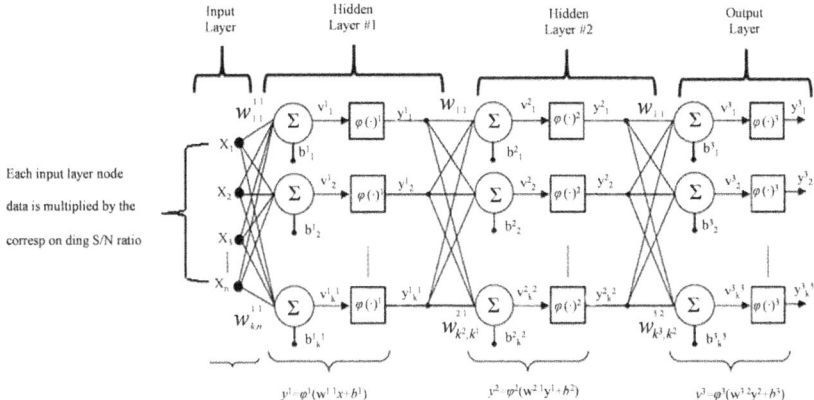

Figure 5: Schamtic of a multilayer perceptron ANN (adapted from [33]), weighted by the proposed S/N ratios at the input layer.

Table 5: Summary of obtained signal-to-noise (SN) ratios

S/N	Sensor 1	Sensor 2	Sensor 3	Sensor 4	Sensor 5	Sensor 6	Sensor 7	Sensor 8	Sensor 9	Sensor 10	Sensor 11	Sensor 12	Sensor 13	Sensor 14	Sensor 15	Sensor 16	Sensor 17
SN-M	-34.8	-34.0	-31.6	-32.8	-32.9	-33.1	-32.7	-33.1	-32.8	-32.8	-32.7	-32.5	-32.3	-32.2	-31.6	-26.9	-34.5
SN-D	30.4	32.2	43.6	37.1	34.3	32.6	32.5	31.9	31.5	31.1	31.3	31.2	31.2	31.3	32.6	41.3	30.8

expressing the vulnerability of each individual sensory point under the simultaneous presence of damage and thickness variation scenarios. Regardless of the type, a higher SN coefficient indicates a stronger average signal compared to the noise level, which implies the robustness of the corresponding sensor in reporting the abnormalities caused by the damage status.

Examples of the application of the concept of signal-to-noise ratio in other engineering and medical applications include the quantification of noise in MRI (Magnetic Resonance Imaging) systems [22], diffusion weig- hted images of spinal cords [23], electrocardiograms [24], ultrasonic non-destructive testing of highly scattering materials [25], etc. The idea of weighting the input layer of a neural network to demonstrate the importance of individual measurement points has also been the topic of some past studies. Zou et al. [26] used back-propag- ation neural networks with a Levenberg-Marquardt learning algorithm to study the protein combinations of both amino acid and amino pairs. The overall accuracy of prediction reached 88.4%, which was a notable enhancement compared to the conventional neural network with an accuracy as low as 66.1%. Chen et al. [27] applied the weighted input layer (adapted from the information entropy theory) to the Elman NN to predict gas turbines performance. The application of the entropy theory to the conventional network resulted in a decreased mean square error of the training/prediction process by about 29%, for real-time predictions.

DAMAGE PREDICTION RESULTS AND DISCUSSION

The developed DSD in Section 2 was used in this section for two complementary analyses. First, an ANOVA analysis was established where the statistical significance of the uncertainty factors on the airfoil SHM predictions was proven statistically. Next, a series of Artificial Neural Network (ANN) SHMs were trained and compared for their prediction quality on the damage location and size, under the presence of manufacturing uncertainty factor and using the analytical framework presented in Section 3.

ANOVA ANALYSIS

Table 5 shows the summary of ANOVA analysis of all 17 sensors in Figure 4, after running the FE model for the entire (166) scenarios of the damage signature database (DSD). The resulting p-values in Table 6 show that no sensor is able to merely focus on predicting the damage status, as all sensor values have been affected by the uncertainty factor (noise) with a P-value < 5%. An earlier statistical study [10], which was based on the same numerical data but using only 3 instead of 166 damage scenarios, had concluded that among the distributed sensors, four sensor locations showed significantly higher sensitivities (i.e., P-values < 5%) to the damage factor than the thickness variation (noise). Hence, one would mistakenly conclude that those few sensor locations may be chosen for the subsequent DSD training purposes during robust SHM development. However, as seen in Table 6, when a large damage database used (here 166 scenarios), in fact all sensors can be significantly affected by the noise against prediction of damage at different locations and with different sizes. Hence, two conclusions may be made here when comparing the present ANOVA results and those of [10] :

- First, it is critical to ensure that a given DSD contains sufficient number of damage scenarios to accurately represent the reality.
- Second, for a sufficiently large damage database, and with uncertainty propagated throughout the structure, caution must be taken not to pre-define/prefer limited sensor locations to be used for training proposes.

Table 6. ANOVA analyses for 166 damage scenarios under uncertainty factor (note that a P-value of nearly zero indicates a very high statistical significance/sensitivity).

Source of variation	SS	df	MS	F	P-value	F crit
Sensor 1						
Damage size/position	865078	165	8318	116	0	1.28
Thickness variation	779133	4	194783	2714	0	2.39
Sensor 2						
Damage size/position	4472109	165	43001	1161	0	1.28
Thickness variation	17496	4	4374	118	0	2.39
Sensor 3						
Damage size/position	11370307	165	109330	436	0	1.28
Thickness variation	7637	4	1909	7.62	6.20508E-06	2.39
Sensor 4						
Damage size/position	9377200	165	90165	748	0	1.28
Thickness variation	10662	4	2665	22.1	0	2.39
Sensor 5						
Damage size/position	3250375	165	31254	564	0	1.28
Thickness variation	9096	4	2274	41.1	0	2.39
Sensor 6						
Damage size/position	2196479	165	21120	380	0	1.28
Thickness variation	11622	4	2906	52.3	0	2.39

Sensor 7						
Source of variation	SS	df	MS	F	P-value	F crit
Damage size/position	1372512	165	13197	414	0	1.28
Thickness variation	8419	4	2105	66.1	0	2.39
Sensor 8						
Source of variation	SS	df	MS	F	P-value	F crit
Damage size/position	864235	165	8310	289	0	1.28
Thickness variation	11254	4	2813	97.7	0	2.39
Sensor 9						
Source of variation	SS	df	MS	F	P-value	F crit
Damage size/position	399288	165	3839	431	0	1.28
Thickness variation	8686	4	2172	244	0	2.39
Sensor 10						
Source of variation	SS	df	MS	F	P-value	F crit
Damage size/position	351589	165	3381	417	0	1.28
Thickness variation	8557	4	2139	264	0	2.39
Sensor 11						
Source of variation	SS	df	MS	F	P-value	F crit
Damage size/position	206701	165	1988	425	0	1.28
Thickness variation	6976	4	1744	373	0	2.39
Sensor 12						
Source of variation	SS	df	MS	F	P-value	F crit
Damage size/position	134266	165	1291	502	0	1.28
Thickness variation	5119	4	1280	498	0	2.39
Sensor 13						
Source of variation	SS	df	MS	F	P-value	F crit
Damage size/position	130149	165	1251	842	0	1.28
Thickness variation	4122	4	1030	694	0	2.39
Sensor 14						
Source of variation	SS	df	MS	F	P-value	F crit
Damage size/position	111839	165	1075	544	0	1.28
Thickness variation	4255	4	1064	538	0	2.39
Sensor 15						
Source of variation	SS	df	MS	F	P-value	F crit
Damage size/position	318843	165	3066	1454	0	1.28
Thickness variation	4725	4	1181	560	0	2.39
Sensor 16						
Source of variation	SS	df	MS	F	P-value	F crit
Damage size/position	304479	165	2928	407	0	1.28
Thickness variation	41060	4	10265	1426	0	2.39
Sensor 17						
Source of variation	SS	df	MS	F	P-value	F crit
Damage size/position	1238033	165	11904	2472	0	1.28
Thickness variation	94708	4	23677	4916	0	2.39

Weighted and Unweighted MLP ANN's

As addressed in Section 3, MLP Artificial Neural Networks are powerful techniques used for pattern recognition purposes in SHM applications [28]-[33] . ANNs are provided with the measured sensory information such as displacement, acceleration, stress/strain, damping ratio, and mode shapes, in time or frequency domains, and are expected to correlate these data to the state of damage (location and size of defect) in the structure.

In the current example, the well-known technique of k-fold cross validation [33] was used to arrive at architecturally optimized ANN architectures. This technique helps the analyst select the number of hidden layers, number of neurons in each hidden layer, activation function, learning algorithm and learning rule. This was initially done for the airfoil's SHM using the nominal DSD only (i.e., without considering noise). The summary of the obtained cross validation results for the nominal damage cases are presented in Table 7. During the cross- validation, in each iteration 70% of the dataset was considered to be the training set and the 30% for validation and testing; i.e. for instance when the network was optimized on the original 166 dataset, 116 sets were used for training and 50 sets for validation and testing. The best trained network for this SHM had 17 neurons in the first layer (input layer), 24 neurons in the second layer, 20 neurons in the third layer, 8 neurons in the fourth layer and 2 neurons in the last layer (also known as output layer). This network (24-20-8-2 NN in Table 7) was then used to predict damage in all simulated airfoils, both with and without uncertainty (i.e., versus the total DSD). In addition, for comparison purposes, the same network was trained with both nominal and noisy damage scenarios and again used to predict the total DSD. Results of the aforementioned approaches are provided in Table 8.

Concerning the low prediction% values in Table 8, it is clear that the conventional (unweighted) ANN has not been capable of predicting the thickness varying scenarios when it is only trained by the nominal DSD. When the same network is trained with the total DSD, however, it has greatly become capable of predicting

Table 7: Optimized neural networks for the nominal DSD without thickness variation.

Architecture	Min. damage location prediction error %	Min. damage size prediction error %
2 hidden layers	14.4% (70-2 neurons)	17.1% (70-2 neurons)
3 hidden layers	1.8% (25-7-2 neurons)	5.2% (46-11-2 neurons)
4 hidden layers	1.1% (24-20-8-2 neurons)	4.0% (16-18-11-2 neurons)

Table 8: Comparison of the neural networks for nominal and thickness varying damage scenarios.

NN type	Training set	Predicting set	% Accuracy of size prediction	% Accuracy of location prediction
Conventional NN	Original	Original + thickness varying	35.5%	57.2%
Conventional NN	Original + thickness varying	Original + thickness varying	95.8%	97.0%
SN-M NN with original architecture	Original	Original + thickness varying	79.0%	70.6
SN-M NN with re-optimized architecture	Original	Original + thickness varying	86.8%	79.0%
SN-D NN with nominal optimal architecture	Original	Original + thickness varying	87.3%	93.1%
SN-D NN with re-optimized architecture	Original	Original + thickness varying	92.2%	91.2%

damage location (97.0%) and damage extent (95.8%) under uncertain input data. This shows a desirable generalizability of the ANN for robust SHM applications, but it can also pose an important challenge. Namely, in practice the second unweighted but robust ANN will require testing all damage scenarios under several random repeats to arrive at a sufficiently large DSD to be included in the training pool. Running such large DSDs with several noise scenarios would normally pose high cost and time limitations in industrial settings.

To address this challenge, we next attempted to use SN ratios to weigh the input layer of the MLP ANN. Table 8 illustrates the prediction results where the conventional neural network has been weighted by the SN ratios of Table 5 and predicted the noisy (total) DSD. Note that the noisy DSD scenarios have not been directly used in the latter training process, but their impacts have been condensed in the SN ratios. Both the SN-M and SN-D weighting methods have resulted in an improved performance of SHM when used on the original optimum ANN architecture with or without re-optimization. By implementing the "dynamic" signal to noise ratio (SN-D), how- ever, the accuracy of damage location prediction was found to be the highest among all tested ANN models (as high as ~92%).

CONCLUDING REMARKS

MLP Artificial Neural Networks have proven to be powerful tools for pattern recognition in the structural health monitoring applications, yet its efficiency can reduce when dealing with unseen uncertainty in the input layers of a system such as ply thickness variation in the composite laminate. Using a case study on a composite airfoil and ANOVA analysis over a DSD with 166 damage scenarios along with five random repeats of each due to manufacturing error, it was found that all the sensory points exhibited almost the same delamination detection sensitivity as the uncertainty effect itself, hence the clear need for a robust SHM development. Also by comparing the above results to the earlier study [10], it was found that the small-size DSDs could be misleading in terms of preselecting specific sensory points to be immune against noise, especially when the uncertainty had been propagated throughout the structure.

One way to deal with uncertainty in the input of SHM data is to include the uncertainty scenarios in the training set, but this approach would involve developing a vast damage database which often requires considerable time and budget for industries. Another approach discussed here would be weighing the conventional neural network by signal-to-noise coefficients. In the performed case study, the noisy DSD was analyzed for each input neuron to calculate an appropriate SN ratio. The approach dramatically increased the efficiency of the ANN- based SHM, even though it was only trained with the original damage scenarios (nominal DSD) and predicted the noisy DSD. Table 8 is a complete summary of all different training and prediction scenarios using the unweighted ANN and weighted ANNs. Among different types of SN, the dynamic SN weighted neural network showed a superior accuracy above 90% in damage prediction. A practical problem with this approach, however, is the reliable estimation of SN weights. This estimation may come from past experience, expert knowledge or by developing a sub-set of initial DSD encompassing uncertainty cases.

This project is a part of a larger research program aiming at identifying practically and economically most viable, yet accurate and robust, algorithms for implementation in real-time SHM systems of composite structures. The assessment of other powerful non-parametric methods such as the Gaussian Processes (GPs) is also of high interest in developing and comparing future robust SHMs.

ACKNOWLEDGEMENTS

This research was funded by the Natural Science and Engineering Research Council of Canada (NSERC), under the Discovery Grants Program.

REFERENCES

1. Boller, C. and Meyendorf, N. (2008) State-of-the-Art in Structural Health Monitoring for Aeronautics. Proceedings of International Symposium on NDT in Aerospace, Fürth/Bavaria, Germany, 3-5 December 2008.

2. Balageas, D., Fritzen, C.P. and Gumes, A. (2006) Structural Health Monitoring. Antony Rowe Ltd., Chippenham, Wiltshire.
3. Perez, I., DiUlio, M., Maley, S. and Phan, N. (2010) Structural Health Monitoring in the Navy. International Journal of Structural Health Monitoring, 9, 199-209.http://dx.doi.org/10.1177/1475921710366498
4. Lopez-Higuera, J.M. (2002) Introduction to Optical Fiber Sensor Technology. In: Lopez-Higuera, J.M., Ed., Handbook of Optical Fibre Sensing Technology, Wiley, New York, 1-21.
5. Teimouri, H., Milani, A.S. and Seethaler, R. (2013) On the Effect of Fabrication and Testing Uncertainties in Structural Health Monitoring. In: Silva, M., Ed., Design of Experiments Applications, InTech, Croatia. http://dx.doi.org/10.5772/56530
6. Kesavan, A., John, S. and Herszberg, I. (2008) Strain Based Structural Health Monitoring of Complex Composite Structures. Structural Health Monitoring, 7, 1-13.
7. Kachlakev, D.I. (1998) Finite Element Method (FEM) Modeling for Composite Strengthening/Retrofit of Bridges. Research Project Work Plan, Civil, Construction and Environmental Engineering Department, Oregon State University, Corvallis, Oregon.
8. Kachlakev, D.I. (1998) Strengthening Bridges Using Composite Materials. FHWA-OR-RD-98-08, Oregon Department of Transportation, Salem, Oregon.
9. Kachlakev, D.I. and McCurry Jr., D. (2000) Simulated Full Scale Testing of Reinforced Concrete Beams Strengthened with FRP Composites: Experimental Results and Design Model Verification. Oregon Department of Transportation, Salem, Oregon.
10. Teimouri, H., Milani, A.S., Seethaler, R., Abedian, A., Heidarzadeh, A. and Teimouri, B. (2013) Towards Strain-Based Structural Health Monitoring of Composite Airfoil under Uncertainty. 19th International Conference on Composite Materials, Montreal, Canada, July-August 2013, 1-8.
11. Gurney. K. (1997) An Introduction to Neural Networks. University of Sheffield, UK.http://dx.doi.org/10.4324/9780203451519
12. Chu, F., Yuan, S. and Peng, Z. (2009) Machine Learning Techniques. Encyclopedia of Structural Health Monitoring.
13. Azzam, H. (1997) A Practical Approach for the Indirect Prediction of Structural Fatigue from Measured Flight Parameters. Proceeding of the Institution of Mechanical Engineering, Part G: Journal of Aerospace Engineering, 211, 29-38.http://dx.doi.org/10.1243/0954410971532479

14. Azzam, H. (1997) The Use of Mathematical Models and Artificial Intelligent Techniques to improve Hums Prediction Capabilities. Proceedings of the Royal Aeronautical Society, Innovation in Rotorcraft Technology Conference, London, 24-25 June 1997, 16.1-16.14.

15. Azzam, H., Hebden, I., Gill, M., Beavan, F. and Wallace, M. (2005) Fusion and Decision Making Techniques for Structural Prognosis Health Management. IEEE Aerospace Conference, Montana, MT, 5-12 March 2005, Paper #1535.

16. Wallace, M., Azzam, H. and Newman, S. (2004) Indirect Approaches to Individual Aircraft Structural Monitoring. Proceeding of the Institution of Mechanical Engineering, Part G: Journal of Aerospace Engineering, 218, 329-346.http://dx.doi.org/10.1243/0954410042467059

17. Reed, S.C. and Cole, D.G. (2003) Development of a Parametric Aircraft Fatigue Monitoring System Using Artificial Neural Network. Proceedings of the 22nd Symposium of the International Committee on Aeronautical Fatigue, Lucern, 9 May 2003, 47-63.

18. Scallonila, G., Cracia, J., Cabrejas, J. and Armijo, J.I. (2007) A Full-Scale Parametric Based Fatigue Monitoring System Using Neural Networks. Proceedings of the 24th Symposium of the International Committee on Aeronautical Fatigue, Naples, 16-18 May 2007.

19. Levinski, O. (2001) Australian Defense Science and Technology Organization. Prediction of Buffet Loads Using Artificial Neural Network, Document DSTO-RR-0218.

20. Teimouri, H. (2015) A New Statistical Approach to Strain-Based Structural Health Monitoring of Composites under Uncertainty. PhD Dissertation in Mechanical Engineering, University of British Columbia, BC, Canada.

21. Fowlkes, W.Y. and Creveling, C.M. (2013) Engineering Methods for Robust Product Design: Using Taguchi Methods in Technology and Product Development. Prentice Hall, Englewood Cliffs.

22. Welvaert, M. and Rosseel, Y. (2013) On the Definition of Signal-to-Noise Ratio and Contrast-to-Noise Ratio for fMRI Data. PLoS ONE, 8, e77089. http://dx.doi.org/10.1371/journal.pone.0077089

23. Griffantia, L., Baglioa, F., Pretia, M.G., Cecconic, P., Rovarisd, M., Basellib, G. and Laganà, M.M. (2012) Signal-to- Noise Ratio of Diffusion Weighted Magnetic Resonance Imaging: Estimation Methods and in Vivo Application to Spinal Cord. Biomedical Signal Processing and Control, 7, 285-294. http://dx.doi.org/10.1016/j.bspc.2011.06.003

24. Poungponsri, S. and Yu, X.H. (2013) An Adaptive Filtering Approach for Electrocardiogram (ECG) Signal Noise Reduction Using Neural Networks. Neurocomputing, 117, 206-213. http://dx.doi.org/10.1016/j.neucom.2013.02.010

25. Liu, A., Lu, M. and Wei, M. (1997) Structure Noise Reduction of Ultrasonic Signals Using Artificial Neural Network Adaptive Filtering. Ultrasonics, 35, 325-328.http://dx.doi.org/10.1016/S0041-624X(97)00009-7

26. Zou, L., Wang, Z. and Huang, J. (2007) Prediction of Subcellular Localization of Eukaryotic Proteins Using Position- Specific Profiles and Neural Network with Weighted Inputs. Journal of Genetics and Genomics, 34, 1080-1087.http://dx.doi.org/10.1016/S1673-8527(07)60123-4

27. Chen, T., Xu, X. and Wang, S. (2011) An Intelligent Prediction Method Based on Information Entropy Weighted Elman Neural Network. Proceedings of the Intelligent Computing and Information Science: International Conference, Chongqing, 8-9 January 2011, Part II, 142-147. http://dx.doi.org/10.1007/978-3-642-18134-4_23

28. Fang, X., Luo, H. and Tang, J. (2005) Structural Damage Detection Using Neural Network with Learning Rate Improvement. Computers and Structures, 83, 2150-2161.http://dx.doi.org/10.1016/j.compstruc.2005.02.029

29. Kesavan, A., John, A. and Herszberg, I. (2008) Strain-Based Structural Health Monitoring of Complex Composite Structures. Structural Health Monitoring, 7, 203-213.http://dx.doi.org/10.1177/1475921708090559

30. Al-Haik, M.S., Hussaini, M.Y. and Garmestani, M. (2006) Prediction of Nonlinear Viscoelastic Behaviour of Polymeric Composites Using an Artificial Neural Network. International Journal of Plasticity, 22, 1367-1392.http://dx.doi.org/10.1016/j.ijplas.2005.09.002

31. Koker, R., Altinkok, N. and Demir, A. (2006) Neural Network Based Prediction of Mechanical Properties of Particulate Reinforced Metal Matrix Composites Using Various Training Algorithms. Materials & Design, 28, 616-627.

32. Singh, A.P., Kamal, T.S. and Kumar, S. (2005) Virtual Curve Tracer for Estimation of Static Response Characteristics of Transducers. Measurement, 38, 166-175.http://dx.doi.org/10.1016/j.measurement.2005.04.005

Chapter 7

ENHANCEMENT IN THE ELECTRICAL AND THERMAL PROPERTIES OF ETHYLENE VINYL ACETATE (EVA) CO-POLYMER BY ZINC OXIDE NANOPARTICLES

Jose Sebastian[1], Eby T. Thachil[1], Jobin Job Mathen[2], Joseph Madhavan[2], Prince Thomas[3], Jacob Philip[4], M. S. Jayalakshmy[4], Shahrom Mahmud[5], Ginson P. Joseph[3]

[1]Department of Polymer Science and Rubber Technology, Cochin University of Science and Technology, Kerala, India

[2]Department of Physics, Loyola College, Chennai, India

[3]Department of Physics, St. Thomas College, Palai, India

[4]Department of Instrumentation, Cochin University of Science and Technology, Kerala, India

[5]School of Physics, Universiti Sains Malaysia, Minden, Malaysia

ABSTRACT

EVA/ZnO nanocomposites of 1%, 2% and 4% ZnO were fabricated by direct probe sonicator method. The ZnO nanopowders were prepared by solvothermal method. As the particle size of the filler incorporated to the polymer matrix decreases, the properties of the polymer-filler interface show dominance over its bulk properties. The dielectric constant and dielectric loss of the composites at ambient temperatures are found to decrease with increasing frequency. The thermal analysis using TGA-DTA is also performed and it is found that the thermal stability of the nanocomposites increases with increasing the filler concentrations. The thermal parameters such as thermal diffusivity (α) and thermal effusivity (e), the thermal conductivity (k) and heat capacity (C_p) were studied using photopyroelectric technique. The band gap of the samples was also determined and found to decrease with increasing filler concentrations. The tensile strength and peel strength of the samples were also investigated and it is found to increase with small inclusion of filler material.

INTRODUCTION

Composites have attracted attention of material scientists as it can combine advantages of different materials. In recent years, material scientists are looking for nano-composites based on polymer matrix due to several added advantages. The advantages include balanced physical and mechanical properties, ease of processability and low production cost [1]. Many previous works have been carried out to improve the optical and electrical properties of polymers through suitable doping [2] [3]. Polymer based dielectric materials can give flexible and light weight electrical devices. It is discovered that nano-particles like Al_2O_3, TiO_2, SiO_2 etc. heterogeneously distributed within the polymer matrix can enhance dielectric properties [4]. Murugaraj and co-workers [5] have fabricated polymer-alumina nano-composites with improved dielectric characteristics. Carbon nanotubes (CNTs), carbon black, carbon nanofibres (CNF) as well as single and multi wall carbon nano tubes (SWNTs & MWNTs) have been incorporated to polymer matrix to use as antistatic coatings [6].

During recent years, colloidal and semiconducting nano particles have attracted a great deal of attention for both researchers and industrialists. Different types of group II-VI nano particles including ZnSe, CdS, CdSe, CdTe are found to be used extensively for light-emitting diode [7], solar cell [8], biomedical tag [9] and laser [10] applications. Nanocrystals (NCs) of semi conducting materials are used in optoelectronic devices like lasers and transistors [11] [12]. The ZnO used in this study is one such type of nanopowder having excellent ultraviolet and visible photoluminescence [13], and it is a semiconductor having large exciton binding energy (60 meV). It has got diverse applications in photovoltaic cells, variable resistors, as fully transparent thin film transistors and in short wavelength light emitting diodes.

Wide variety of polymers are found application in the synthesis of nano composites. They form the continuous phase termed as the matrix of the composite. Polymer matrix composites (PMC) with ceramic and metals as fillers have been developed to improve electrical properties like dielectric permittivity [14] [15]. Poly methylmethacrylate [16] [17], epoxy [18], poly(vinyl alcohol) [19], polyaniline [20] [21] are extensively used as the matrix for composites.

Poly (ethylene-co-vinyl acetate), EVA, was used as the base polymer in this experiment as they are compatible even with inert fillers. EVA is noted for its rubbery nature along with gloss, permeability and good impact strength. EVA-TiO_2 nanocomposites were investigated for the effect of TiO_2 particle size on the co-efficient of thermal expansion [22]. EVA copolymer irradiated with gamma rays can cause modification in its electronic structure [23]. Ethylene vinyl acetate is particularly used in electrical industry as cable insulating material due to good stress cracking resistance.

The II-VI semiconducting materials show significant properties from the optoelectronic point of view [24]. In the bulk form and in the quantum dot form these materials exhibit high density and quantum confinement. This paper deals with the effect of ZnO nanopowder on the electrical, optical, mechanical properties EVA polymer matrix and reported for the first time.

SYNTHESIS OF ZNO NANOPARTICLE

A solution of 0.2M-$(CH_3COO)_2Zn \cdot 2H_2O$, Zinc Acetate, was prepared by dissolving 4.39 gm of $(CH_3COO)_2Zn \cdot 2H_2O$ in 100 ml of methanol in a beaker and the mixture was kept stirred for 15 minutes. Another mixture of 0.5M-NaOH and methanol was prepared by dissolving 0.5 gm of NaOH in 25 ml of methanol and was kept for stirring for 15 minutes. Then the NaOH-methanol mixture was added to the basic solution and the reaction mixture was kept stirred for 30 minutes. The prepared solution was kept in autoclave for drying at 180°C for 5 hours to obtain nano-sized ZnO particles. Dried ZnO nanoparticles, white in colour was obtained. The size of the nanopowder was determined using TEM and it as confirmed 20 nm (Figure 1).

Synthesis of ZnO/EVA Polymer Nanocomposite

Poly (ethylene-co-vinyl acetate), EVA copolymer used for the experiment was obtained from ExxonMobil Chemicals, Singapore. The vinyl acetate content of the copolymer used was 9.4 wt% (Density—0.931 g/cm^3, Melt Flow Index—2.1 g/10min @190°C, 2.16 kg). Ethylene Vinyl Acetate-ZnO nanocomposites were prepared for different weight percentage (1%, 2%, and 4%) of ZnO by direct probe sonicator method. Initially pure EVA film was made in a glass mould by solvent casting method using toluene. Then 1% by weight of ZnO nanoparticles was added to EVA- toluene mixture taken in a beaker and was subjected to direct probe sonication. Finally

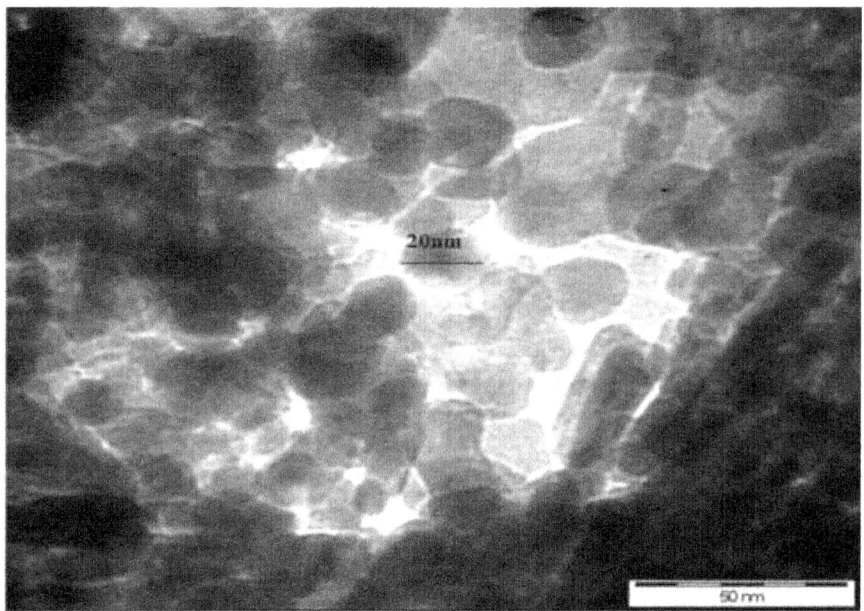

Figure 1: TEM image of ZnO nanopowder.

the polymer was dried in a glass mould for 3 - 4 hours at 50°C and thus polymer nanocomposite with 1% ZnO was formed. Similarly nanocomposites for remaining weight percentage, (2% and 4%) of ZnO were prepared.

RESULTS AND DISCUSSION

Morphological Studies

TEM micrograph shown in Figure 1 confirmed that the synthesized ZnO powder are solid in nature and obtained particle size is about 20 nm. SEM micrographs were used to identify the relative differences in surface characteristics of EVA and its nanocomposites. The ZnO particle distribution and its influence on the EVA copolymer morphology were also investigated. The interaction between ZnO with host matrix is strong and the surface micelles are homogeneous. Micrographs at 10 μm and 1 μm showed the developed shape of the EVA particles and the filler distribution in EVA as shown in Figure 2(b) & Figure 2(c). A homogeneous dispersion of ZnO in the EVA matrix was observed in Figure 2(a). The nano ZnO dispersion in the EVA surface was evident with higher magnification at 100 nm.

Electrical and Optical Properties

The ZnO/EVA polymer nanocompsite of area 15 mm^2 having silver coating on the opposite faces was introduced between two copper electrodes and then connected to HIOKI 3532-50 LCR Impedance analyzer for dielectric measurement. The dielectric constant of the sample is calculated using the relation $\epsilon_r = Cd/\epsilon_0 A$; where the nanocomposite acts as a dielectric with ϵ_0 the absolute permittivity, C is the capacitance, d is the thickness and A is the area (mm^2) of the ZnO/EVA composite. Figure 3 shows the variation of dielectric constant of ZnO/EVA with different filler concentrations. From the figure one can easily examine the behaviour of dielectric constant of nanocomposites with varying frequency from 100 Hz to 5 MHz. It is observed that initially the dielectric constant has larger values at lower frequencies and then decreases with increase in frequency for all films; however there is an increase in the dielectric constant of the nanocomposites as the percentage of the filler concentrations increases. The value of dielectric constant at 1 KHz for 1%, 2% and 4% are around 3.62, 4.1 and 4.87 respectively. The dielectric constant remains almost constant for all samples in the higher frequencies. At low frequencies, all the four polarizations are active. The space charge contribution depends on the purity and perfection of the material and its influence is noticeable in the low frequency region. The orientational effect can sometimes be seen in some materials even up to 10^{10} Hz. Ionic and electronic polarizations always exist below 10^{13} Hz. Hence, the larger values of dielectric constant and dielectric loss exhibited by nanocomposite at low frequencies may be attributed to space charge polarization due to impurities and defects present in the nanocomposites. Figure 4 shows the variation of dielectric loss of nanocomposites as a function of frequency. In the lower frequency region, dielectric loss shows larger values due to the loss associated with ionic mobility. The trend in the variations of both dielectric constant and dielectric loss as a function of frequency is the same. Temperature has a striking effect on the dielectric properties. Interestingly, the variations of both dielectric constant and dielectric loss as a function of frequency are the same for all temperatures. It is observed that the dielectric constant and dielectric loss slightly decrease with the temperature which may due to the reduction in charge carriers. The variation of dielectric constant and loss with temperatures for 1 KHz and 2 KHz frequency

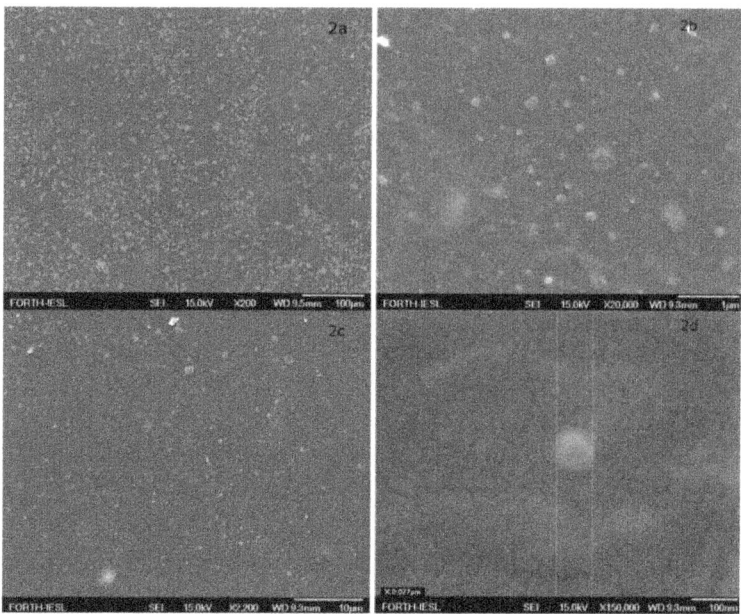

Figure 2: (a)-(d) SEM Micrographs of ZnO/EVA nanocomposite in different magnifications.

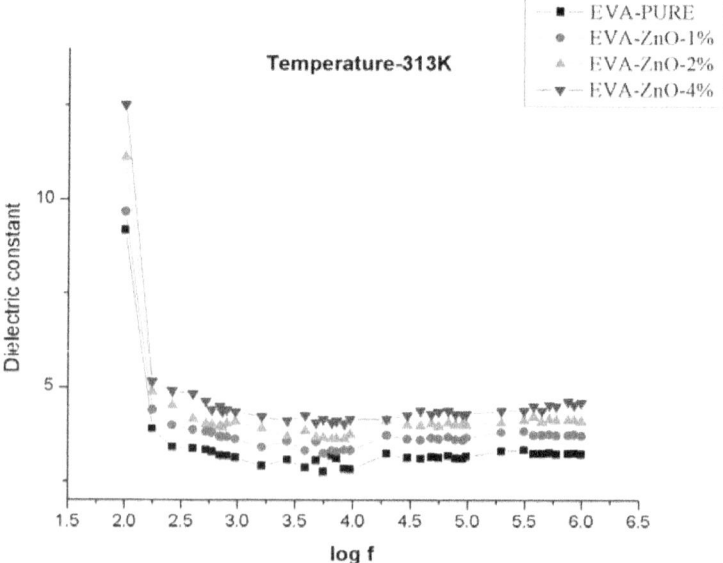

Figure 3: Variation of dielectric constant of ZnO/EVA with different filler concentrations as a function of frequency.

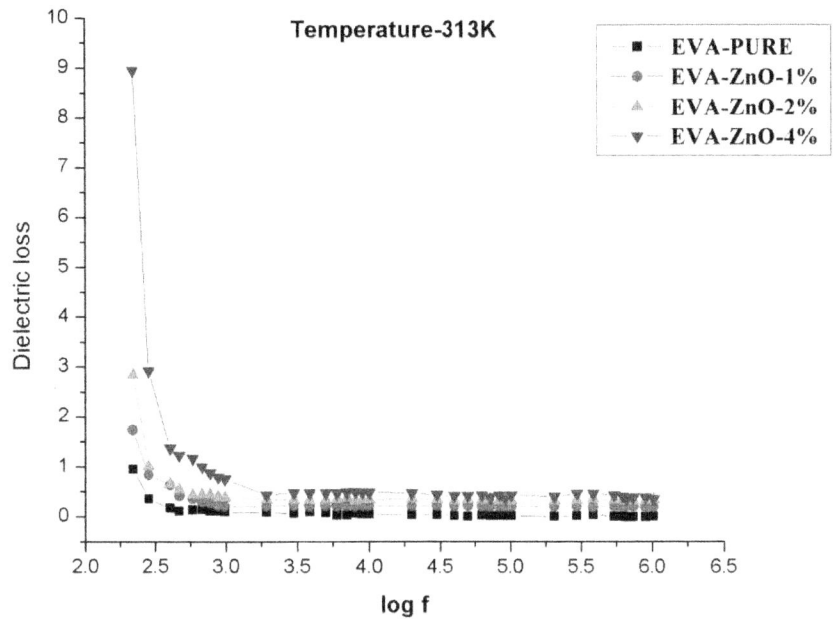

Figure 4. Variation of dielectric loss of ZnO/EVA with different filler concentrations as a function of frequency.

are shown in Figures 5-8.

The optical absorption coefficients of the nanocomposites are done using VARIAN CARY 5000 spectrophotometer in the range of 200 to 2000 nm and are shown in Figure 9. The spectra show large transparency window between 500 nm and 1600 nm. But there is a absorption at 1200 and 1450 nm. It is observed that the intensity of the absorption peak is increased with increasing the filler concentrations. The band gap of the PNC's are calculated using Tauc plotting technique and is shown in Figure 10. The band gaps of 1%, 2% and 4% EVA/ZnO PNC's are found to be 4.56 eV, 4.18 eV and 3.97 eV respectively, which decreases with increasing filler concentrations thus increases the conductivity by increasing the ZnO concentration.

The analysis of Fourier transform infrared (FT-IR) spectra of the samples have been carried out using a Thermo Nicolet Make Avatar 370 FTIR Spectrometer in the wave number range 400 - 4000 cm^{-1}. DTGS detector is used for signal Detection. Figures 11(a)-(c) show the FT-IR spectra of pristine EVA and 2% & 4% ZnO/EVA nanocomposites. The spectrum of the nanocomposites exhibits the characteristics absorption bands

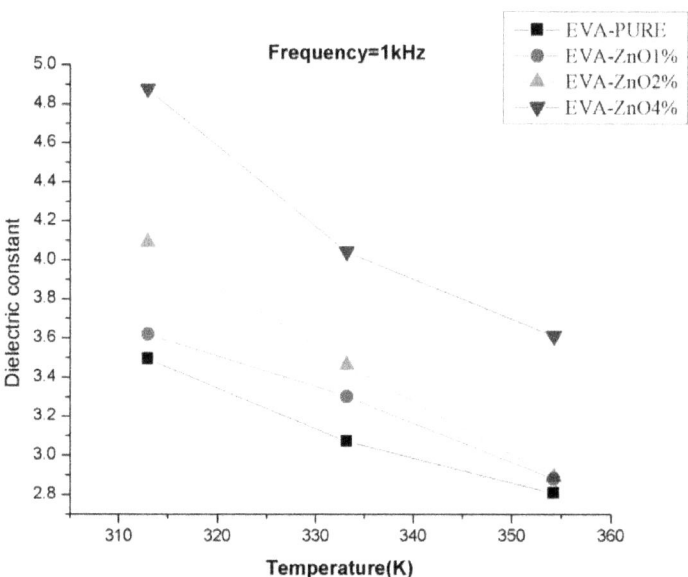

Figure 5: Variation of dielectric constant of ZnO/EVA with different temperature at 1 KHz frequency.

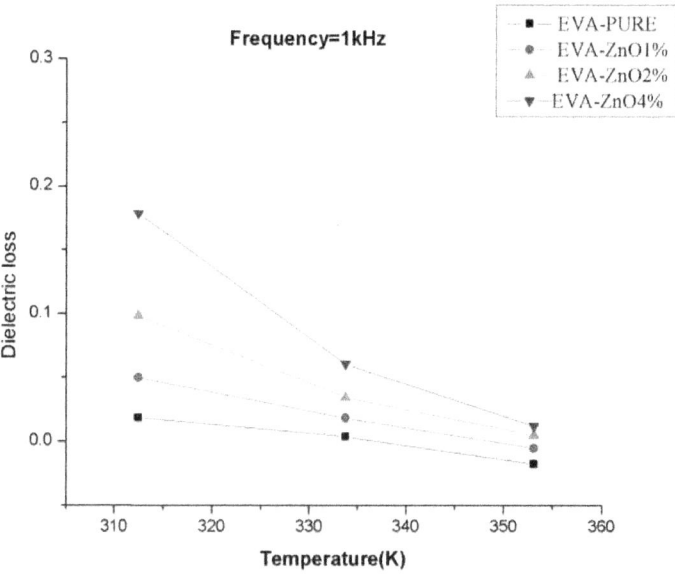

Figure 6: Variation of dielectric loss of ZnO/EVA with different temperature at 1 KHz frequency.

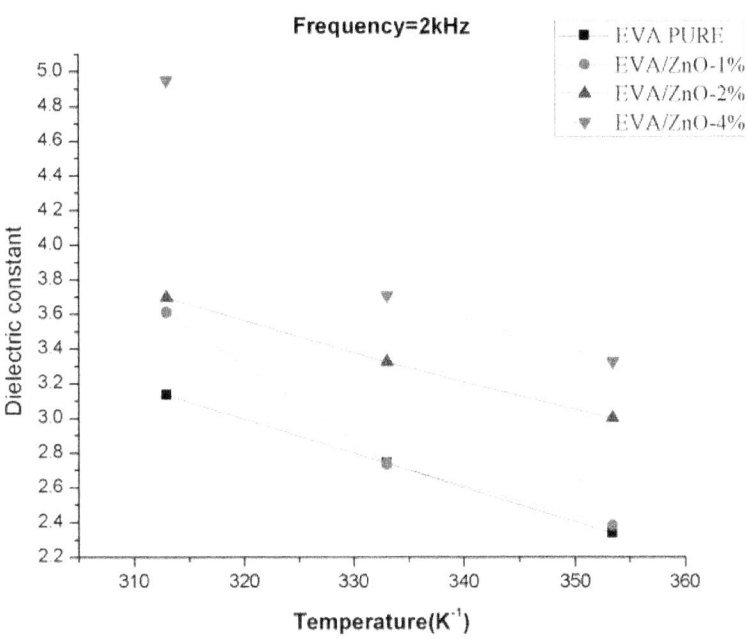

Figure 7. Variation of dielectric constant of ZnO/EVA with different temperature at 2 KHz frequency.

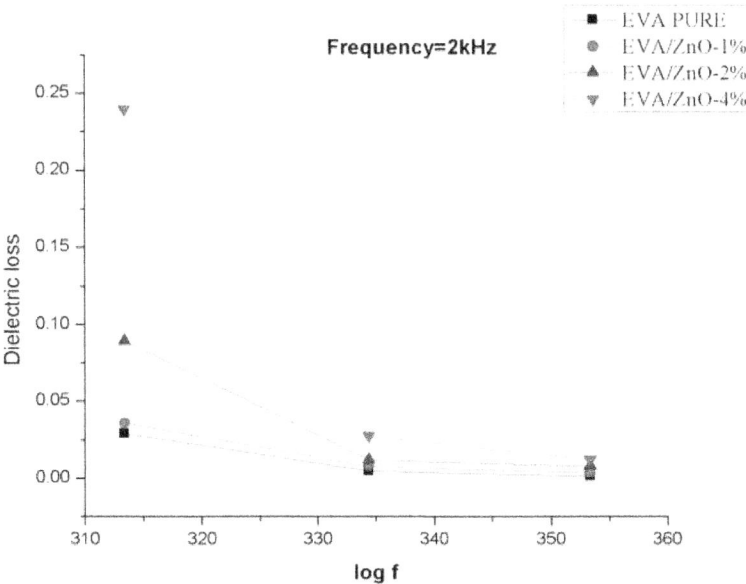

Figure 8. Variation of dielectric loss of ZnO/EVA with different temperature at 2 KHz frequency.

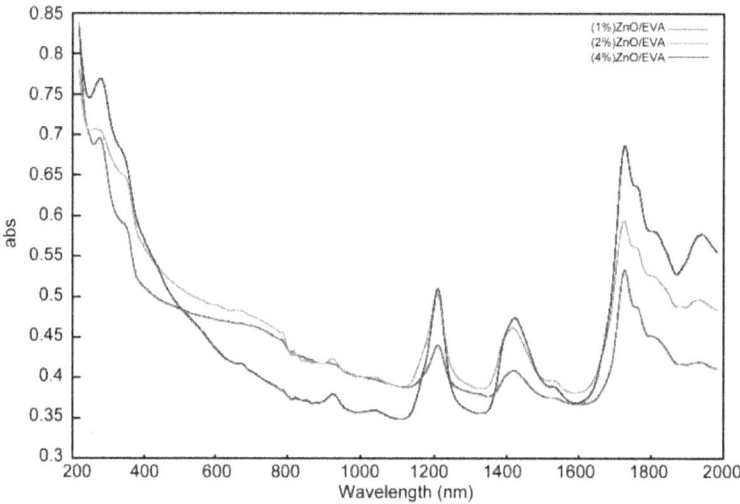

Figure 9. UV-Vis-NIR spectra of EVA/ZnO Nanocomposite with different filler concentrations.

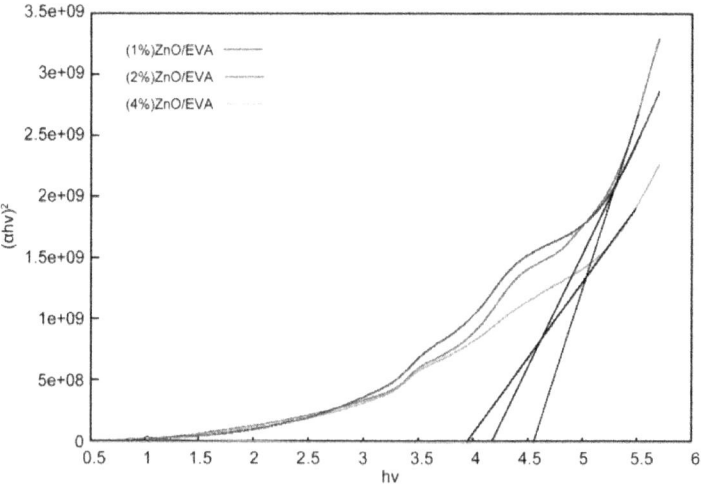

Figure 10: Tauc plot of EVA/ZnO nanocomposite with different filler concentrations.

corresponding to polymer groups and ZnO nanoparticles. In following spectra of ZnO/EVA composites, the transmittance intensity changes at 479.24 cm^{-1}, 444.73 cm^{-1} and 434 cm^{-1} are corresponding to Zn-O vibrations. The bands centered at 1371.96 cm^{-1} and 1163.75 cm^{-1} are indicates the transmittance intensity is increased and is attributed to the stretching vibrations of C=O and C-H groups in acetate species.

The characteristics vibration bands of aromatic C=C units are observed at 1465.70 cm^{-1} in the spectra of EVA and ZnO/EVA has changed as increasing percentages of ZnO nanoparticles and extra peaks are observed at 1566.25 cm^{-1} and 1624.13 cm^{-1}. The band centered at 2920 cm^{-1} and 2850.28 cm^{-1} are observed in EVA and ZnO/EVA spectra are the stretching vibrations of C-H alkanes. There is a specific peak is observed in the spectrum of both 2% and 4% ZnO/EVA at 3431.14 cm^{-1} and 3396.19 cm^{-1} which emphasized the O-H bond stretching of ZnO nanofillers. The changes in the relative intensities of the bands in the region 1420.31 cm^{-1}, 1115 cm^{-1} and 902.61 cm^{-1} in composite can be due to the presence of absorbed species on the surface of the ZnO nanoparticles. The characteristics bands of EVA and ZnO are both observed in the ZnO/EVA spectrum which confirmed that ZnO is well dispersed in the EVA matrix.

Thermal Studies

The TGA and DTA analyses of polymer nanocomposites were carried out between 28°C and 1300°C at a heating rate of 20 K/min using the instrument NETSZCH STA 409C. The TGA-DTA curves are shown in Figures 12(a)-(d), which confirms the decomposition of the nanocomposites occurs in two steps. 28.9% and 28.5% of weight is lost for Pure EVA and 1% added composite respectively in the first stage while that of 14% and 13% respectively for 2% and 4% of filler concentrations. It is also observed that the thermal stability of the polymer nanocomposites increases with increasing the filler concentrations and the onset decomposition temperature of the pure EVA, 1%, 2% and 4% are around 290°C, 295°C, 318°C and 337°C respectively. There is no much difference in the thermogram of Pure and 1% doped ZnO. The increment in the thermal stability may be due to the increase in the strength of the nanocomposite by increase of interfacial area.

The thermal parameters such as thermal diffussivity (α) and thermal efffusivity (e), the thermal conductivity (k) and heat capacity (C_p) were determined by the technique developed Preethy C menon et al. [25] . During the measurement the sample, the pyroelectric detector and the backing should be thermally thick. The sample was illuminated by an intensity-modulated beam of light, which gives rise to periodic temperature variation by optical absorption. The thermal waves so generated propagate through the sample and were detected by the pyroelectric detector.

A He-Cd laser of (wavelength λ = 442 nm KIMMON) output power 120 mW was used as the optical heating source. A polyvinylidene difluoride (PVDF) film of thickness 28 μm was used as the pyroelectric detector. The sample was attached to the pyroelectric detector by means of a thermally thin layer of a compound whose contribution to the signal was negligible. The signal output was measured using a lock-in amplifier (SR830). The frequency of modulation of the light was kept above 40 Hz to ensure that the detector, the sample and the backing

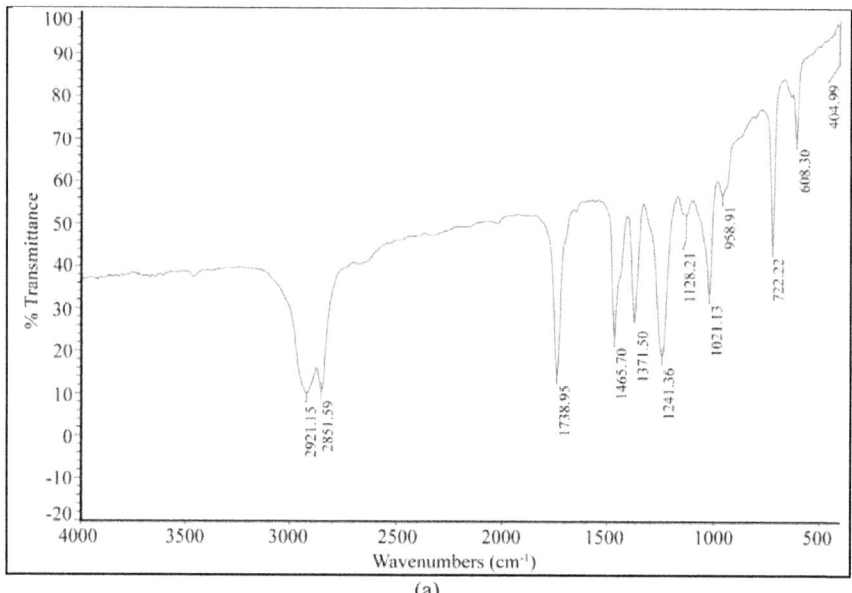
(a)

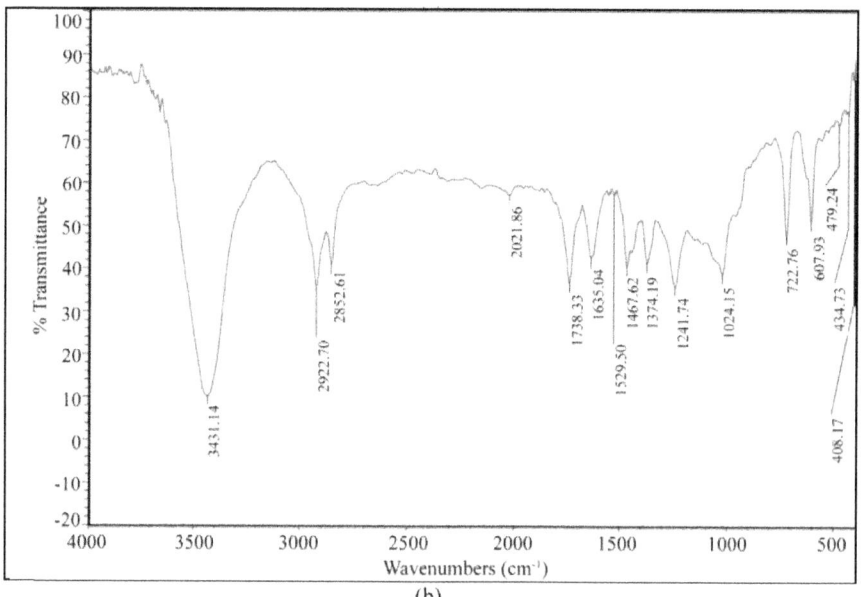
(b)

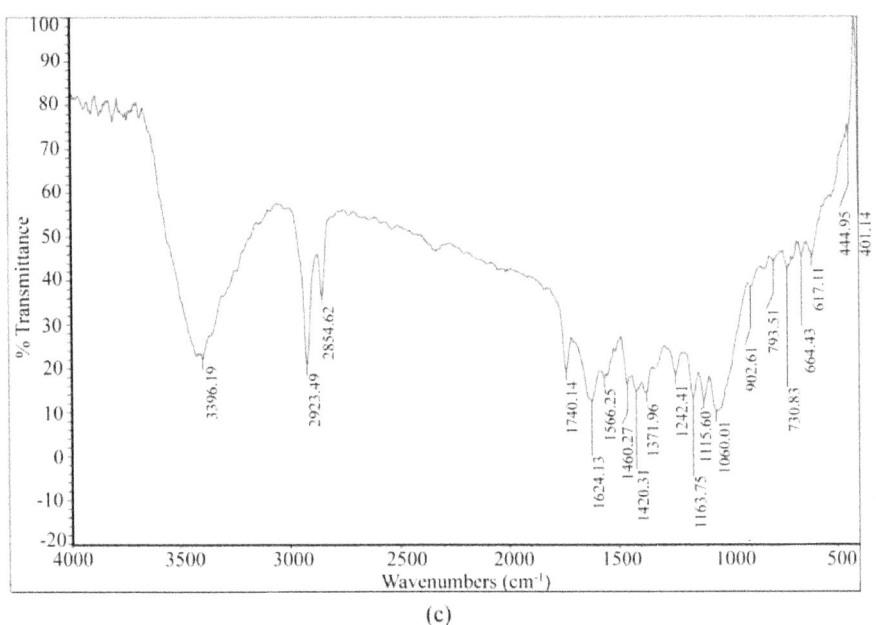

(c)

Figure 11: (a)-(c) The comparison of FT-IR Spectra of pure EVA, 2%ZnO/EVA and 4%ZnO/EVA polymer nanocomposites.

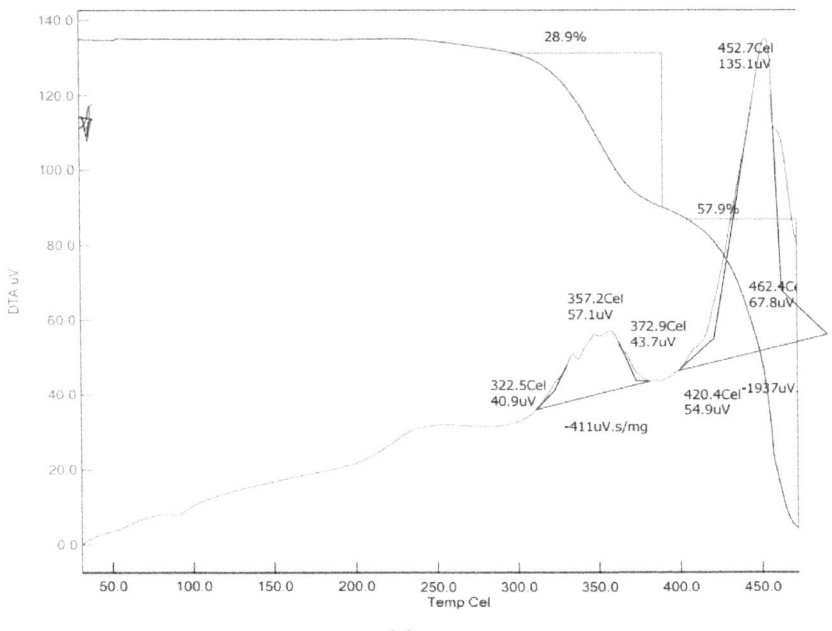

(a)

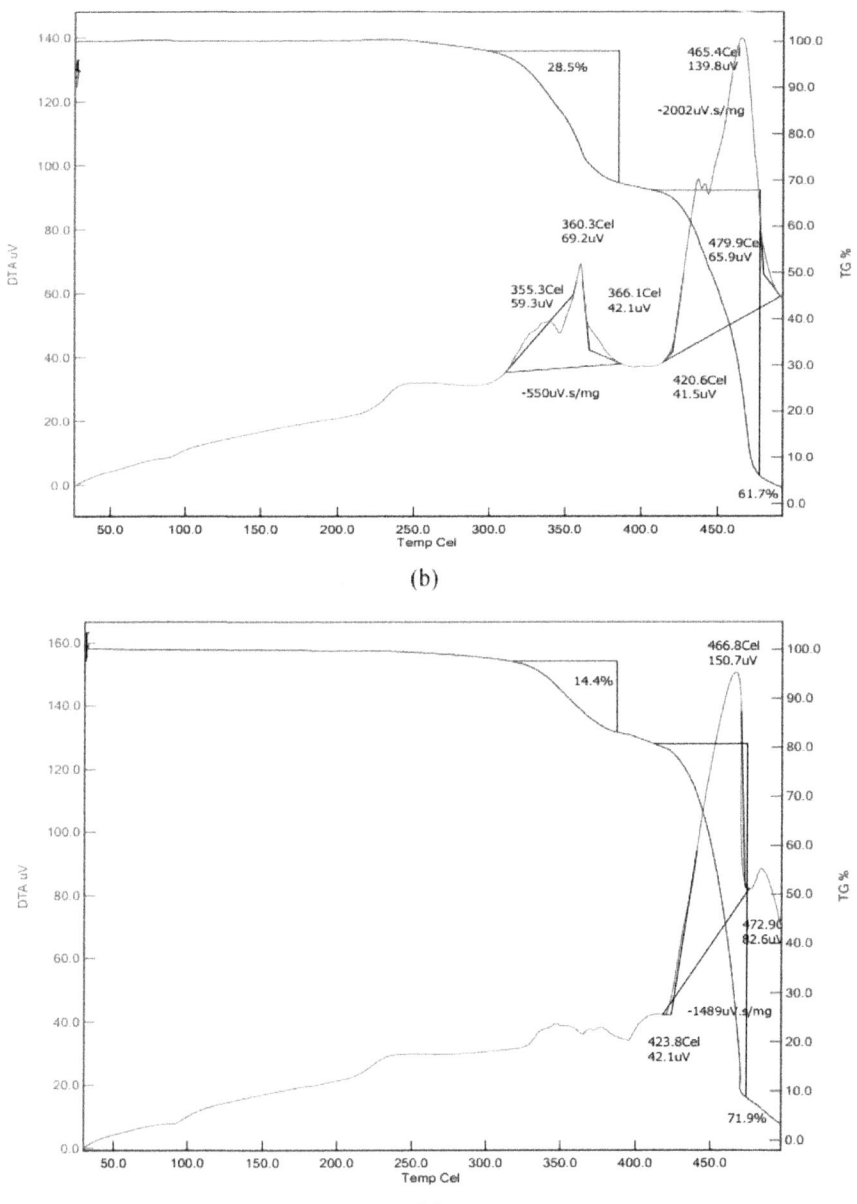

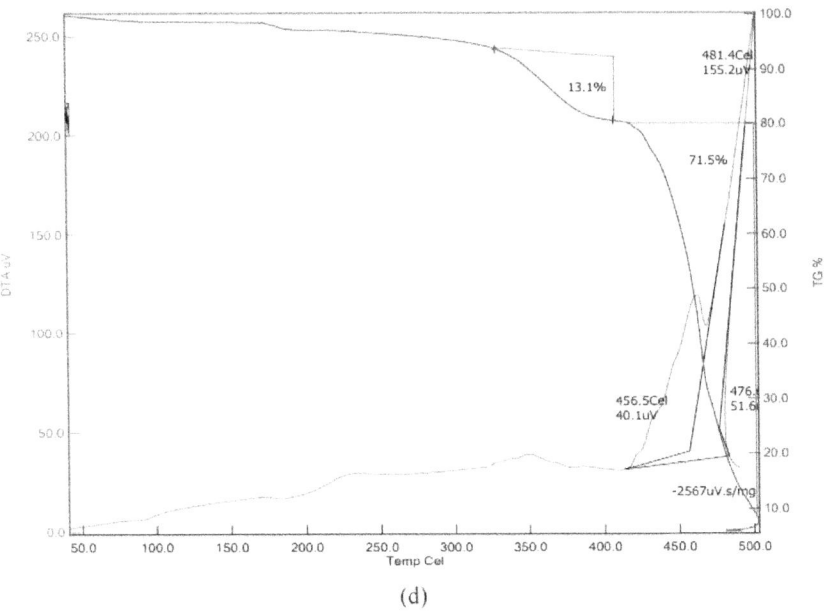

(d)

Figure 12: (a) TGA-DTA Curve of pure EVA; (b) TGA-DTA Curve of EVA-ZnO 1%; (c) TGA-DTA Curve of EVA-ZnO 2%; (d) TGA-DTA Curve of EVA-ZnO 4%.

medium were thermally thick during measurements. The values of the thermal parameters such as thermal effussivity (e) and thermal diffusivity (α), the thermal conductivity (k) and heat capacity for 2% ZnO are 1637 ± 18 $Ws^{1/2}/m^2K$, 2.4154 ± 0.14 × 10^{-6} m^2/s, 2.63 ± 0.10 W/mK and 1431 ± 33 J/kgK respectively while that of 4% are 2758 ± 35 $Ws^{1/2}/m^2K$, 8.5113 ± 0.18 × 10^{-6} m^2/s, 7.98 ± 0.15 W/mK and 1300 ± 28 J/kgK respectively. When we doubled the filler concentration the value of thermal parameters has got enhanced whereas the specific heat capacity decreased.

Mechanical Properties

The tensile strength and elongation at break of the virgin EVA and EVA nanocomposites samples were performed using an Instron 3366 testing machine according to ASTM D882. Each sample had a width of 6.4 mm. The average thickness of the samples was about 0.060 mm. The tensile test was conducted using a cross head speed of 1.3 mm/min. The stress-strain graph and elongation at various stages of the test was recorded and shown in Figure 13. Ethylene vinyl acetate (EVA) films shows good tensile strength (30 MPa) and stretches 430% to its original dimension before break. Thin films of EVA-ZnO nanocomposites have got comparable tensile and deforming properties to that of the virgin polymer. The tensile strength improves to 33 MPa on addition of 1% of ZnO nanoparticles. The nano sized particles tend to tie up the EVA molecules, leading to greater resistance to the tensile deformations. Further increase in ZnO, reduces the tensile strength and the strain deformations. The chain flexibility of the macromolecules might have been reduced with the incorporation of ZnO nanoparticles, leading to reduction of elongation.

Peel strength of the nanaocomposites were performed using an Instron tensile testing machine at a peel speed of 50 mm/min. Peel test with 180° stripping is carried out as per ASTM D 1876. Peel test involves stripping away of substrate joined by the adhesive. The substrates (glass paper, cotton and polyester) were flexible enough to permit a 180° turn near the point of loading. Peel strength values was recorded in Newton per millimeter (N/mm) of width of the bonded specimen. Peel strength of the EVA nanocomposites on various substrates are shown in Figure 14. The inclusion of the ZnO nanoparticles to the EVA matrix improves its peel strength on 2% of nano ZnO loading on all substrates. When ZnO is added more, it is found that the peel adhesion properties get reduced compared to the virgin compound. The surface finish and smoothness of the glass paper might have attributed to its inferior adhesion compared cotton and polyester fabric. Cotton fabric gives maximum peel adhesion (2.4 N/mm) as the EVA copolymer impregnate on its porous surface.

CONCLUSION

The EVA/ZnO nanocomposites materials of different concentrations of ZnO nanoparticle (0%, 1%, 2% and 4%)

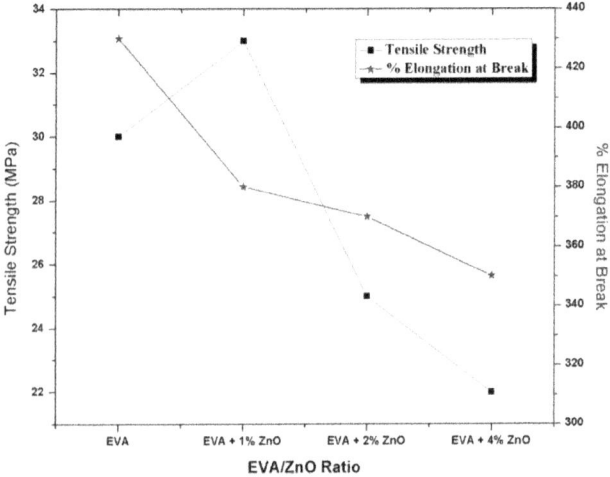

Figure 13: Tensile strength and Elongation at break of EVA/ZnO nanocomposite.

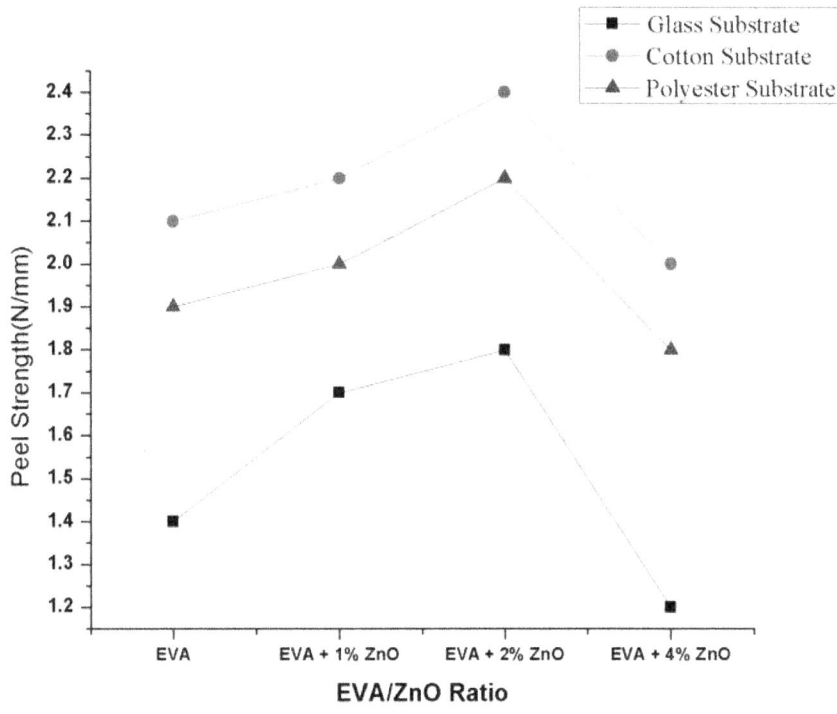

Figure 14: Peel strength of EVA/ZnO nanocomposite.

have been prepared successfully by ultrasonic probe method. The ZnO nanoparticles used in the synthesis of polymer nanocomposite were synthesized using solvothermal method. The investigations on the electrical properties of the nanocomposites revealed that the conductivity increases with increasing the filler concentrations and resistivity increases with increasing the temperature. The band gap of the nanocomposites found to be decreased with increasing the filler concentration. The thermal stability of the nanocomposites was also found to be increased. Photopyroelectric Technique is used to find the thermal parameters such as such as thermal diffussivity (α) and thermal efffusivity (e), the thermal conductivity (k) and heat capacity (C_p) and except heat capacity all others found to be increased with increasing the filler concentrations. The tensile strength and peel strength of the nanocomposites found to increase initially with the addition of ZnO nanomaterials. On further addition of ZnO, the mechanical properties drop down.

REFERENCES

1. Ray, S., Easteal, A.J., Cooney, R.P. and Edmonds, N.R. (2009) Structure and Properties of Melt-Processed PVDF/ PMMA/Polyaniline Blends. Materials Chemistry and Physics, 113, 829-838. http://dx.doi.org/10.1016/j.matchemphys.2008.08.034
2. Blom, P.W.M., Schoo, H.F.M. and Matters, M. (1998) Electrical Characterization of Electroluminescent Polymer/ Nanoparticle Composite Devices. Applied Physics Letters, 73, 3914. http://dx.doi.org/10.1063/1.122934
3. Kiesow, A., Morris, J.E., Radehaus, C. and Heilmann, A. (2003) Switching Behavior of Plasma Polymer Films Containing silver Nanoparticles. Journal of Applied Physics, 94, 6988. http://dx.doi.org/10.1063/1.1622990
4. Siegel, R.W., Schadler-Feist, L., Ma, D., Hong, J.I., Martensson, E. and Onneby, C. (2005) Nanocomposites with controlled Electrical Properties. Publication No: WO2005036563, 28 p.
5. Murugaraj, P., Mainwaring, D. and Mora-Huertas, N. (2005) Dielectric Enhancement in Polymer-Nanoparticle Composites through Interphase Polarizability. Journal of Applied Physics, 98, 054304. http://dx.doi.org/10.1063/1.2034654
6. Tishkova, V., Raynal, P.-I., Puech, P., Lonjon, A., Le Fournier, M., Demont, P., Flahaut, E. and Basca, W. (2011) Electrical Conductivity and Raman Imaging of Double Wall Carbon Nanotubes in a Polymer Matrix. Composites Science and Technology, 71, 1326-1330. http://dx.doi.org/10.1016/j.compscitech.2011.05.001
7. Lee, J., Sunder, V.C., Heine, J.R., Bawendi, M.G. and Jensen, K.F. (2000) Full Color Emission from II-VI Semiconductor Quantum Dot-Polymer Composites. Advanced Materials, 12, 1102. 3.0.CO;2-J>http://dx.doi.org/10.1002/1521-4095(200008)12:15<1102::AID-ADMA1102>3.0.CO;2-J
8. Shaheen, S.E., Brabec, C.J., Sariciftci, N., et al. (2001) 2.5% Efficient Organic Plastic Solar Cells. Applied Physics Letters, 78, 841-843. http://dx.doi.org/10.1063/1.1345834
9. Wang, H. and Branton, D. (2001) Nanopores with a Spark for Single-Molecule Detection. Nature Biotechnology, 19, 622-623. http://dx.doi.org/10.1038/90216
10. Artemyeu, M., Woggon, U. and Langbein, W. (2002) Quantum Dot Emission Confined by a Spherical Photonic Dot. Physica Status

Solidi (b), 229, 423-426. 3.0.CO;2-#>http://dx.doi.org/10.1002/1521-3951(200201)229:1<423::AID-PSSB423>3.0.CO;2-#

11. Klein, D., Roth, R., Lim, A.K.L., Alivisatos, A.P. and McEuen, P.L. (1997) A Single-Electron Transistor Made from a Cadmium Selenide Nanocrystal. Nature, 389, 699-701. http://dx.doi.org/10.1038/39535

12. Klimov, V.I., Milkhailovsky, A.A., Xu, S. Leatherdale, C.A., Eisler, H.J. and Bawendi, M.G. (2000) Optical Gain and Stimulated Emission in Nanocrystal Quantum Dots. Science, 290, 314-317.http://dx.doi.org/10.1126/science.290.5490.314

13. Konenkamp, R., Word, R. and Schlegel, C. (2004) Vertical Nanowire Light-Emitting Diode. Applied Physics Letters, 85, 6004. http://dx.doi.org/10.1063/1.1836873

14. Dang, Z.M., Zhou, T., Yao, S.H., Yuan, J.K., Zha, J.W. and Song, H.T. (2009) Advanced Calcium Copper Titanate/ Polyimide Functional Hybrid Films with High Dielectric Permittivity. Advanced Materials, 21, 2077-2082. http://dx.doi.org/10.1002/adma.200803427

15. Dang, Z.M., Yuan, J.K., Zha, J.W., Zhou, T., Li, S.T. and Hu, G.H. (2012) Fundamentals, Processes and Applications of High-Permittivity Polymer-Matrix Composites. Progress in Materials Science, 57, 660-723. http://dx.doi.org/10.1016/j.pmatsci.2011.08.001

16. Monti, O.L.A., Fourkas, J.T. and Nesbitt, D.J. (2004) Diffraction-Limited Photogeneration and Characterization of Silver Nanoparticles. The Journal of Physical Chemistry B, 108, 1604-1612. http://dx.doi.org/10.1021/jp030492c

17. Yuen, S.M., Ma, C.C.M., Chuang, C.Y., Yu, K.C., Wu, S.Y., Yang, C.C. and Wei, M.H. (2008) Effect of Processing Method on the Shielding Effectiveness of Electromagnetic Interference of MWCNT/PMMA Composites. Composites Science and Technology, 68, 963-968. http://dx.doi.org/10.1016/j.compscitech.2007.08.004

18. Stankovich, S., Dikin, D.A., Dommett, G.H.B., Kohlhaas, K.M., Zimney, E.J., Stach, E.A., Piner, R.D., Nguyen, S.T. and Ruoff, R.S. (2006) Graphene-Based Composite Materials. Nature, 442, 282-286. http://dx.doi.org/10.1038/nature04969

19. Khanna, P.K., Singh, N., Charan, S. and Mulik, U.P. (2005) Synthesis and Characterization of Ag/PVA Nanocomposite by Chemical Reduction Method. Materials Chemistry and Physics, 93, 117-121.

20. Bhadra, S., Khastgir, D., Singha, N.K. and Lee, J.H. (2009) Progress in Preparation, Processing and Applications of Polyaniline. Progress

in Polymer Science, 34, 783-810. http://dx.doi.org/10.1016/j.progpolymsci.2009.04.003

21. Subramanian, S. and Pathinettam Padiyan, D. (2008) Effect of Structural, Electrical and Optical Properties of Electrodeposited Bismuth Selenide Thin Films in Polyaniline Aqueous Medium. Materials Chemistry and Physics, 107, 392- 398. http://dx.doi.org/10.1016/j.matchemphys.2007.08.005

22. Gonzalex-Benito, J., Castillo, E. and Caldito, J.F. (2013) Coefficient of Thermal Expansion of TiO_2 Filled EVA Based Nanocomposites. A New Insight about the Influence of Filler Particle Size in Composites. European Polymer Journal, 49, 1747-1752. http://dx.doi.org/10.1016/j.eurpolymj.2013.04.023

23. Chu, K.C., Jordan, K.J., Battista, J.J., Van-Dyk, J. and Rutt, B.K. (2000) Polyvinyl Alcohol-Fricke Hydrogel and Cryogel: Two New Gel Dosimetry Systems with Low Fe^{3+} Diffusion. Physics in Medicine and Biology, 45, 955. http://dx.doi.org/10.1088/0031-9155/45/4/311

24. Kasap, S. (2007) Springer Handbook of Electronic and Photonic Materials. Springer, Berlin. http://dx.doi.org/10.1007/978-0-387-29185-7

25. Preethy Menon, C. and Philip, S. (2000) Simultaneous Determination of Thermal Conductivity and Heat Capacity Near Solid State Phase Transitions by a Photopyroelectric Technique. Measurement Science and Technology, 11, 1744. http://dx.doi.org/10.1088/0957-0233/11/12/314

Chapter 8

GRAPHENE-BORON NITRIDE COMPOSITE: A MATERIAL WITH ADVANCED FUNCTIONALITIES

Sumanta Bhandary[1] and Biplab Sanyal[1]
[1]Department of Physics and Astronomy, Uppsala University, Uppsala, Sweden

INTRODUCTION

The discovery of two dimensional materials is extremely exciting due to their unique properties, resulting from the lowering of dimensionality. Physics in 2D is quite rich (e.g., high temperature superconductivity, fractional quantum Hall effect etc.) and is different from its other dimensional counterparts. A 2D material acts as the bridge between bulk 3D systems and 0D quantum dots or 1D chain materials. This can well be the building block for materials with other dimensions. The discovery of graphene, the 2D allotrope of carbon by Geim and Novoselov [1] made an enormous sensation owing to a plethora of exciting properties. They were awarded the Nobel prize in Physics in 2010. Graphene, an atomically thick C layer, has broken the jinx of impossibility of the formation of a 2D structure at a finite temperature, as argued by Landau *et al.*[2, 3]. The argument that a 2D material is thermodynamically unstable due to the out-of-plane thermal distortion, which is comparable to its bond length, was proven invalid with this discovery. One of the recent interests is to understand this apparent discrepancy by considering rippled structures of graphene at finite temperatures.

Graphene, with its exciting appearance, has won the crowns of the thinnest, the strongest, the most stretchable material along with extremely high electron mobility and thermal conductivity [4]. The linear dispersion curve at the Dirac point gives rise to exciting elementary electronic properties. Electrons in graphene behave like massless Dirac fermions, similar to the relativistic particles in quantum electrodynamics and hence has brought different branches of science together under a truly interdisciplinary platform. At low temperature and high magnetic field, a fascinating phenomenon, called the half integer quantum hall effect, is observed. The relativistic nature of carriers in graphene shows 100% tunneling through a potential barrier by changing its chirality. The phenomenon is known as "Klein-Paradox". Minimum conductivity of a value of conductivity quantum (e^2/h per spin per valley) is measured at zero field, which makes graphene unique. "The CERN on table top" is thus a significant naming of the experiments performed with this fascinating material [5].

An infinite pristine graphene is a semi metal, i. e., a metal with zero band gap [6]. Inversion symmetry provided by $P6/mmm$ space group results in a band degeneracy at the Dirac points (K and K') in the hexagonal Brillouin zone (BZ). This limits its most anticipated application in electronics as the on-off current ratio becomes too small to be employed in a device. The opening of a band gap is thus essential from electronics point of view retaining a high carrier mobility. Several approaches have been already made by modifying graphene, either chemically [7, 8] or by structural confinement [9–11] to improve its application possibilities, both from theory and experiment. It should be noted that in theoretical studies, the use of density functional theory (DFT) [12] has always played an instrumental role in understanding and predicting the properties of materials, often in a quantitative way.

Boron Nitride (BN), on the other hand, can have different forms of structures like bulk hexagonal BN with sp^2 bond, cubic BN with sp^3 bond, analogous to graphite and diamond respectively. A 2D sheet with strong sp^2 bonds can also be derived from it, which resembles its carbon counterpart, graphene. But two different chemical species in the two sublattices of BN forbid the inversion symmetry, which results in the degeneracy lifting at Dirac points in the BZ. Hexagonal BN sheet thus turns out to be an insulator with a band gap of 5.97 eV.

This opens up a possibility of alloying these neighboring elements in the periodic table to form another interesting class of materials. Possibilities are bright and so are the promises. B-N bond length is just 1.7% larger than the C-C bond, which makes them perfect for alloying with minimal internal stress. At the same time, introduction of BN in graphene, breaks the inversion symmetry, which can result in the opening up of a band gap in graphene. On top of that, the electronegativities of B, C, and N are respectively 2.0, 2.5 and 3.0 [13], which means that the charge transfer in different kinds of BCN structures is going to play an interesting role both in stability and electronic properties.

Hexagonal BNC (h-BNC) films have been recently synthesized [14] on a Cu substrate by thermal catalytic chemical vapor deposition method. For the synthesis, ammonia borane (NH_3-BH_3) and methane were used as precursors for BN and C respectively. In the experimental situation, it is possible to control the relative percentage of C and BN. The interesting point is that the h-BNC films can be lithographically patterned for fabrication of devices. The atomic force microscopy images indicated the formation of 2-3 layers of h-BNC. The structures and compositions of the films were characterized by atomic high resolution transmission electron microscopy and electron energy-loss spectroscopy. Electrical measurements in a four-probe device showed that the electrical conductivity of h-BNC ribbons increased with an increase in the percentage of graphene. The h-BNC field effect transistor showed ambipolar behavior similar to graphene but with reduced carrier mobility of 5-20 cm^{2}V^{-1}s^{-1}. From all these detailed analysis, one could conclude that in h-BNC films, hybridized h-BN and graphene domains were formed with unique electronic properties. Therefore, one can imagine the h-BN domains as extended impurities in the graphene lattice.

The structure and composition of BN-graphene composite are important issues to consider. As mentioned before, substitution of C in graphene by B and N can give an alloyed BCN configuration. Considering the possibilities of thermodynamic non-equilibrium at the time of growth process, one can think of several ways of alloying. The potential barrier among those individual structures can be quite high and that can keep these relatively high energetic structures stable at room temperature. For example, a huge potential barrier has to be crossed to reach a phase segregated alloy from a normal alloy, which makes normal alloy stable at room temperature. Now, depending on the growth process, different types of alloying are possible. Firstly, one can think of an even mixture of boron nitride and carbon, where one C_2 block is replaced by B-N. In this case, the formula unit will be BC_2N. Secondly, a whole area of graphene can be replaced by boron nitride, which makes them phase separated. This we call as phase segregated alloy. The formula unit of phase segregated alloy can change depending on the percentage of doping. The final part of the following section will be devoted to the phase segregated BCN alloys. Apart from those, a distributed alloying is possible with different BN: graphene ratios.

The substitution of C_2 with B-N introduces several interesting features. Firstly, B-N bond length is 1.7% bigger that C-C bond but C-B bond is 15% bigger than C-N bond. So, this is going to create intra-layer strain, which is going to affect its stability. Secondly, the difference in electro negativity in B (2.0) and N (3.0) will definitely cause a charge transfer. The orientation of charged pair B-N do have a major contribution in cohesive energy. Thirdly, as mentioned earlier, this will break inversion symmetry in graphene, which brings a significant change in electronic properties. Keeping these in mind, we are now going to discuss stability and electronic structure of BC_2N.

STABILITY OF BC_2N

In this section we will mainly focus on the stability issues for various BC_2N structures [13, 15]. To demonstrate the factors for structural stability, we have chosen five different structures of BC_2N (Fig. 1). Let's first have a closer look at structure I. Every C atom has one C, N and B as nearest neighbors while B(N) has two Cs and one N(B) as their nearest neighbors. There is a possibility of all bonds to be relaxed, retaining the hexagonal structure. Stress is thus minimized in this structure, which helps obviously in the stability. In structure II, each C has two C and either one B or N as nearest neighbors. C_2 and BN form own striped regions, which lie parallel to each other in this structure. Now C-B bond is much larger that C-C. So this is definitely going to put some internal stress. From the point of view of intra-layer stress, this structure is definitely less stable than the previous one. Looking at structure III, one can see this structure looks similar to structure I but B-N bond orientations are different. Each C atom now has either two N or B and one C in its neighboring position while N(B) has two C and one B (N) as neighbors. This obviously adds some uncompensated strain in the structure. Structure III, thus consists of two parallel C-N and C-B chains and as C-B bond length is much larger than C-N (15%), this mismatch is going to introduce a large strain in the interface. On the other hand in structure I, C-N and C-B are lined up making the structural energy lower compared to structure III. Structure IV does not contain any C-C bond. C-B and C-N chains are lined parallel to each other. Finally in structure V, B-N bonds are placed in such a way that they make 60° angle to each other. Both of the last two structures thus have uncompensated strain, which increases their structural energies.

Bond energy is another key factor in stability. When the bond energies are counted, the ordering of the bonds is the following [13]:

$B - N(4.00 \ eV) > C - C(3.71 \ eV) > N - C(2.83 \ eV) >$
$B - C(2.59 \ eV) > B - B(2.32 \ eV) > N - N(2.11 \ eV)$

The maximization of stable bonds like B-N and C-C will thus stabilize the structure as a whole. Now, a structure like II, with a striped pattern of C and B-N chains has maximum number of such bonds. This makes it most stable even though a structural strain is present. In this case bond energy wins over structural stress. For the structures like I and III, number of such bonds is equal. In that case, intra-layer stress acts as the deciding factor. Structure IV, on the other hand does not have any C-C or B-N bond but only C-B or C-N. Therefore, the issue of stability is the most prominent here. The number of strong bonds is sufficiently large in structure V for the stabilization despite of 60° arrangement of B-N bonds.

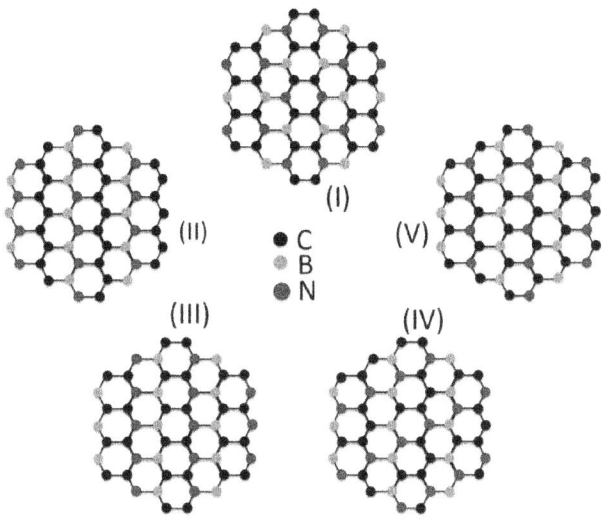

Figure 1: Crystal structure of different isomers of BC_2N. Filled black, pink, and blue circles represent carbon, boron and nitrogen respectively.

Another important issue is the charge transfer as there is a difference between the electronegativities of B, C, and N. As mentioned earlier, N is the most electronegative and B is the least one while C behaves as a neutral atom. This also adds an ionic character in the bond formation. B (N) always gains some +ve (-ve) charges. So, the gain in the electrostatic energy only happens if these +/- charges are situated in an alternative manner. Otherwise the electrostatic repulsion makes the structure unstable. From this point of view, structures II, III, and V are more stable than the other two following the trend shown from bond energies. Thus, as reported by Itoh et al.[13], ordering of the structures will be: II>V>(I,III)>IV. Stability of other possible isomers can as well be anticipated with the same arguments.

So far we have talked about the substitutional alloying of BN and graphene, where BN to C_2 ratio is 1:1. Another group of structures, which can be formed by alloying BN and grapheme is the phase segregated BCN. In this kind of structure, BN (graphene) retains its own phase, separated by graphene (BN). The experimental evidence of these kinds of structures have been shown [14]. The size of the graphene or BN phase has an impact on the stability and electronic properties. Here, the BN:C_2 ratio is thus not only 1:1 but can be varied and if varied controllably, one can control the electronic properties such as band gap [16]. Lam et al.[16] have shown that, by controlling the graphene phase, one can control the band gap according to the desired values for technological applications. The phase segregated $(BN)m(C_2)n$ alloys are also found to be stable over the first kind of alloying, which indicates a transformation due to thermal vibration. Yuge et al.[17] with DFT studies and Monte Carlo simulations have shown a tendency of phase separation between BN and graphene.

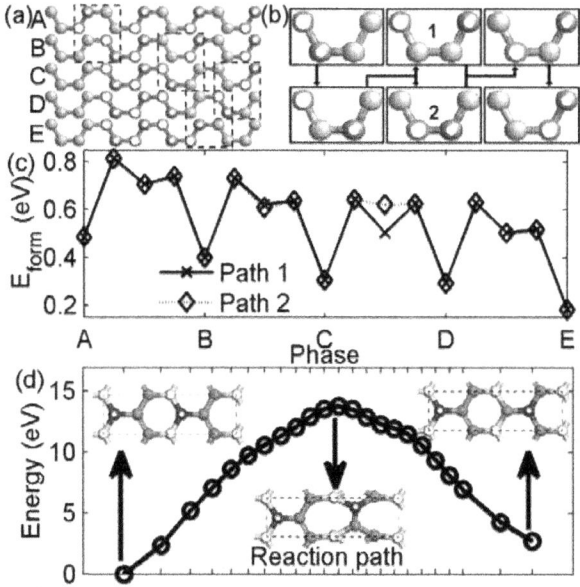

Figure 2: a) Different steps (A-E) of phase separation process, (b) Swapping of BN and C dimers, (c) Formation energies for different steps shown in (a) for two different paths demonstrated in (b), (d)

Activation energy in going from left to right configuration in the initial step of phase segregation. Reprinted with permission from Appl. Phys. Lett. 98, 022101 (2011). Copyright (2011) American Institute of Physics.

Even though a tendency is indicated, a recent calculation by Lam et al., have shown that this possibility is hindered as the activation energy required for phase segregation is extremely high. As shown in Fig. 2, they have chosen a possible path for phase separation by swapping B-N bond to C-C bonds. This kind of swapping can also happen in two ways (Fig. 2(b)). Calculated formation energies for these two process are shown in Fig. 2(c), which basically demonstrates that the intermediate structures are quite high in energy compared to evenly distributed and phase separated structures. The authors also performed nudged elastic band (NEB) calculations to determine the activation barriers for the first step to occur, i. e. to change a B-N bond to B-C and N-C bonds (Fig. 2(d)). Activation energy required is 1.63eV/atom suggesting that this process can happen only at elevated temperatures. At room temperature that is why the pristine (BN)m(C$_2$)n should be stable and so are the phase separated ones.

There can be two different patterns for phase segregated BCN alloys. One is the phase separated island-like and the other one is a striped pattern. The island-like pattern consists of larger graphene-BN interface region than that in the striped pattern. This means that the number of B-C and N-C bonds are less in striped pattern than in an island form. As we have discussed earlier, the maximization of C-C and B-N bonds thus favors a striped pattern [13].

Till now, we have discussed mainly the stability issues of (BN)m(C$_2$)n with 1:1 ratio and phase separated BCN alloys. A distributed mixture of BN and graphene with different $m : n$ ratios can also form depending on the growth condition. Different isomeric structures are also possible for a particular $m : n$ ratio. In the following section we are going to present a DFT study to analyze the stability and electronic properties of (BN)m(C$_2$)n with different $m : n$ ratios. Utilizing the concept of aromaticity, the aim is to find out stable isomers for a particular $m : n$ ratio and also to explore the possibilities of achieving desired electronic properties.

Aromaticity, as extensively used to determine the stability of organic molecules, can provide us a working principle for determining stability of the structures as well. Benzene (C$_6$H$_6$) is the prototype for the organic molecules, which are stabilized by aromaticity. Borazine (B$_3$N$_3$H$_6$), an isoelectric BN analogue of benzene on the other hand has one-third stability of benzene from the point of view of aromaticity [18–21]. This is particularly interesting in (BN)m(C$_2$)n, as the admixture of two not only changes the electronic property but also affects its stability. To investigate a stable isomer, our first working principle thus is to maximize the carbon hexagons, which essentially mimic benzene rings. A carbon-hexagon again can be surrounded by BN and each hexagon can be kept aloof or all hexagons can form a carbon-pathway. In a carbon pathway, ϖ-conjugation is allowed whereas it is hindered in isolated C-hexagons.

To look for reasonable isomers, we consider that the following structural possibilities will not occur. Firstly, a hexagon will not contain B and N in 1 and 3 positions with respect to each other. These kind of structures are described by zwitterionic and biradical resonance structures, which basically result in an odd number of ϖ-electrons on two of the Cs in the hexagon (Fig. 3).

Hence, a B-N pair should be placed either in 1,4 or 1,2 position in the hexagon with respect to each other. ϖ-electrons will thus be distributed over a C-C bond and form a resonance structure. Second kind of structural constraint, that we consider, is the absence of B-B or N-N bonds. As discussed earlier, these kind of bonds result in the lowering of ϖ-bonds and thus decreased relative stability of an isomer.

The relative positions of B and N around an all C hexagon is also a key factor that controls the electronic properties. To illustrate the phenomenon, let's consider the following two isomers. As in Fig. 3, the isomer I and isomer II, both have similar chemical configuration. But in Isomer I, B and N are connected to C at position 1 and 4 in the hexagon, which we can call B-ring-N para-arrangement. A donor- acceptor (D-A) interaction is thus established in this kind of structural arrangement. On the other hand in isomer II, B and N are connected to 1st & 2nd (4th & 5th) positioned C atoms in the hexagon. Although a D-A interaction occurs between neighboring B and N, B-ring-N interaction is forbidden. The local D-A interaction around a C-hexagon, as shown in Fig. 4, increases the HOMO-LUMO gap whereas N-ring-N (or B-ring-B) para arrangement results in the lowering of the HOMO-LUMO gap.

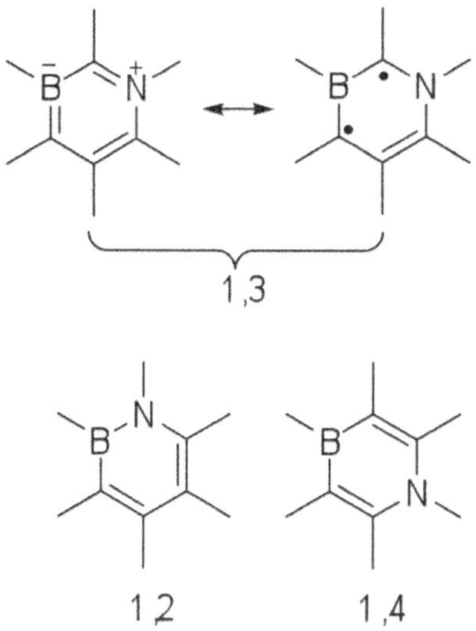

Figure 3: Schematic representation of zwitterionic and biradical resonance structures. Reprinted with permission from J. Phys. Chem. C 115, 10264 (2011). Copyright (2011) American Chemical Society.

We have performed density functional calculations to investigate the isomers of (BN)m(C$_2$)n [22]. All the structures are optimized with both (Perdew-Burke-Ernzerhof) PBE [23] and (Heyd-Scuseria-Ernzerhof) HSE [24] functionals. The functionals based on local spin density approximation or generalized gradient approximations reproduce the structural parameters reasonably well, whereas the band gaps come out to be much smaller compared to experiments. The reason behind this is the self interaction error. HSE, with a better description of exchange and correlation within hybrid DFT, yields a band gap, which is much closer to the experimental value.

The degree of aromaticity is calculated quantitatively, with a harmonic oscillator model of aromaticity (HOMA) prescribed by Krygowski *et al.*[25]. The HOMA value of an ideal aromatic compound (Benzene) will be 1, whereas the value will be close to zero for non aromatic compounds. Anti-aromatic compound with the least stability will have a negative HOMA value. As mentioned earlier, the aim is to find the important isomers with relatively high stability and reasonable band gaps among $(BN)m(C_2)n$ compounds with $m:n$ ratios 1:1, 2:1, 1:3 and 2:3. Let's focus on each type separately.

1:1 H-BN:Graphene (BC_2N)

We have considered six isomers for BC_2N, among which two structures BC_2N-I and BC_2N-II consist of all C-hexagon pathways. In the third one, BC_2N-III, all C-hexagons are connected linearly as in polyacenes where as the fourth one, BC_2N-IV, has disconnected all-C-hexagons. The other two structures, BC_2N-V & BC_2N-VI do not have any all-C hexagon but BC_2N-VI has at least polyacetylene paths whereas BC_2N-V has only isolated C-C bonds. Although there are several other isomers possible, we limit ourselves with these and try to understand the properties with the knowledge of aromaticity and conjugation. Firstly, the first three isomers, among all six are most stable and the relative energies differ by at most 0.15 eV (PBE) and 0.07 eV (HSE). The presence of all C-hexagons connected to each other not only increases the stable C-C and B-N bonds but also helps in the ϖ-conjugation. The result is reflected in the HOMA values of first two structures, which are 0.842 and 0.888 respectively. This suggests the formation of aromatic benzene like all-C hexagons. The HOMA value of BC_2N-III is little less (0.642) but this structure in particular is not stable due to aromaticity rather due to the formation of polyacetylene paths. A slightly lower HOMA value observed in BC_2N-I compared to BC_2N-II is due to the difference in B-C bond (0.02Å), which leads to a change in D-A interaction.

If we look at the formation energies of BC_2N-IV & BC_2N-VI, the values are quite close. BC_2N-IV consists of completely isolated all-C hexagons. This is the reason of having high aromaticity of 0.88. But at the same time this increases N-C & B-C bonds and restricts ϖ-conjugation. Therefore, this structure is less probable thermodynamically. BC_2N-VI, which was suggested to be the most stable BC_2N structure by Liu *et al.*[15], on the other hand has no aromatic all C-hexagon. But this structure contains all-C polyacetylene paths with C-C bond length 1.42 Å, which explains its low formation energy. BC_2N-V is the least stable among all, which has neither all-C hexagon nor polyacetylene C-paths. Obviously most unstable B-C and B-N bonds are maximized here creating an enormous strain in the structure. The presence of only C-C bond of 1.327 Å explains that. These factors make this compound thermodynamically most unstable among all five structures.

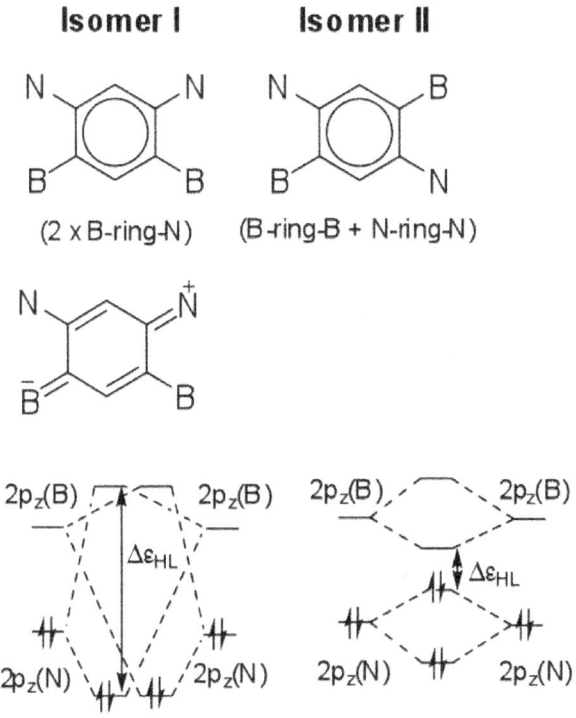

Figure 4: Qualitative representation of opening up a band gap and D-A interaction in isomer I and reduction of band gap in isomer II, with molecular orbital diagrams and valence bond representation. Reprinted with permission from J. Phys. Chem. C 115, 10264 (2011). Copyright (2011) American Chemical Society.

All these results give us a stand point from where we can judge the thermodynamic stability of other $(BN)m(C_2)n$ structures with the following working principles in hand:

a. ϖ-Conjugation within all C path increases stability.

b. The formation of aromatic all C-hexagons also does the same, while this is more effective when hexagons are connected.

c. There is not much contribution of B-ring-B or B-ring-N arrangement of B and N around poly(para-phenylene) (PPP) path in total energies. But indeed these, as discussed earlier, will affect the band gap, which we will present in the following section.

Coming to the band gap issue, the first three structures, which are close in energy, have band gaps ranging from 1.6 to 2.3 eV in HSE calculations (0.7 to 1.7 eV in PBE). The difference in BC_2N-I & BC_2N-II comes from the arrangement of B and N around all-C hexagon. As discussed in Fig. 3, D-A interaction increases for para-positions (i. e.1,4 or 2,6), which is observed in BC_2N-I. The band gap is 0.5 eV (0.65 eV in PBE), higher than that in BC_2N-II, where B and N are oriented in ortho-position (i. e.1,2 or 4,5). Quite obviously, BC_2N-III has all-C chain, which resembles a graphene nanoribbon and has the least value of the band gap.

2:1 H-BN:Graphene (BCN)

We have investigated three structures of BCN, which have recently been synthesized [26]. The first one (BCN-I) has aromatic all-C hexagons connected in PPP path whereas the second one (BCN-III) contains all-C hexagon but connected in zigzag polyacene bonds. The final one (BCN-IV) consists of neither all-C hexagon nor a stripe of all-C region. BCN-I & BCN-II are iso-energetic, which is expected and is ~ 0.5 eV lower than BCN-IV. This again explains the importance of aromatic all-C hexagon and ϖ-conjugation. The absence of these and also the increased B-C, N-C bonds make BCN-IV relatively unstable. Another key point in BCN-I & BCN-I is the position of B and N around the hexagon. The stability may not be affected but the band gap is definitely changed by this. As expected, from the discussion in Fig. 4, BCN-I has a quite high value of the band gap. Aromaticity is higher in BCN-I (0.846) compared to BCN-III (0.557), which is also seen in BC_2N structures. Iso-energetic BCN-I & BCN-II are equally probable during growth process but with a band gap range 1.3 to 2.7 eV.

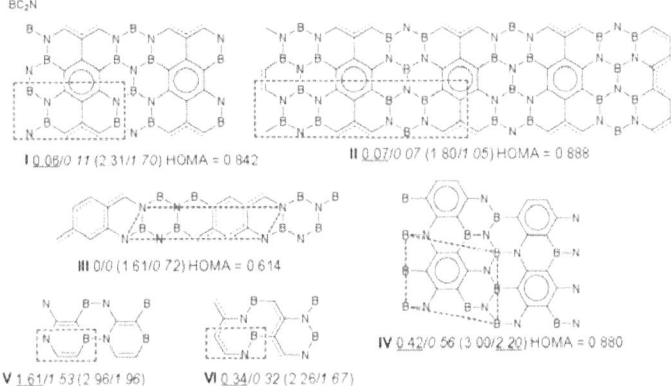

Figure 5: Six isomers of BC_2N are considered for this study. Relative energy (per formula unit) with respect to most stable structures and band gaps (in parentheses) are

shown from HSE (normal print) and PBE (italic) calculations. HOMA values of all-C hexagons, obtained from HSE calculations are also provided. Reprinted with permission from J. Phys. Chem. C 115, 10264 (2011). Copyright (2011) American Chemical Society.

2:3 & 1:3 H-BN:Graphene (BC3N & BC6N)

We now gradually increase the C-percentage with the anticipation of lowering the band gap because of increased graphene region. Two structures of BC_3N and three structures of BC_6N have been examined. Both the structures of BC_3N consist of aromatic hexagonal all-C rings connected in PPP path. The similarity in HOMA values depicts that picture. The position of B and N around C-ring is different though in BC_3N-I and BC_3N-II. The para arrangement of B-ring-N results in a large band gap in BC_3N-I (2.14 eV) while the ortho arrangement of the same in BC_3N-II lowers the value of the band gap. The stability is not affected by that fact as aromaticity and ϖ-band formation are quite similar, which make those structures iso-energetic. A similar situation is also seen in BC_6N structures. BC_6N-I and BC_6N-II are iso-energetic mainly due to a similarity in structures. Both of them contain all-C rings connected in PPP path. But in the third one (BC_6N-III), all-C rings are separated, which makes this structure relatively unstable due to restricted ϖ-conjugation. The HOMA value is maximum (0.93) in BC_6N-III, whereas the values are 0.78 and 0.86 respectively for BC_6N-I and BC_6N-II. The para arrangement of B-ring-N in BC_6N-I and BC_6N-III leads to large band gaps (1.58 and 1.34 eV respectively) while ortho-positioning of the same results in a reduced band gap in BC_6N-II. Finally, we have summarized the calculated (HSE and PBE) band gaps for all $(BN)m(C_2)n$ isomers (Fig. 7A and Fig. 7B). As expected, the band gaps are increased for HSE functional. Apart from that, both HSE and PBE -level calculations show a similar trend. A general trend is observed that an increase in the graphene region reduces the band gap. All the lowest energy structures for different compositions of $(BN)m(C_2)n$ have band gaps around 1 eV, which is a desired value for technological applications. One very important thing that we learnt from this study is that the position of B and N around C-ring controls the band gap without affecting the stability.

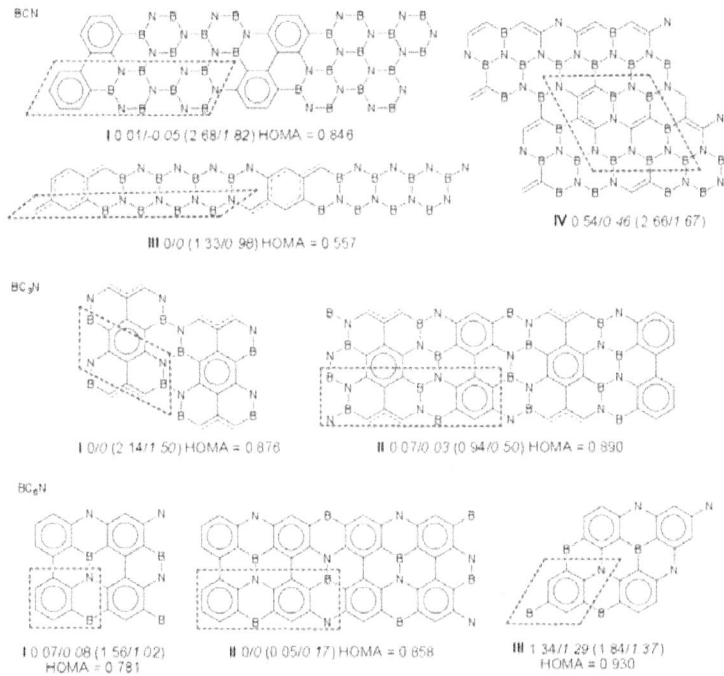

Figure 6: The isomers of BCN, BC$_3$N, BC$_6$N. Relative energy (per formula unit) with respect to most stable structures and band gaps (in parentheses) are shown from HSE (normal print) and PBE (italic) calculations. HOMA values of all-C hexagons, obtained from HSE calculations are also provided. Reprinted with permission from J. Phys. Chem. C 115, 10264 (2011). Copyright (2011) American Chemical Society.

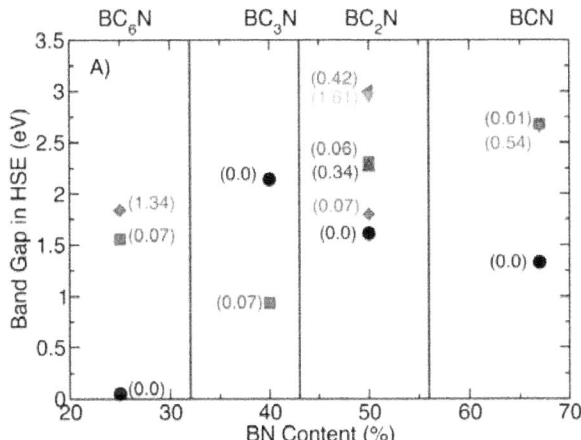

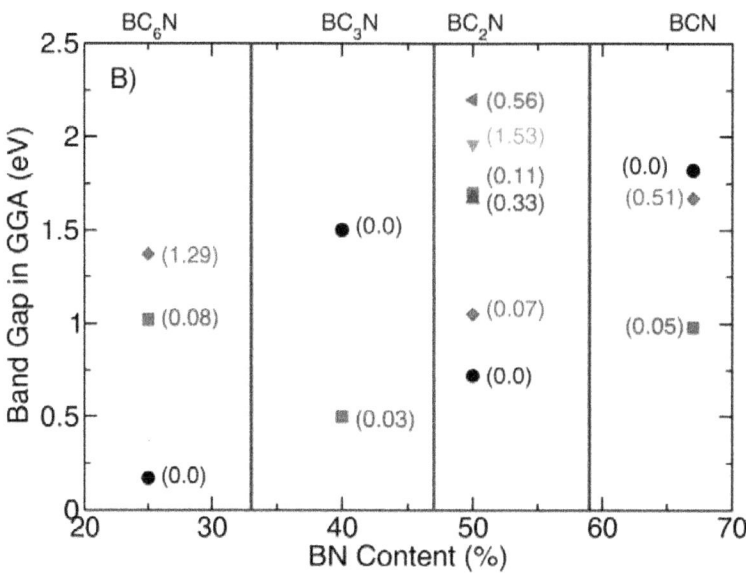

Figure 7: Band gaps of different isomers of $(BN)m(C_2)n$ are plotted against BN density, obtained from both (A) HSE and (B) PBE calculations. Relative energies are given in parentheses.

FUNCTIONALIZATION

Incorporation of magnetism in 2D sp-materials has been an important point of discussion in recent times. The combination of localized moments of $3d$ transition metal atoms and the sp electrons of the host 2D lattice can give rise to interesting magnetic properties relevant to nano devices based on the principle of magnetoresistance, for example. Ferromagnetic long-ranged order, half metallicity, large magnetic anisotropy, electric field driven switching of magnetization etc. are being studied for transition metal atoms adsorbed on graphene and 2D BN sheets. Another important point is the adsorption of these species at the interface between BN and graphene. A recent study [27] based on first principles electronic structure calculations has revealed some interesting electronic and magnetic properties of Fe, Co and Ni adatoms adsorbed on a h-BC_2N sheet. A hexagonal site at the interface between BN and graphene turns out to be a favorable site for adsorption. The presence of Fe, Co and Ni makes the system a magnetic semiconductor, a magnetic semi-metal and a non magnetic semiconductor respectively. Another interesting observation was that the adatoms are highly mobile on the surface and hence have the possibility to have a clustered configuration among themselves. It is interesting to note that the properties of this hybrid system are tunable in the sense that they can be modified by having different combinations of the width of each subsystem, BN and graphene.

Not only by magnetic adatoms, but by intrinsic edge properties, one can render magnetism in these 2D sheets. By DFT calculations, Dutta *et al.* [28] predicted interesting magnetic properties of H-passivated zigzag nanoribbons (ZGNRs) of various widths, doped by boron and nitrogen, keeping the whole system isoelectronic with C atoms in graphene. In the extreme case, all C atoms of ZGNRs are replaced by B and N atoms and zigzag BN nanoribbons are formed. In the ground state, the two edges are antiferromagnetically coupled and remain so for all dopings. However, the application of an external electric field affects the electronic structure of the nanoribbon giving rise to semiconducting and half-metallic properties. Electric-field induced changes in the magnetic properties are very interesting from a technological point of view. Other related studies [29] based on DFT revealed energetics, electronic structure and magnetism of quantum dots and nanorods of graphene embedded in BN sheet. It was showed that the formation energies and the HOMO-LUMO gaps of quantum dots vary as $1/n\sqrt{1/n}$, where n is the number of carbon atoms in the dots.

Adsorption of gases on 2D materials is an important topic from the technological and environmental points of view. Materials for clean energy are always sought for and in this respect, efficient hydrogen uptake of suitable materials is an important issue. Raidongia *et al.* [26] have studied H_2 adsorption on BCN at 77 K and 1 atm. pressure. From their experiments, H_2 uptake of 2.6 wt % was observed. Also, CO_2 adsorption is very important for environmental issues. In their study, BCN was found to have a very high CO_2 uptake of 100 wt % at 195 K and 1 atm. pressure. It should be noted that the uptake is only 58 % by the activated charcoal under identical conditions. At room temperature and 40 bar pressure, the CO_2 uptake was found to be 44 wt %.

In a recent theoretical study, Cao *et al.* [30] showed that a zigzag interface between BN and graphene can have a strong capability of adsorbing hydrogen, much stronger than pure graphene, BN or the armchair interface between them. Moreover, the adsorption of hydrogen induces a semiconductor to metal transition. As the mobility of hydrogen on the surface is rather high, the hydrogen atoms can migrate to the zigzag interface and hence will increase the density of hydrogen storage with the added functionality of band gap engineering.

SUMMARY AND OUTLOOK

Graphene-BN nanocomposites offer a huge potential in various technological sectors, e.g., nano electronics, gas sensing, hydrogen storage, nanomagnetic storage devices, to name a few. The unique combination of these two materials with different electronic properties, forming a 2D network, offers many possibilities for studying fundamental science and applications for nanotechnology. However, many challenges, both in the domains of experiment and theory, will come in the way. Experimental synthesis of samples of good quality and state-of-the-art characterization techniques to reveal the atomic scale physics will be the issues. From the point of view of theory, one faces difficulties in having a correct description of the band gaps and electronic structures in standard approximations of materials-specific theories. However, with the availability of powerful supercomputing facilities, it is nowadays possible to treat large systems by sophisticated many body theories to have a much better quantitative descriptions. Nevertheless, one may envisage many interesting directions for the applications of these nanocomposites to utilize the interface properties of BN and graphene. One of them is the spin switching properties of organometallics adsorbed at the interface, similar to what has been studied recently [31] for a 2D graphene sheet. The other application can be the adsorption of amino acids [32] at the interface to increase the activity by their immobilization. Hopefully, in near future, we will observe many applications of these nanocomposites, useful for the human society.

REFERENCES

1. K. S. Novoselov, A. K. Geim, S. V. Morozov, D. Jiang, Y. Zhang, S. V. Dubonos, I. V. Grigorieva, A. A. Firsov, 30. Science, 2004
2. L. D. Landau, Z. Phys, 1. Sowjetunion, 1937
3. R. E. Peierls, I. H. Ann, . Poincare, 1935
4. A. K. Geim, K. S. Novoselov, Mater. . Nat, 1. , 2007A.H. Castro Neto et al., Rev. Mod. Phys. 81, 109 (2009); A.K. Geim, Science 324, 1530 (2009).
5. M. I. Katsnelson, Today. 1. Materials, 2007
6. P. R. Wallace, Rev. . Phys, 1947
7. J. O. Sofo, A. S. Chaudhari, G. D. Barber, Rev. B. 7. Phys, 1534, (2007, D. C. Elias, et al. 32, 6. , 2009O. Leenaerts, H. Peelaers, A. D. Hernandez-Nieves, B. Partoens, and F. M. Peeters, Phys. Rev. B 82, 195436 (2010).
8. M. Klintenberg, S. Lebegue, M. I. Katsnelson, O. Eriksson, Rev. B. Phys, 2010

9. M. Y. Han, B. Özyilmaz, Y. Zhang, P. Kim, Rev. Phys, 9. Lett, 2007
10. Y. , W. Son, M. L. Cohen, S. G. Louie, 44. Nature, 2006
11. S. Bhandary, O. Eriksson, B. Sanyal, M. I. Katsnelson, Rev. B. Phys, 2010
12. P. Hohenberg, W. Kohn, Rev. B. Phys, 13, 8. , 1964W. Kohn and L. J. Sham, Phys. Rev. A 140, 1133 (1965).
13. H. Nozaki, S. Itoh, J. Phys, Solids. 5. Chem, 1996
14. L. Ci, L. Song, C. Jin, D. Jariwala, D. Wu, Y. Li, A. Srivastava, Z. F. Wang, K. Storr, L. Balicas, F. Liu, P. M. Ajayan, Mater. . Nat, 2010
15. A. Y. Liu, R. M. Wentzcovitch, M. L. Cohen, Rev. B. Phys, 1989
16. K. , T. Lam, Y. Lu, Y. P. Feng, G. Liang, Phys. Appl, 9. Lett, 2011
17. K. Yuge, Rev. B. 7. Phys, 2009
18. P. v. R. Schleyer, H. Jiao, Appl. Pure, 6. Chem, 1996
19. E. D. Jemmis, B. Kiran, Chem. 3. Inorg, 1998
20. P. W. Fowler, E. J. Steiner, J. Phys, A. Chem, 10, 1997
21. W. H. Fink, J. C. Richards, J. Am, Soc. Chem, 11, 1991
22. J. Zhu, S. Bhandary, B. Sanyal, H. Ottosson, J. Phys, C. Chem, 11, 2011
23. J. P. Perdew, Y. Wang, Rev. B. 4. Phys, 1992
24. J. Heyd, G. E. Scuseria, J. Chem, 12. Phys, 72, 2004J. Heyd, G. E. Scuseria and M. Ernzerhof, J. Chem. Phys. 118, 8207 (2003).
25. T. M. Krygowsky, J. Chem, Comput. Inf, 3. Sci, 1993
26. K. Raidongia, A. Nag, K. P. S. S. Hembram, U. V. Waghmare, R. Datta, C. N. R. Rao, Eur. J. 1. Chem, 2010
27. P. Srivastava, M. Deshpande, P. Sen, Chem. Phys, Phys. 1. Chem, 2011
28. S. Dutta, A. K. Manna, S. K. Pati, Rev. Phys, 10. Lett, 2009
29. S. Bhowmick, A. K. Singh, B. I. Yakobson, J. Phys, C. Chem, 11, 2011
30. T. Cao, J. Feng, E. G. Wang, Rev. B. 8. Phys, 2011
31. S. Bhandary, S. Ghosh, H. Herper, H. Wende, O. Eriksson, B. Sanyal, Rev. Phys, 10. Lett, 2011
32. S. Mukhopadhyay, R. H. Scheicher, R. Pandey, S. P. Karna, J. Phys, Lett. . Chem, 2011

Chapter 9

PROPERTIES OF MWNT-CONTAINING POLYMER COMPOSITE MATERIALS DEPENDING ON THEIR STRUCTURE

Ilya Mazov[1], Vladimir Kuznetsov[2], Anatoly Romanenko[3] and Valentin Suslyaev[4]

[1]Boreskov Institute of Catalysis, Novosibirsk, RussiaNational Research Technical University "MISIS", Moscow,, Russia

[2]Boreskov Institute of Catalysis, Novosibirsk,, Russia

[3]Nikolaev Institute of Inorganic Chemistry, Novosibirsk,, Russia

[4]National Research Tomsk State University, Tomsk,, Russia

INTRODUCTION

Carbon nanotubes (CNTs) are tubular structures composed of curved graphene sheets with diameter up to several tens of nanometers with typical length up to several micrometers. Single- and doublewall CNTs have diameters from 1.2 to ca. 3 nm and are usually packed in relatively dense structures ("ropes"). Multiwall carbon nanotubes can contain up to tens of concentrically aligned tubules and have diameter from 3-4 to tens of nanometers. Carbon nanotubes, both single- and multiwall, show outstanding mechanical and electrical properties [1, 2]. Nowadays CNTs are regarded as one of the key materials for development of various nanotechnology applications – new materials, sensors, actuators, field emitters etc. [3, 4, 5, 6]. In the last decade great effort was done in this field by many research groups, investigating structural, physical, mechanical, and electrical properties of CNTs.

Among all types of nanotubes single-wall nanotubes (SWNTs) were widely recognized as most perspective in regard of their predicted properties. Depending on chirality and diameter SWNTs can show significantly different electronic structure thus revealing metallic of semiconducting properties [7,8]. Mechanical properties of SWNTs were investigated both theoretically and experimentally and were shown to outstand all other construction materials such as steel, carbon fibers etc [9].

Extremely remarkable properties of SWNTs are strongly limited in usage by their high cost and low yield of production methods. Commonly used methods of SWNT synthesis include arc discharge [10], laser evaporation of carbon targets [11], or catalytic decomposition of gaseous carbonaceous species (carbon monoxide [12], alcohols [13], various hydrocarbons [14]) by CVD technique [15]. As-produced SWNTs need purification from amorphous carbon and other graphene-like species (fullerenes, multiwall CNTs etc.) which is usually performed by strong oxidative media and/or by selective surfactants (followed by ultra-centrifugation etc.). Involvement of such complex techniques results in high cost of resulting material, especially in the case of production of CNTs with tailored mechanical and electronic properties. According to market analysis the average price of highly purified SWNTs (90-99 wt. %) lays in range 200-600 $ per gram depending on purity, chirality and surface composition. Note that high amounts of SWNTs are still less available.

Multiwall carbon nanotubes (MWNTs) were firstly described in 1953 and now are one of the most common and widely used nanotubes allotrope. Multiwall nanotubes are composed with several concentrically aligned tubular graphene sheets, with typical diameter in range 8-30 nm. Physical and mechanical properties of MWNTs are significantly lower than that for SWNTs but still are higher than properties of commonly used construction materials and reinforcement additives.

Multiwall carbon nanotubes can be synthesized in the same way as SWNTs by arc discharge, graphite evaporation by laser irradiation or by catalytic decomposition of gaseous carbon-containing species [16] by CVD. The last one is the most perspective due to possibility to regulate CNT diameter and length; due to high yield and high selectivity of the process less or even no purification by aggressive oxidation is needed to achieve MWNT with high purity (higher than 90 wt. %) [17]. CVD process has high scaling potential, e.g. by realization of the fluidized bed technique [18, 19,20].

In the last few years significant progress was achieved in the scaling of synthesis of MWNTs by catalytic CVD route. Several companies have demonstrated large-scale facilities for the process, for example, Bayer AG (Germany), Nanocyl (Belgium), Arkema (France), Hyperion Catalyst (USA), CheapTubes Inc. (China). Development of the large-scale synthesis route for MWNTs with high purity and relatively low defectiveness allowed to significantly lower market price for such product which is in range 1-15 $ per one gram depending on purity, mean diameter and surface functionalization.

Relatively high availability of MWNTs and their remarkable properties result in great interest of their usage in various nanotechnology applications. At the present time high amount of research work was done in the field of

MWNT investigation and application. Multiwall carbon nanotubes can be used as components of composite materials with polymer, metal or ceramic matrices [21]; as chemical sensors [22, 23]; as components of catalytic systems [24, 25]; as electromagnetic shielding materials [26]; for biomedical applications such as selective drug delivery [27, 28] etc.

One of the most perspective approaches of usage of MWNT's superior properties is development of new multi-functional composite materials with improved and tailorable properties. Such composites can be used in various applications, for example as construction materials [29], anti-static coatings [30], low-weight electromagnetic shielding [31], conductive polymers [32] etc. Polymer matrices are mostly used for development of such composite due to their light weight, low price, good processability and controllable chemical, physical and mechanical properties as well as good scaling perspectives.

To date several tens of polymer matrices were investigated for the synthesis of MWNT-and SWNT-loaded polymeric composites. These are epoxy resins [33], polyurethanes [34], polyolefines [35, 36], polymethylmethacrylate [37], polystyrene [31], and others.

Systematic investigation of properties of novel multifunctional composite materials, containing carbon nanotubes is of essential importance to understand and improve their properties.

It is known [24, 38] that CNTs of same type but produced by different vendors often show significant difference in chemical and physical properties depending on their diameter, length, defectiveness, agglomeration state, surface chemistry etc. Variation in properties of the filler may result in non-linear changes in properties of resulting composite. Thus investigation of properties of CNT-containing composites depending on properties of incorporated nanotubes and polymer host matrix is an important task.

In this chapter we describe an attempt to systematic investigation of structural, physical-chemical, and electrophysical and electromagnetic properties of thermoplastic composite materials, comprising multiwall carbon nanotubes with different mean diameters and morphology.

MWNT-CONTAINING POLYMER COMPOSITES

Approaches to Synthesis, Main Problems

As it was mentioned above, CNT (and, first of all, MWNT)-containing polymer-based composite materials attract great interest in the last decade. Great work was done in this field and tens of various host matrices were investigated.

Several main problems in the area of design and preparation of nanotube-filled composite materials can be outlined according to literature analysis.

1. As-synthesized carbon nanotubes are usually arranged either in dense aligned arrays ("ropes") or tangled "furballs", composed of several tens of closely matted CNTs. First type is more typical for SWNTs, and the second one is most typical for CVD-produced MWNTs. Dense entangled MWNT arrays should be destructed during composite synthesis in order to achieve maximum dispersion state of nanotubes and subsequent maximum increase in properties of the composite. Also in this case it is possible to reach electrical percolation threshold at relatively low concentration of the filler due to intensive linking of high-dispersed nanotubes between each other. See review [39] for details of dispersion of MWNTs in various liquids using different technique.

2. Carbon nanotube fillers in the polymer matrix (and, in common sense, in all types of composite materials) can act as reinforcement material in several ways.

The first, incorporated nanotubes with high mechanical properties and high electrical conductivity, may act in the same way as macroscopic fillers (carbon fiber, glass wool etc.) providing stress transfer from the low-strength matrix in the case of mechanical load or charge transfer through continuous linked conductive network in the case of electrical load [40, 41].

The second, CNT filler is providing nucleation sites for the growth of polymer nanosized crystallites. Introduction of high amount of nanotubes with high surface chemical potential results in significant reduction of the grain size of resulting composite [42, 43] thus leading to increase of its mechanical properties.

Thus it is of crucial importance to obtain high dispersion degree of nanotube filler in the bulk volume of the polymer matrix and provide intensive interaction between CNTs and polymer. These two tasks should be resolved during composite synthesis in order to obtain material with increased properties. This can be done in several ways depending on the type of the polymer used as matrix material.

All polymer materials can be roughly divided in two parts – thermoplastic and thermoreactive polymers. Both types were used for synthesis of MWNT-containing composites with certain success.

Thermoreactive matrices, such as epoxy resins and polyurethanes, were one of the first used for preparation of MWNT-loaded composites. The main route of preparation of such type of materials involves mixing of carbon filler with resin or with chosen intermediate solvent (acetone, dimethylformamide etc. [44]) which is later mixed with the resin. The mixing process is often assisted with ultrasonic treatment which results in higher dispersion degree of the nanotube filler in the matrix [45]. Obtained mixture is molded to form necessary shape and cured. The technique described is quite experimentally simple and scalable. However, due to high viscosity of epoxy resin it is hard to disperse entangled nanotubes uniformly in whole volume of the polymer. In the case of usage of intermediate solvents such as acetone, the last must be evaporated before curing process. Carbon nanotubes tend to spontaneous agglomeration while staying in suspension, thus destroying achieved dispersive state and resulting in lowering of composite's properties as compared with theoretically predicted.

Several special procedures can improve the process, such as chemical surface functionalization [46, 47] or usage of short aligned CNT arrays as starting material [48, 49], allowing to obtain good distribution of CNTs with low electrical percolation threshold and increased mechanical properties.

Thermoplastic polymers can be used as matrix materials for design and synthesis of nanotube-filled composites. These matrices can be processed by variety of techniques, such as solution casting, extrusion, pressure molding, hot pressing etc. Processing methods can be divided in two main parts – *temperature*-assisted and *solution*-assisted technique.

Twin-screw extrusion, pressure molding, hot pressing, liquid casting and similar methods can be described as *temperature-assisted*. The main step of these techniques involves melting of mixture of the polymer matrix material and CNT filler with subsequent processing of the melt blend [35]. Such approaches are relatively cheap, scalable and experimentally simple, but still have several disadvantages.

Usually the polymer blend has high viscosity in molten state thus preventing disaggregation of entangled agglomerates of CNTs. Usage of ultrasonic treatment is this case is complicated by high temperatures and closed volume of experimental setup. Moreover, increase of CNT loading results in further sharp increase of viscosity of polymer-nanotube blend, preventing achieving high dispersion of incorporated nanotubes.

These problems can be partially solved by using of additional mixing procedure, such as mechanical activation of the solid polymer powder with CNTs, or using of high CNT shear flow mixers for the molten mixture providing high dispersion degree. However, these procedures can result in breaking of nanotubes, especially in the case of multiwall CNTs with defective walls, with corresponding CNT shortening and decrease in mechanical and electrophysical properties of the composite.

Solution-assisted technique of preparation of nanotube-filled polymer composites includes dissolving of polymer material in appropriate solvent, mixing of the resulting solution with CNTs and subsequent evaporation of the solvent with formation of the polymer-nanotube film [50]. High disaggregation state of CNTs can be achieved by using ultrasonication and/or high-intensive mixing [51] of the nanotube-polymer suspension due to reasonably low viscosity of the solution. Thin films can be produced by this technique, allowing one to design functionally grade materials, electrostatic coatings, polarizing films etc.

However, this technique has some disadvantages – it is hard to obtain massive samples by solution casting and also special precautions must be applied in order to avoid CNT agglomeration during drying of the composite film (for example, surface functionalization, shortening of nanotubes etc. [52, 53]).

There are some specific methods of synthesis of CNT-loaded composite materials, which cannot be ascribed to abovementioned types, for example *in situ* polymerization and coagulation precipitation techniques. The first one includes deposition of the catalyst on the surface of CNTs with further polymerization [54] or radical polymerization of the monomer (e.g. polystyrene, polymethylmethacrylate) in presence of carbon nanotubes or other nano-sized fillers or monomers [37,55, 56, 57]. High dispersion degree of nanotubes can be achieved by such methods, with following processing using conventional techniques, mentioned above.

The second method of CNT/polymer composite synthesis, coagulation precipitation, was firstly developed by Du et al. [58] for SWNT/PMMA composites. The coagulation precipitation (CP) technique involves dissolution of the polymer in appropriate solvent, mixing of this solution with carbon nanotubes (or other filler materials). The resulting slurry is mixed with the second solution in which the first solvent is soluble, but the polymer is not. As a result the polymer/CNT mixture immediately precipitates, forming disperse composite material which can be later processed in usual ways.

Coagulation precipitation technique has several remarkable advantages as compared with abovementioned methods. By right choice of the first solvent it is possible to obtain both dissolution of the polymer and good wetting of nanotubes. For example, in the case of MWNT/PMMA or MWNT/PS composites dimethylformamide of N-methylpyrrolidone can be used for this task. These solvents are known to provide very stable CNT suspensions [59] and can dissolve corresponding polymers in high concentrations. The second solvent for this system is water.

High CNT dispersion state is produced by ultrasonic treatment of the polymer-solvent mixture and is quite stable during minutes. It is of crucial importance that such high dispersion can be "frozen" on the second step by mixing with the second solvent, thus there is no reason to obtain super-stable CNT suspension.

The precipitation of polymer starts immediately after mixing with the second solvent and this process proceeds at high rate, producing small particles. Dispersed carbon nanotubes act as nucleation sites allowing reaching intimate interaction between polymer and individual CNTs.

CP technique still has certain inconveniences – for example, lack of scalability potential. Solution pair should be chosen carefully, providing both mutual solubility, and partial solubility of the polymer matrix material.

In the present chapter we describe preparation of MWNT-containing composite materials using coagulation precipitation technique and polymethylmethacrylate (PMMA) and polystyrene (PS) as matrices. All these polymers are thermoplastics and can be processed using common pressing and extrusion techniques.

Experimental: Synthesis of Mwnts AND MWNT-Loaded Composites

Multiwall carbon nanotubes were synthesized *in-lab* by ethylene decomposition over bimetallic FeCo catalyst in hot-wall CVD reactor at 680 °C. Details of preparation technique can be found elsewhere [60]. As-prepared MWNTs were additionally purified by reflux in HCl (15 wt. %) during 3 hours, washed with distilled water and dried in air at 80 °C for 24 hours.

Preparation of MWNT-loaded composites **was performed using coagulation precipitation technique.**

MWNT/PMMA and MWNT/PS composites were synthesized in similar way as follows. Polymer powder (PMMA, m.w. ~ 100000; Polystyrene, m.w. ~ 120000) was dissolved in dimethylformamide with concentration 0.1 g/ml. Calculated amount of air-dried MWNTs was loaded in water-cooled glass reactor and poured with polymer/DMF solution (40 ml) diluted with pure DMF (40 ml).

Resulting mixture was sonicated by Ti horn ultrasonicator (output power 8.5 W/cm^2) during 15 minutes under constant water cooling (the temperature of mixture was not higher than 50 °C). Resulting slurry was poured in ~ 1.5 liters of distilled water (t = 60-70 °C) under vigorous stirring immediately after US treatment. Precipitation of the polymer-MWNT composite proceeds immediately, resulting in formation of spongy-like deposit, which was left to stay overnight to complete coagulation.

The precipitate was filtered using Buchner funnel, washed with water (3×500 ml) and dried in air at 60 °C overnight. Residual water was removed by drying in vacuum (10^{-2} torr) at 60 °C for 2 hours.

Composite powder can be processed using common technique. In our case hot pressing was chosen as one of the most simple ways to make polymer films. Powder was placed between two polished steel plates, which were heated up to melting temperature of the polymer and pressed with hydraulic press with pressure ca. 400 kg/cm^2. Copper ring with 0.5 mm thickness was used as spaces. Produced composite films were Ø 60×0.5 mm^3 in dimensions.

Scanning and transmission electron microscopy was used for investigation of the structure of composite powder and films (JSM6460LV and JEM 2010 electron microscopes were used). Powder samples were placed on the conductive carbon adhesive tape, films were cut in plates with size ca. 8×3×0.5 mm^3, which were broken and glued to the copper stand with breaks upwards using silver glue. In order to avoid surface charging during SEM investigations all samples were additionally covered with 5-10 nm gold layer.

Electrophysical properties of composite films were investigated using four-probe technique with silver wires connected to sample surface with silver glue.

Electromagnetic response properties of free-standing polymer composite films were investigated in frequency range 0.01-12 GHz, 26-37 GHz. Array of experimental setups was used: quasi-optical setup based on panoramic meter KSvN R2-65 (Russia), Mach-Zehnder interferometer based on backward wave oscillator (Russia), HP Agilent PNA 8363B network analyzer (Agilent, USA) with multimode resonators (for dielectric permittivity spectra), and R2M-04 reflectance/transmission meter (Mikran, Russia). For ε spectra measurement samples were cut in pieces with size 0.5×2.5×30 mm^3 and placed in antinode of electric field parallel to electric field lines. Transmission coefficient was measured using polyfoam gasket, samples were cut in rings with an outer diameter Ø16.5 mm and inner diameter Ø6.95 mm and glued to the gasket. The measuring setup was calibrated using pure polyfoam gasket covered with glue.

Properties of Pure Mwnts

Macro-scale properties of MWNTs and MWNT-based composites depend strongly on their nanoscale parameters, such as diameter, particle size distribution, morphology of agglomerates. In this work we have investigated three types of MWNTs differencing in main diameter (and diameter distribution), defectiveness and morphology of agglomerates. On Figure 1 TEM data and diameter distribution for all types of MWNTs is shown.

Mean diameter of MWNT (as obtained by statistical analysis of TEM images) makes the value of ~7-9 nm for MWNT[8], ~10-12 nm for MWNT[12], and ~22-24 nm for MWNT[22]. From TEM images one can roughly evaluate defectiveness of MWNT (amount of amorphous species on the surface of CNT, opened walls etc.) which is increasing with decreasing of their mean diameter.

According to SEM data the morphology of MWNT secondary agglomerates varies, changing from rope-like structure for "thin" tubes (MWNT[8], MWNT[12]) to tangled furball-like structure for "thick" tubes (MWNT[22]). On figure 2 SEM images of typical MWNT agglomerates are presented.

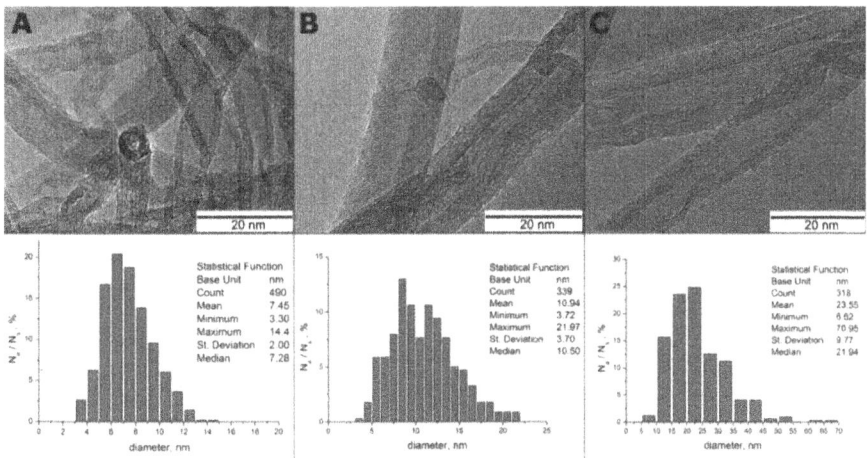

Figure 1: HRTEM images of MWNT samples used for preparation of MWNT/PMMA composites and their statistically calculated diameter distribution. A – MWNT[8], B – MWNT[12], C – MWNT[22].

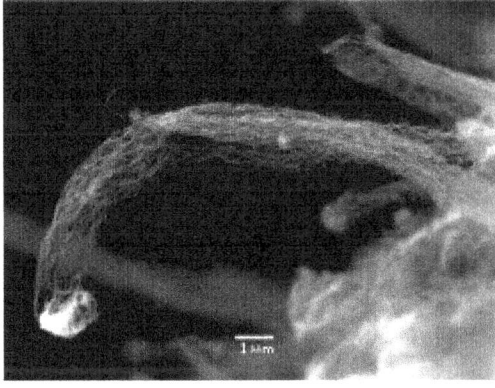

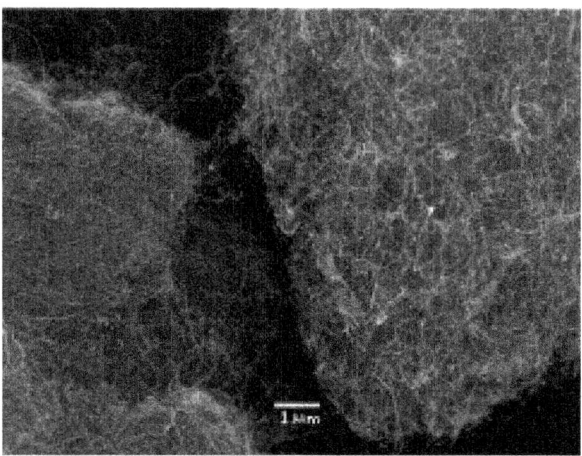

Figure 2: SEM micrographs of secondary aggregates of MWNTs. Left – MWNT[8], "rope-like" structure; right – MWNT[12], MWNT[22], "furball" structure.

Entangled nanotubes, which are forming "furball" structure, are hard to disperse in liquid media. MWNTs of all types tend to agglomerate in liquid dispersion, thus it is necessary to achieve certain conditions, allowing to "freeze" dispersed state of nanotubes. In our case usage of the coagulation precipitation technique allows to do this, as coagulation proceeds almost immediately after sonication and dispergation.

Pmma and PS Composites – Influence of Matrix Type

Polymethylmethacrylate (PMMA) and polystyrene (PS) both are cheap, large-scale and easily processable polymer matrices, which are often used as model systems for synthesis of polymer composites. Polymers with similar average molecular weight were used for synthesis of composites, both PMMA and PS are soluble in DMF and precipitate using water as secondary solvent.

Principal physical-chemical properties of polymethylmethacrylate and polystyrene polymers are quite similar (see Table 1 for details) [61].

From the Table 1 it can be clearly seen that PMMA and PS have very similar physical and chemical properties and the main difference is their surface composition and polarity of the surface [62, 63]. Carboxylic functions are present in PMMA structure thus resulting in increased oxygen content and increased polarity of the surface. Polystyrene has no oxygen-containing groups, showing lower polarity and surface tension [64].

Table 1: Physical-chemical properties of polymethylmethacrylate (PMMA) and polystyrene (PS)

Polymer	Formula	Surface free energy	Surface oxygen (XPS)	Polarity	Density, g/cm^3	Thermal conductivity, W/m·K	Dielectric constant, ε'@1MHz@25°C
PMMA		41.2 mJ/m^2 [65]	38.6 at. % [66]	0.28 [6"]	1.17	0.19	2.80 [65]
PS		40.1 mJ/m^2	0 at. %	0.17	1.04	0.22	2.55

Thus investigation of these model polymer systems allows one to reveal basic principles of the influence of the matrix type on the properties of MWNT-loaded composites.

Transmission electron microscopy analysis of MWNT/PMMA and MWNT/PS composites was performed to investigate internal structure of materials. Corresponding micrographs are shown on figure 3. Interaction of carbon nanotubes with polymer matrix results in surface wetting of MWNTs with polymer, which depends strongly both on the surface composition of CNTs and on surface properties of the polymer. As-prepared nanotubes have low amount of polar (oxygen-containing) groups, thus surface is mostly hydrophobic. Chemical functionalization (e.g. oxidation) can modify surface composition of the materials, making it hydrophilic. According to our previous study [65], as-prepared CNTs have ca. 0.3-0.5 oxygen-containing groups per nm^2, so their surface is mostly hydrophobic, but still can show several hydrophilic behavior.

It is reasonable to suggest that hydrophilic surface would have higher affinity to polar polymer matrices, and pristine nanotubes would be better wetted by low-polar polymers.

Such behavior was observed for both compared MWNT/PMMA and MWNT/PS samples. As it can be seen from Figure 3, the surface of CNTs is covered with polymer. In the case of PMMA as a matrix, nanotubes are not fully covered with polymer. Both covered and naked surface areas are observed (marked with arrows on Fig. 3 A, B) for PMMA-treated composite. It is important to note that observed contact angles between nanotube surface and PMMA layer are still lower than 35°.

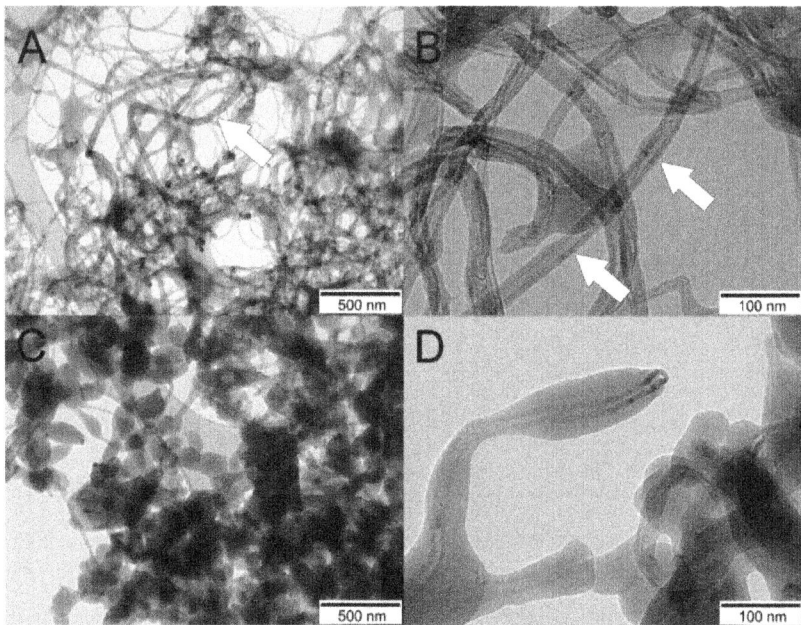

Figure 3: TEM images of MWNT-containing composites. A, B – MWNT^{22}/PMMA (5 wt. %), C, D – MWNT^{22}/PS (10 wt. %). Arrows are indicating naked parts of MWNTs (not covered with PMMA).

The possible reason for partial surface coverage with PMMA is certain irregularity of the functional composition of MWNT surface, for example, PMMA-covered areas are observed due to higher polarity of that nanotube part (e.g. due to partial oxidation), providing better wetting and lowering of the contact angle.

In contrast, PS-covered composite sample reveal significantly different structure where the surface of MWNTs is completely covered with thin continuous layer of the polymer (with thickness up to several nanometers). In the case of low CNT loading in composite one can observe disaggregated nanotubes, completely wrapped and wetted by PS (fig. 3, D). For higher nanotube content polystyrene "drops" are observed settled on the surface of nanotubes covered still with undestructured PS layer. As soon as polystyrene shows lower polarity as compared with PMMA and moreover has aromatic rings in structure, which have high affinity to the conjugated π-system of pristine (low-oxidized) nanotube, the wetting is more ease and coverage is higher in this case.

The contact angle between PS drops and CNT surface is in range 22-25° indicating high wetting ability of PS towards untreated surface of nanotubes.

Interfacial interaction between polymer matrix and incorporated nanotubes significantly affects structural and physical properties of composite films. Formation of dispersed nanotube array or network in the volume of the polymer matrix is essential to drastical change of material properties. Carbon nanotubes tend to agglomerate during synthesis of composites, thus decreasing dispersion degree.

Formation of CNT agglomerates in composite powder was observed for both PMMA- and PS-based materials using transmission electron microscopy. TEM micrographs of different nanotubes agglomerates are shown on Figure 4.

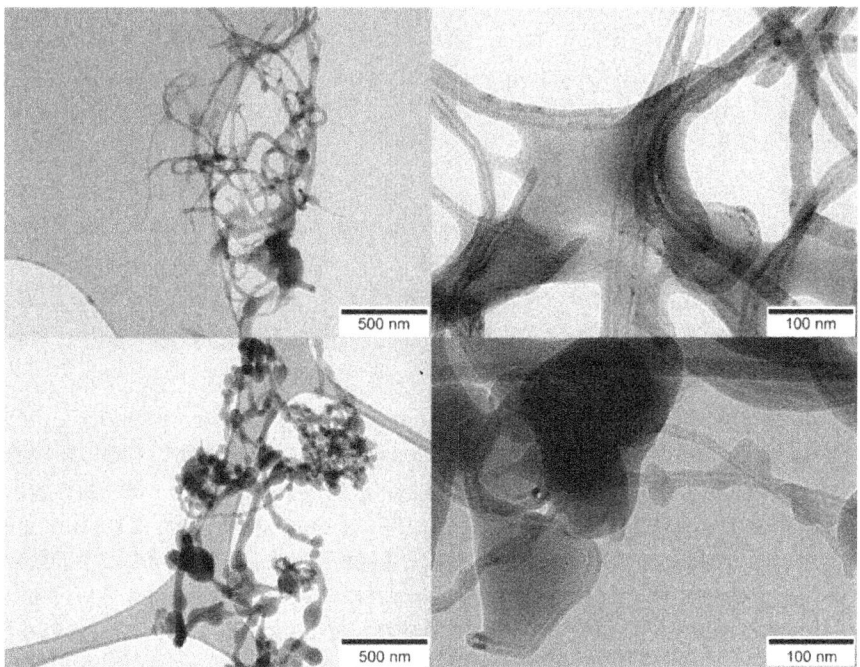

Figure 4: TEM images of agglomerated nanotubes in composite powders. A, B – MWNT[22]/PMMA (5 wt. %), C, D – MWNT[22]/PS (10 wt. %).

Two types of nanotube agglomerates can be distinguished in polymer-based composite materials. These agglomerates have been identified as "primary" and "secondary", according to their origin. *Primary agglomerates* are formed mainly during synthesis of carbon nanotubes themselves and can be avoided in growth-aligned nanotube arrays [66]. These aggregates must be destroyed during composite synthesis by ultrasonic treatment, high-shear flow mixing etc. Destruction of such particles in necessary to achieve high dispersion degree and, for example, correspondingly low percolation threshold for composites. Nevertheless, for CVD-grown nanotubes these aggregates are hard to destroy and they are still occurring in composites, covered with polymer layer. Such aggregates are typical for systems with low wetting of CNTs with polymer, e.g. for MWNT-PMMA (fig. 3-A & 4-A).

Agglomerates of secondary type are formed during the process of composite synthesis due to re-aggregation of dispersed nanotubes in solution due to van der Walls forces. In this case individual nanotubes are separated from each other in solution and linked by the polymer particle in solid composite (fig. 4-B&D). Strong wetting of carbon nanotubes with polymer assists destruction of primary agglomerates with subsequent formation of secondary-type agglomerates. Such phenomenon is observed for PS-based composites and is clearly seen on TEM micrographs – even for high MWNT loading (up to 10 wt. %) almost none primary agglomerates were observed. MWNTs are dispersed and separated from each other, forming secondary agglomerates, clued by polymer particles (fig. 4-C&D).

Investigation of composite films using scanning electron microscopy (SEM) was performed in order to elucidate influence of CNT-polymer interface on structural properties of composites. Corresponding SEM images of fresh breaks of composite films with different MWNT loading are shown on fig. 5.

For both types of polymer matrices separated nanotubes can be clearly seen on the surface of film breaks.

Difference in polymer-nanotube interface properties results in significant changes of the dispersion state of nanotubes in PMMA- and PS-based composites. For PMMA-based samples only single nanotubes can be seen for samples with 1 wt. % of nanotubes (fig. 5-A). Increase of MWNT content to 4 wt. % results in more uniform and dense dispersion of nanotubes in the volume of the polymer (fig. 5-C). Nevertheless, low-filled areas can be observed indicating relatively low dispersion degree of nanotubes. Certain amount of MWNTs can still occur in the agglomerated form, preventing formation of highly-disaggregated nanotube "network" in the polymer matrix.

Such phenomenon can be observed for the highest investigated MWNT loading in the composite (10 wt. %, fig. 5-E). Dense and uniform nanotube array can be clearly seen on the surface of composite film break. Note that the length of MWNT residues on the surface of the film is 1-2 μm, moreover, the surface of nanotubes is not covered with the polymer. This may be attributed to stretching of CNTs from the volume of the polymer during breakage of the film, which can be easily assumed taking into consideration low wetting ability of PMMA towards hydrophobic surface of untreated carbon nanotubes.

Polystyrene-based composite materials show opposite phenomenon of high wetting and high dispersion degree even at low MWNT loadings. Nanotubes in all MWNT/PS composite film samples are well-dispersed and covered with polymer layer. Uniform CNT distribution can be observed for samples with both low and high nanotube loading. According to SEM data nanotubes are randomly and evenly distributed on whole breakage area, which can be clearly seen on fig. 5-B, D, F.

Note that protruding parts of MWNT for PS-based composites are shorter as compared with similar PMMA-based samples and are wrapped with polymer layer. Such phenomenon can be attributed to higher wetting of CNTs with polystyrene and higher adhesion of filler to the material of the polymer matrix.

Thereby according to TEM and SEM data significant difference in wetting ability of investigated polymers towards MWNTs results in drastic changes of structural properties of composite powders and films. Highly-wetting aromatic polystyrene matrix allows one to reach higher dispersion degree at lower filler concentrations as compared with more polar polymethylmethacrylate due to mainly hydrophobic surface character of untreated nanotubes.

Area of application of novel materials is strongly dependent on their physical properties. As it was shown above, MWNT-based composites are perspective as tailorable materials for electrical and electromagnetic applications. Structure and dispersion state of nanotubes in composite, as well as their interconnections (i.e. formation of connected array of CNTs) strongly affect physical properties of composite materials. It is well known that introduction of continuous carbon nanotubes in the dielectric polymer matrix allows to increase conductivity of the resulting composite by several orders of magnitude. Conductive composite materials can be characterized by the percolation threshold which for multiwall carbon nanotubes lays in range from ca. 0.005 wt. % [67] to 3-4 wt. % [68] depending on the electrophysical properties of initial nanotubes and peculiarities of the composites' preparation (alignment and dispersion of nanotubes etc.).

Electrical conductivity measurements reveal significant difference between PMMA- and PS-based composites with similar MWNT loading. Conductivity data for PMMA-based composites with different types of MWNTs is shown on figure 6.

Electrical percolation threshold for PMMA-based composites was estimated as ~ 1-2 wt. %, reaching maximum value of 10^{-1}-10^{-2} S/m for 3-5 wt. % of CNT content.

PS-based composite materials show surprisingly low electrical conductivity. Significant conductivity (~ 4.5×10^{-4} S/cm) was observed for composite sample with 10.0 wt. % loading of CNTs, all other samples show electrical resistivity higher than 10^9 Ohm/cm which was the sensitivity threshold for the setup used for measurements.

As it was shown by electron microscopy investigations, PS-based composites reveal higher wetting and higher dispersion degree of nanotubes in the volume of the polymer as compared with PMMA-based materials. Unusually high percolation threshold and low conductivity value can be explained taking into consideration insulation of conductive nanotubes with wrapped polystyrene layer. In this case perfectly dispersed three-dimensional nanotube network in the bulk volume of the polymer is formed by insulating objects, preventing current flow through composite even at high CNT loadings.

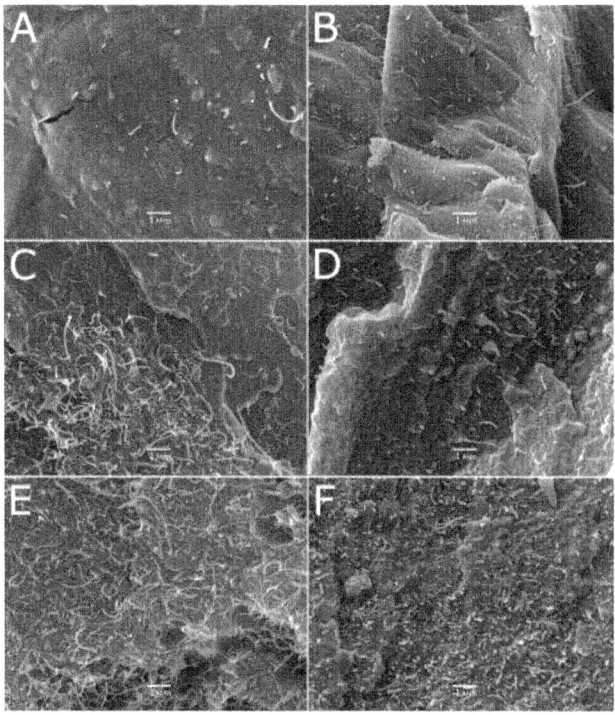

Figure 5: SEM images of MWNT/PMMA (left column) and MWNT/PS (right column) composite films. A, B – 1 wt. %, C, D – 4 wt. %, E, F – 10 wt. %.

In the case of poorly-wetted PMMA-based composites the surface of nanotubes is not completely insulated, thus allowing reaching of electrical saturation at relatively low CNT loadings. Moreover, destruction of primary agglomerates of nanotubes, observed for PS and not for PMMA-based samples, results in diminishing of electrical contacts between nanotubes due to their insulation. Residual CNT agglomerates in the case of PMMA-based composites facilitate formation of conductive paths in the volume of the polymer matrix. As a result overall electrical conductivity of MWNT/PMMA materials is higher as compared with PS-based samples.

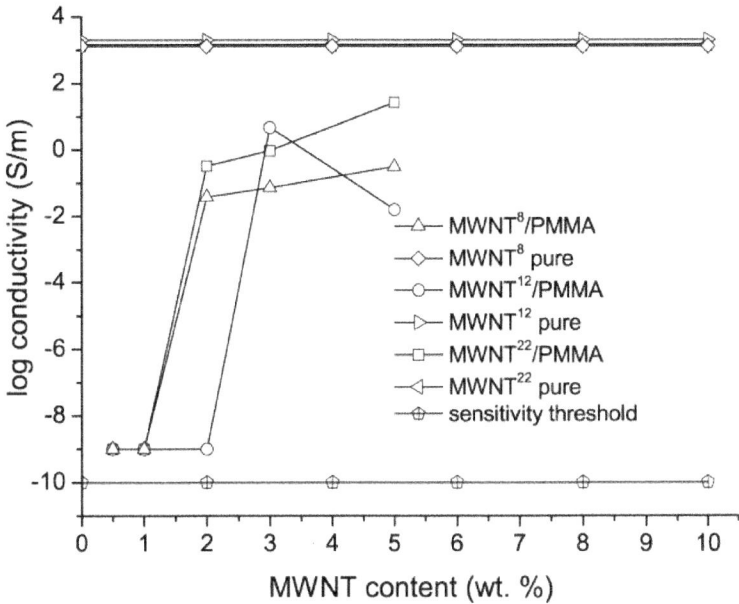

Figure 6: Electrical conductivity of MWNT/PMMA composites with various types of nanotubes versus filler concentration.

Electromagnetic response properties of composite materials play significant role in their application area. Incorporation of conductive media, such as carbon nanotubes, affects strongly the way of interaction of certain material with electromagnetic irradiation (EMI). Conjugated π-system, occurring in carbon nanotubes, allows both to dissipate and reflect electromagnetic wave, thus providing two possible mechanisms of EM response – reflectance and absorbance.

Electromagnetic response properties of MWNT-containing composites were investigated in broadband region (2-36 GHz) and was found to be strongly dependent on MWNT type and diameter as well as on polymer matrix type.

Influence of structural properties of MWNTs on EM shielding properties of composites was investigated for MWNT/PMMA materials due to their high electrical conductivity. Structure of initial CNT affects electrical and electromagnetic properties of CNT-based composites. We have investigated EM response properties of MWNT-based composites in frequency range 3-11 GHz (complex dielectric permittivity for MWNT/PMMA samples). Transmission (T) and reflection (R) coefficients can be easily calculated using following equations. Absorption coefficient of the EM radiation in the sample can be calculated as $A = 1 - R - T$.

$$R = \left| \rho \frac{1-e^{-i2kd}}{1-\rho^2 e^{-i2kd}} \right|; \quad T = \left| \frac{(1-\rho^2)e^{-ikd}}{1-\rho^2 e^{-i2kd}} \right|,$$

where

$$\rho = \frac{Z-1}{Z+1}, \quad Z = \sqrt{\mu^*/\varepsilon^*},$$

is wave impedance, and

$$k = \frac{2\pi f \sqrt{\varepsilon^* \mu^*}}{c}$$

is the wavenumber, f is the frequency of the EM wave, c is the speed of light, d is the thickness of the sample, $\varepsilon^* = \varepsilon' - i\varepsilon''$ and $\mu^* = \mu' - i\mu''$ are complex permittivity and magnetic permeability of the investigated material correspondingly. Dielectric losses (ε' and ε'') are measured experimentally, and $\mu'=1, \mu''=0$ (no magnetic losses are observed in the sample). Observed values of ε' and ε'' are growing with increase of the CNT loading in the composite and lay in range 20-70 for MWNT[22]/PMMA and 5-40 for MWNT[12] and MWNT[8]/PMMA composites.

On the figure 7 data on measured transmission and reflections coefficients are shown.

Increase of the conductivity of the composites with MWNT loading in all samples leads to growth of *R* and diminishing of *T* parameters. For example for the MWNT[8]/PMMA composite one can see higher *R* and lower *T* values for the sample with 3 wt.% of MWNT whereas for the sample with 5 wt. % these values are lower and higher, correspondingly, correlating with its conductivity. It should be mentioned that even composites with the lowest electrical conductivity (with CNT loading 0.5-2 wt.%) show high values of the *R*, which may be due to formation of isolated conductive MWNT structures in the volume of the polymer that cannot be registered by macroscopic measurements of the electrical conductivity but still can interact with electromagnetic field. Polarization of such isolated structures gives contribution in EM response of non-conductive materials with subsequent growth of the permittivity.

Note that transmission and reflection coefficients do not depend directly on the electrical conductivity for different MWNT types. The highest values of R are observed for the samples MWNT[22]/PMMA which show the lowest conductivity among other composites. This phenomenon may be described taking into account significant difference between number of individual nanotubes of each type incorporated into polymer matrix. Relationship between

CNT number can be roughly estimated as relationship between $r_{oj}^2-r_{ij}^2$, where r_{oj} is the outer diameter of CNT type j, r_{ij} is the inner diameter of CNT type j

Simple math gives rough approximate of relationship between CNT number in the composite with the same weight loading as $N(MWNT^{22}):N(MWNT^{12}):N(MWNT^8) \approx 8:4:1$.

Thus it is possible to propose that despite of the lower macroscopic electrical conductivity, composite materials comprising MWNT with lower diameter possess higher amount of polarizable species, giving higher values of dielectric permittivity and reflection coefficient.

All samples with low CNT content (0.5 wt. %) are almost transparent in all frequency range. Increase of MWNT loading leads to reduction of transmission coefficient. Transmission is strongly affected by MWNT type and is changing unidirectionally with the dependence of permittivity – the lowest transmission is observed for MWNT^{22}-containing composites which have higher ε value as compared with MWNT^{12}/PMMA, MWNT^8/PMMA and composites.

For both matrices percolation-like concentration dependence can be observed with sharp decrease of EM transmission coefficient (and corresponding increase of EM shielding of composites) after 1-2 wt. % of MWNTs in materials.

Nevertheless, the most significant difference for investigated matrices is that in case of PS matrix EM shielding is provided mainly by absorption, growing with increase of MWNT loading, and for the case of PMMA matrix main part of EM shielding is constituted by EM reflectance. Principal changes in mechanisms of EM attenuation in polymer composites filled with same type MWNTs can be explained taking into consideration abovementioned differences in their structural and electrophysical properties.

PMMA-based composites show high electrical conductivity and relatively low percolation threshold therefore above percolation threshold such materials interact with incident EM radiation as typical conductor, reflecting most part of electromagnetic wave.

Sharp increase in EM absorbance for PMMA-based composites at MWNT loading lower than 2 wt. % may be attributed to absence of reflectance allowing propagation of incident EM wave through the sample and its interaction with conjugated π-system of incorporated carbon nanotubes.

The value of EM absorbance reaches saturation for MWNT loadings higher than 2 wt. %, corresponding to electrical percolation threshold value. This phenomenon indicates formation of interconnected array (or cluster) of carbon nanotubes in the volume of the polymer, providing conductive paths for electrical current, decreasing electrical resistivity of the composite with subsequent increase in EM reflectance. Further increase of EM reflectance with growing MWNT loading is caused by formation of new conductive paths in linked nanotube network, increasing its electrical conductivity and reflective properties, however not affecting or diminishing absorption of EM radiation.

Polystyrene-based composites with same MWNTs as filler show surprisingly high shielding efficiency, as they are almost insulating at low nanotube loadings. In contrast to PMMA-based materials main part of shielding efficiency of PS-based composites is provided by absorption, growing with increase of MWNT concentration, especially for higher frequencies. At the same time reflection of EM radiation is increasing slowly, showing similar behavior as compared with EM absorbance for MWNT/PMMA composites.

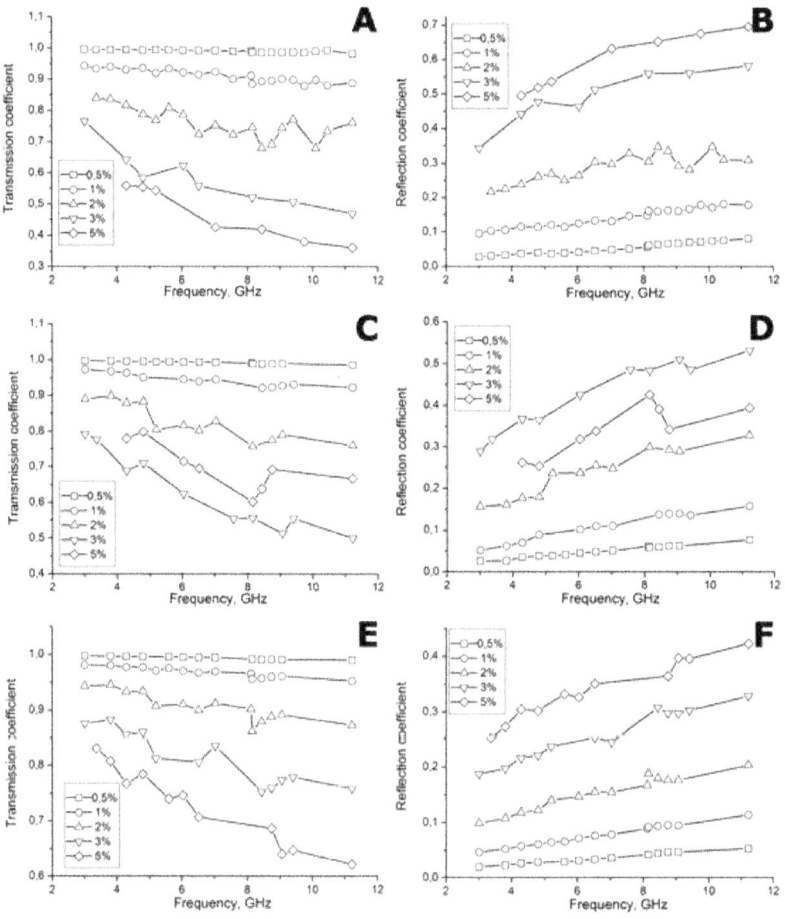

Figure 7: Transmission (A, C, E) and reflection (B, D, F) coefficients for MWNT#1, #2, #3/PMMA composites respectively.

Internal structure of composite materials and interaction between MWNTs and polystyrene matrix show *i)* strong wetting of MWNT surface with polymer, providing certain insulation of nanotubes from each other; *ii)* high dispersion degree of MWNT without primary agglomerates; *iii)* random uniform distribution state of individual nanotubes in the volume of the polymer even for high filler concentrations. All these factors affect changes in EMI shielding mechanism of MWNT/PS composites as compared with relatively poorly-dispersed highly-conductive PMMA-based materials.

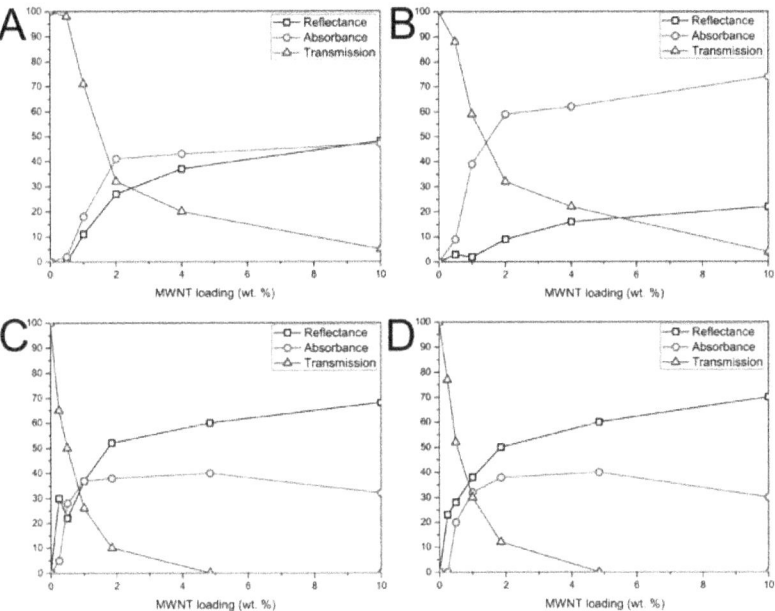

Figure 8: Concentration dependence of electromagnetic response properties of PS-based (A, B) and PMMA-based (C, D) composites in Ka-band. A, C – 29 GHz, B, D – 34 GHz.

Low-conductive PS composites possess correspondingly low EMI reflectance from the surface, allowing electromagnetic radiation to propagate in-depth of the sample. The characteristic wavelength for the EM radiation in microwave region is 1-10 mm, which is higher than the size of the individual nanotube and MWNT agglomerate and is comparable with the macroscopic size of the composite sample. Thus incident EM wave may interact with whole volume of the composite sample and this interaction is strongly facilitated due to the absence of surface conductivity and reflectance due to skin-effects.

Propagating EM wave interacts with individual nanotubes uniformly dispersed in the volume of the composite with corresponding attenuation.

High dispersion degree of MWNTs in polymer results in uniform density of attenuating particles, providing correspondingly high "extinction coefficient" of composite material.

CONCLUSION

Properties of multiwall carbon nanotubes and MWNT-containing composite materials are strongly dependent on their structure and morphology and on the type of polymer matrix. Variations of mean diameter distribution and morphology of agglomerates results in significant changes of their dispersive and electrical properties, affecting electromagnetic response of composites.

MWNT-filled composites with PMMA matrix show higher EM reflectance, higher electrical conductivity with low percolation threshold due to relatively low wetting of hydrophobic MWNTs with polar PMMA molecules and corresponding low dispersion degree of individual nanotubes.

In contrast for low-polar polystyrene matrix high wetting was observed resulting in formation of uniformly dispersed array of PS-covered MWNTs in the matrix. Coverage of MWNT surface with polystyrene results in insulation of nanotubes, thus resulting in high electrical resistivity with high percolation threshold. High shielding efficiency was observed for such composites, provided mainly by absorbance of EM radiation with low-conductive media.

Thus variation of main properties of polymer matrices and incorporated nanotubes allows to obtain composite materials with predictable tailorable properties, which may be used in various applications.

ACKNOWLEDGEMENT

This work was partially supported by RFBR grant #11-03-00351, ISTC project B-1708, and RNP projects #2.1.1/10256, #16.740.11.0016 & #16.740.11.0146.

REFERENCES

1. J. , P. Salvetat, J. , M. Bonard, N. H. Thomson, A. J. Kulik, L. Forro, W. Benoit, L. Zuppiroli, properties. Mechanical, carbon. of, Appl. nanotubes, Phys. A 69, 255-260 (1999

2. M. Terrones, And. Science, Of. Technology, Twenty. The-First, Synthesis. Century, And. Properties, Of. Applications, Nanotubes. Carbon, Annu, Rev. Mater.Res. (2003419 EOF

3. P. M. Ajayan, O. Z. Zhou, of. Applications, Nanotubes. Carbon, M. in, S. Dresselhaus, G. Dresselhaus, Ph. Avouris (Eds.): Carbon Nanotubes, Topics Appl. Phys. 80, 391-425 (2001

4. D. Srivastava, C. Wei, K. Cho, of. Nanomechanics, nanotubes. carbon, Appl. composites, Mech. Rev. 5620032003215230
5. J. R. Stetter, G. J. , Maclay Carbon nanotubes and sensors: a review, Advanced Micro and Nanosystems, 1edited by: H. Baltes, O. Brand, G.K. Fedder, C. Hierold, J. Korvink, O. Tabata 357-382 (2004
6. J. Wang, Based. Carbon-Nanotube, Biosensors. A. Electrochemical, Electroanalysis. . Review, 20057 EOF14 EOF
7. J. W. G. Wilder, L. C. Venema, A. G. Rinzler, R. E. Smalley, C. , Dekker Electronic structure of atomically resolved carbon nanotubes Nature 391199819985962
8. Q. Liu, W. Ren, Z. , G. Chen, L. Yin, F. Li, H. Cong, H. , M. Cheng, properties. Semiconducting, cup-stacked. of, nanotubes. carbon, . Carbon, 2009731 EOF736 EOF
9. R. S. Ruoff, D. Qian, W. K. , Liu Mechanical properties of carbon nanotubes: theoretical predictions and experimental measurements C.R. Physique 4200320039931008
10. T. W. Ebbesen, P. M. Ajayan, Large scale synthesis of carbon nanotubes, Nature, 358, 220 (1992).
11. A. Thess, R. Lee, P. Nikolaev, H. Dai, P. Petit, J. Robert, C. Xu, Y. H. Lee, S. G. Kim, A. G. Rinzler, D. T. Colbert, G. E. Scuseria, D. Tománek, J. E. Fischer, R. E. , Smalley Crystalline Ropes of Metallic Carbon Nanotubes, Science 27319961996483487
12. B. Kitiyanan, W. E. Alvarez, J. H. Harwell, D. E. Resasco, production. Controlled, single-wall. of, nanotubes. carbon, catalytic. by, of. C. O. decomposition, bimetallic. on, catalysts. Co-Mo, Physics. Chemical, 3. Letters, 2000497 EOF503 EOF
13. Y. Murakami, S. Chiashi, Y. Miyauchi, M. Hu, M. Ogura, T. Okubo, S. Maruyama, of. Growth, aligned. vertically, carbon. single-walled, films. nanotube, quartz. on, substrates, optical. their, Chemical. anisotropy, Letters. 3. Physics, 2004298 EOF
14. G. Zhang, D. Mann, L. Zhang, A. Javey, Y. Li, E. Yenilmez, Q. Wang, J. P. Mc Vittie, Y. Nishi, J. Gibbons, H. , Dai Ultra-high-yield growth of vertical single-walled carbon nanotubes: Hidden roles of hydrogen and oxygen PNAS 1022005200516141116145
15. M. Kumar, Y. Ando, Vapor. Chemical, of. Deposition, Nanotubes. A. Carbon, on. Review, Mechanism. Growth, Production. Mass, of. Journal, Nanoscience, . Nanotechnology, 2010

16. Y. Li, X. B. Zhang, X. Y. Tao, J. M. Xu, W. Z. Huang, J. H. Luo, Z. Q. Luo, T. Li, F. Liu, Y. Bao, H. J. Geise, production. Mass, high-quality. of, carbon. multi-walled, bundles. nanotube, a. on, Mo. Ni, O. Mg, Carbon. . catalyst, 2005295 EOF301 EOF

17. K.B. Hong, A.A.B. Ismail, M.E. Mahayuddin, A.R. Mohamed, S.H.S. Zeiri, Production of High Purity Multi-Walled Carbon Nanotubes from Catalytic Decomposition of Methane Journal of Natural Gas Chemistry1520062006266270

18. D. Venegoni, Serp. R. Ph, . Y. Feurer, C. Kihn, . Vahlas, Kalck. Ph, study. Parametric, the. for, of. growth, nanotubes. carbon, chemical. by, deposition. vapor, a. in, reactor. fluidized, . Carbon, 20021799 EOF

19. M. Corrias, B. Caussat, A. Ayral, J. Durand, Y. Kihn, Kalck. Ph, Serp. Ph, nanotubes. Carbon, by. produced, catalytic. C. V. D. fuidized, approach. first, the. of, Chemical. process, Science. . Engineering, 2003

20. S.W. Jeong, S.Y. Son, D.H. Lee, Synthesis of multi-walled carbon nanotubes using Co–Fe–Mo/Al2O3 catalytic powders in a fluidized bed reactor, Advanced Powder Technology 21 (2010) 93–99

21. J. Wang, H. Kou, X. Liu, Y. Pan, J. Guo, of. Reinforcement, matrix. mullite, multi-walled. with, nanotubes. carbon, International. . Ceramics, 2007719 EOF

22. Y. , T. Jang, S. , I. Moon, J. , H. Ahn, Y. , H. Lee, B. , K. Ju, A. simple, in. approach, chemical. fabricating, using. sensor, grown. laterally, carbon. multi-walled, Sensors. nanotubes, B. Actuators, . Chemical, 2004

23. N. Sinha, J. J. T. W. Yeow, Nanotube. Carbon-Based, Journal. Sensors, Nanoscience. of, . Nanotechnology, 2006

24. N. Maksimova, G. Mestl, R. , Schlogl Catalytic activity of carbon nanotubes and other carbon materials for oxidative dehydrogenation of ethylbenzene to styrene, Studies in Surface Science and Catalysis, G.F. Froment and K.C. Waugh (Editors) 13320012001383389

25. M. S. Saha, R. Li, X. Sun, loading. High, Pt. monodispersed, on. nanoparticles, carbon. multiwalled, for. nanotubes, performance. high, exchange. proton, fuel. membrane, Journal. cells, Power. of, 1. Sources, 2008314 EOF322 EOF

26. M. H. Al-Saleh, U. Sundararaj, interference. Electromagnetic, mechanisms. shielding, C. N. of, composites. T/polymer, . Carbon, 20091738 EOF1746 EOF

27. T. A. Hilder, J. M. , Hill Carbon nanotubes as drug delivery nanocapsules, Current Applied Physics 820082008258261

28. C. Tripisciano, K. Kraemer, A. Taylor, E. Borowiak, Borowiak-Palen Single-wall carbon nanotubes based anticancer drug delivery system, Chemical Physics Letters 47820092009200205
29. J. Makar, J. Beaudoin, Carbon nanotubes and their application in the construction industry. Nanotechnology 331-341 (2003
30. M. Sangermano, S. Pegel, P. Potschke, B., Voit Antistatic Epoxy Coatings With Carbon Nanotubes Obtained by Cationic Photopolymerization, Macromol. Rapid Commun. 2920082008396400
31. Y. Yang, M. C. Gupta, K. L. Dudley, R. W. Lawrence, A. Comparative, of. E. M. I. Study, Properties. Shielding, Carbon. of, Nanofiber, Carbon. Multi-Walled, Filled. Nanotube, Composites. Polymer, of. Journal, Nanoscience, . Nanotechnology, 2005
32. X. Zhang, J. Zhang, Z. Liu, polymer/carbon. Conducting, composite. nanotube, made. films, in. by, electropolymerization. situ, an. using, surfactant. ionic, supporting. the, Carbon. . electrolyte, 2005
33. L. Sun, H. Sue, J. , 2010Epoxy/CarbonNanotube Nanocomposites, in Epoxy Polymers: New Materials and Innovations (eds J.-P. Pascault and R. J. J. Williams), Wiley-VCH Verlag GmbH & Co. KGaA, Weinheim, Germany
34. S. Shang, W. Zeng, X. Tao, stretchable. M. W. N. High, conductive. Ts/ polyurethane, J. . nanocomposites, Mater, Chem., 212011201172747280
35. Y. Zou, Y. Feng, L. Wang, X. Liu, Processing, of. M. W. N. T. H. D. P. E. properties, Carbon. . composites, 2004271 EOF
36. W. Tang, M. H. Santare, S. G. Advani, processing. Melt, property. mechanical, of. characterization, carbon. multi-walled, /high. nanotube, polyethylene. . M. W. N. T. H. D. P. E. density, films. composite, . Carbon, 20032779 EOF2785 EOF
37. S. Shang, L. Li, X. Yang, Y. , Wei Polymethylmethacrylate-carbon nanotubes composites prepared by microemulsion polymerization for gas sensor, Composites Science and Technology 692009200911561159
38. J. H. Lehman, M. Terrones, E. Mansfield, K. E. Hurst, V. Meunier, the. Evaluating, of. characteristics, carbon. multiwall, Carbon. . nanotubes, 20112581 EOF2602 EOF
39. J. Hilding, E. A. Grulke, Z. G. Zhang, F. Lockwood, of. Dispersion, Nanotubes. Carbon, Liquids. in, Of. Journal, Science. Dispersion, Technology. . And, 20031 EOF41 EOF
40. L.S. Schadler, S.C. Giannaris,P.M. Ajayan, Load transfer in carbon nanotube epoxy composites, Appl.Phys. Lett. 73, 3842 (1998

41. L. R. Xu, S. Sengupta, stress. Interfacial, transfer, mismatch. property, discontinuous. in, composite. nanofiber/nanotube, J. materials, Nanosci Nanotechnol. 420052005620626

42. K. T. Kim, J. Eckert, S. B. Menzel, T. Gemming, S. H. Hong, refinement. Grain, strengthening. assisted, carbon. of, reinforced. nanotube, matrix. copper, Appl. nanocomposites, Phys. Lett. 92, 121901 (2008

43. X. Zeng, G. Zhou, Q. Xu, Y. Xiong, C. Luo, J. Wu, A. new, for. technique, of. dispersion, nanotube. carbon, a. in, melt. metal, Science. Materials, A. 5. Engineering, 20105335 EOF5340 EOF

44. K. Q. Xiao, L. C. Zhang, separation. Effective, of. alignment, entangled. long, nanotubes. carbon, epoxy. in, of. Journal, Science. . Materials, 20056513 EOF6516 EOF

45. S. Bal, S. S. , Samal Carbon nanotube reinforced polymer composites-A state of the art, Bull.Mater. Sci. 3020072007379386

46. Y. J. Kim, T. S. Shin, H. D. Choi, J. H. Kwon, Y. , C. Chung, H. G. , Yoon Electrical conductivity of chemically modified multiwalled carbon nanotube/epoxy composites, Carbon 43200520052330

47. R.E. Gorga, K.K.S. Lau, K.K. Gleason, R.E. Cohen, The Importance of Interfacial Design at the Carbon Nanotube/Polymer Composite Interface, Journal of Applied Polymer Science,1022006200614131418

48. Z. Yang, Z. Cao, H. Sun, Y. , Li Composite Films Based on Aligned Carbon Nanotube Arrays and a Poly(N-Isopropyl Acrylamide) Hydrogel, Adv. Mater. 200820082022012205

49. E.J. Garcia, A.J. Hart, B.L. Wardle, A.H. Slocum, Fabrication and Nanocompression Testing of Aligned Carbon-Nanotube-Polymer Nanocomposites, Adv.Mater. 200720071921512156

50. W. Yuan, J. M. B. Chan-Park, A. Novel, Dispersing. Polyimide, for. Matrix, Electrically. Highly, Solution. Conductive-Cast, Nanotube. Carbon-Based, Chem. Composite, Mater. 232011201141494157

51. P. Ciselli, R. Zhang, Z. Wang, C. T. Reynolds, M. Baxendale, T. Peijs, U. H. M. W. Oriented-P, E. C. N. T. composite, by. a. tapes, casting-drawing. solution, using. process, European. mixed-solvents, Journal. . Polymer, 2009

52. N.G. Sahoo, Y.C. Jung, H.J. Yoo, J.W. Cho, Effect of Functionalized Carbon Nanotubes on Molecular Interaction and Properties of Polyurethane Composites, Macromol.Chem. Phys. 2072006200617731780

53. K. Zhang, J. Y. Lim, . B. J. Park, H. J. Jin, H. J. Choi, Acid. Carboxylic, Multi. Functionalized-Walled, Nanotube. Carbon-Adsorption,

Poly(methyl. onto, Microspheres. methacrylate, of. Journal, Nanoscience, . Nanotechnology, 2009

54. L. Cui, N.H. Tarte, S.I. Woo, Synthesis and Characterization of PMMA/MWNT Nanocomposites Prepared by in Situ Polymerization with Ni(acac)2 Catalyst, Macromolecules 2009, 42, 8649–8654

55. H. , J. Jin, H. J. Choi, S. H. Yoon, S. J. Myung, S. E. Shim, Nanotube. Carbon-Adsorbed, Polystyrene, methacrylate. Poly(methyl, Chem. Microspheres, Mater. 200520051740344037

56. D. Wang, J. Zhu, Q. Yao, C. A. Wilkie, A. Comparison, Various. of, for. Methods, Preparation. the, Polystyrene. of, methacrylate. Poly(methyl, Nanocomposites. Clay, Chem, Mater.200220021438373843

57. T. Wang, S. Shi, F. Yang, L. M. Zhou, S. Kuroda, methacrylate)/ polystyrene. Poly(methyl, latex. composite, with. a. particles, core/shell. novel, J. morphology, Sci. Mater, 2010

58. F. Du, J. E. Fischer, K. I. Winey, Method. Coagulation, Preparing. for, Carbon. Single-Walled, Poly(methyl. Nanotube, Composites. methacrylate, Modulus. Their, Conductivity. Electrical, Stability. Thermal, of. Journal, Science. Polymer, B. Part, Physics. . Polymer, 2003

59. T. Gabor, D. Aranyi, K. Papp, F. H. Karman, E. Kalman, of. Dispersibility, Nanotubes. Carbon, Science. Materials, 1. Forum, 2007

60. A. Usoltseva, V. Kuznetsov, N. Rudina, E. Moroz, M. Haluska, S. Roth, of. Influence, ctivation. catalysts', their. on, activity, in. selectivity, nanotubes. carbon, Phys. synthesis, Stat. Sol. 112007200739203924

61. M. Sikka, N. N. Pellegrini, E. A. Schmitt, K. I. Winey, a. Modifying, Poly. Polystyrene, methacrylate. (methyl, with. Interface, methacrylate. Poly(styrene-co-methyl, Copolymers. Random, . Macromolecules, 1997

62. S. Varennes, H. P. Schreiber, Origins. On, Time. of-Dependence, Contact. in, Measurements. Angle, Journal. The, Adhesion. of, 76 29330 2001,

63. Y. Li, Y. Yang, . F. Yu, L. Dong, Surface, Morphology. Interface, Polystyrene. of, (methyl. Poly, Thin. methacrylate-Film, Blends, Journal. Bilayers, Polymer. of, Part. B. Science, Physics. Polymer, Vol, 2006

64. Y. T. Sung, W. J. Seo, Y. H. Kim, H. S. Lee, W. N. , Kim, H.S. Lee, W.N. Kim Evaluation of interfacial tension for poly(methyl methacrylate) and polystyrene by rheological measurements and interaction parameter of the two polymers, Korea-Australia Rheology Journal 1620042004135140

65. K. Tanaka, A. Takahara, T. Kajiyama, Macromolecules 19962932329

66. C. Ton-That, A. G. Shard, D. O. H. Teare, R. H. Bradley, X. P. S. , A. F. M. surface, of. studies, P. S. P. M. M. A. solvent-cast, Polymer. . blends, 20011121 EOF1129 EOF
67. S. Wu, Polar, Interactions. Nonpolar, Adhesion. J. in, Adhesion, J. Adhesion, 197353955
68. P. K. C. Pillai, P. Khurana, A. Tripathi, studies. Dielectric, poly(methyl. of, /polystyrene. methacrylate, layer. double, Journal. system, Materials. of, Letters. . Science, 1986
69. I. Mazov, V. L. Kuznetsov, I. A. Simonova, A. I. Stadnichenko, A. V. Ishchenko, A. I. Romanenko, E. N. Tkachev, O. B. , Anikeeva "Oxidation behavior of multiwall carbon nanotubes with different diameters and morphologyApplied Surface Science20126272 EOF6280 EOF
70. L. Ci, J. Suhr, V. Pushparaj, X. Zhang, P. M. Ajayan, Carbon. Continuous, Reinforced. Nanotube, Nano. Composites, . Letters, 20082762 EOF2766 EOF
71. A. Moisala, Q. Li, I. A. Kinloch, A. H. Windle, Thermal, conductivity. electrical, single. of, carbon. multi-walled, composites. nanotube-epoxy, Compos, Sci. Technol., 66, 10 (200612851288
72. W. K. Park, J. H. , Kim Effect of Carbon Nanotube Pre-treatment on Dispersion and Electrical Properties of Melt Mixed Multi-Walled Carbon Nanotubes / Poly(methyl methacrylate) Composites, Macromolecular Research, 1332062112005

Chapter 10

CARBON NANOTUBE REINFORCED ALUMINA COMPOSITE MATERIALS

Go Yamamoto[1] and Toshiyuki Hashida[1]
[1]Fracture and Reliability Research Institute (FRRI), Tohoku University, Japan

INTRODUCTION

Novel materials and processing routes provide opportunities for the production of advanced high performance structures for different applications. Ceramic matrix composites are one of these promising materials. Engineering ceramics such as Al_2O_3, Si_3N_4, SiC and ZrO_2 produced by conventional manufacturing technology have high stiffness, excellent thermostability and relatively low density, but extreme brittle nature restricted them from many structural applications (Mukerji, 1993). Considerable attention has been adopted to improve the fracture toughness. An approach has been paid to the development of nanocrystalline ceramics with improved fracture properties. Decreasing the grain size of ceramics to the sub- and nano-meter scale leads to a marked increase in fracture strength (Miyahara et al., 1994). However, fracture toughness of nanocrystalline ceramics generally displays modest improvement or even deterioration (Miyahara et al., 1994; Rice, 1996; Yao et al., 2011). As one possible approach, incorporation of particulates, flakes and short/long fibers into ceramics matrix, as a second phase, to produce tougher ceramic materials is an eminent practice for decades (Evans, 1990). Recently, researchers have focused on the carbon nanomaterials, in particular carbon nanotubes (CNTs), which are nanometer-sized tubes of single- (SWCNTs) or multi- layer graphene (MWCNTs) with outstanding mechanical, chemical and electrical properties (Dai et al., 1996; Ebbesen et al., 1996; Treacy et al., 1996; Huang et al., 2006; Peng et al., 2008), motivating their use in ceramic composite materials as a fibrous reinforcing agent.

It is well recognized that some difficulties appear to be the major cause for the limited improvement in CNT/ceramic composites prepared to date. The first is the inhomogeneous dispersion of CNTs in the ceramic matrix. Pristine CNTs are well known for poor solubilization, which leads to phase segregation in the composite owing to the van der Waals attractive force (Chen et al., 1998). Such clustering produces a negative effect on the physical and mechanical properties of the resultant composites (Yamamoto et al., 2008). The second is the difficulty in controlling connectivity between CNTs and the ceramics matrix, which leads to a limited stress transfer capability from the matrix to the CNTs (Peigney, 2003; Sheldon & Curtin, 2004; Chen et al., 2011). The strengthening and toughening mechanisms of composites by fibers are now well established (Evans, 1990; Hull & Clyne, 1996); central to an understanding is the concept of interaction between the matrix and reinforcing phase during the fracture of the composite. The fracture properties of such composites are dominated by the fiber bridging force resulting from debonding and sliding resistance, which dictates the major contribution to the strength and toughness. Thus, the adequate connectivity with the matrix, and uniform distribution within the matrix are essential structural requirements for the stronger and tougher CNT/ceramic composites. To overcome these obstacles, various efforts, such as surface modification (De Andrade et al., 2008; Yamamoto et al., 2008; Kita et al., 2010; Gonzalez-Julian et al., 2011), heterocoagulation (Fan et al., 2006a, 2006b), extrusion (Peigney et al., 2002), and their combination, have been made to effectively achieve good dispersion of CNTs in ceramic matrix. Until now, however, most results for strengthening and toughening have been disappointing, and only little or no improvement have been reported in CNT/ceramic composite materials, presumably owing to the difficulties in homogeneous dispersion of CNTs in the matrix and in formation of adequate interfacial connectivity between two phases.

This chapter presents that novel processing approach based on the precursor method. The MWCNTs used in this study are modified with an acid treatment. Combined with a mechanical interlock induced by the chemically modified MWCNTs, this approach leads to improved mechanical properties. Mechanical measurements on the composites revealed that only 0.9 vol.% acid-treated MWCNT addition results in 37% and 36% simultaneous increases in bending strength (689.6 ± 29.1 MPa) and fracture toughness (5.90 ± 0.27 MPa $m^{1/2}$), respectively, compared with a MWCNT-free alumina sample prepared under similar processing conditions. Structure-property relationship of present composites will be explained on the basis of the detailed nano/microstructure and fractographic analysis. We also explain why previous reports indicated only modest improvements in the fracture properties of MWCNT based ceramic composites. Here, the failure mechanism of the MWCNTs during crack opening in a MWCNT/alumina composite is investigated through transmission electron microscope (TEM) observations and single nanotube pullout tests. Achieving tougher ceramic composites with MWCNTs is discussed based on these results.

A NOVEL APPROACH FOR PREPARATION OF MWCNT/ ALUMINA COMPOSITES

To disperse the MWCNTs homogeneously in the matrix and improve the connectivity between MWCNTs and matrix, we developed a novel approach with combination of a precursor method for synthesis of an alumina matrix, an acid treatment of MWCNTs and a spark plasma sintering method. The improvement on the bending strength and fracture toughness was confirmed by the fracture tests.

Materials and Specimen Preparation

Starting Materials

The MWCNT material (Nano Carbon Technologies) used in this research was synthesized by a catalytic chemical vapor deposition method followed by high temperature annealing at 2600°C. The purity was claimed to be 99.5% by the producer. Fig. 1 shows scanning electron microscope (SEM, Hitachi S-4300) and TEM (Hitachi HF-2000) images of the pristine MWCNTs. It can be seen fromFig. 1b that the pristine MWCNTs have a highly crystalline multi-walled structure with a narrow central channel. The corresponding geometrical and mechanical properties of the pristine MWCNT are listed in Table 1. The estimated diameter and length of the pristine MWCNTs from SEM and TEM measurements ranged from 33 to 124 nm (average: 70 nm) and 1.1 to 22.5 µm (average: 8.7 µm), respectively. Tensile-loading experiments with individual MWCNTs using a nanomanipulator tool operated inside SEM revealed that the tensile strengths of 10 pristine MWCNTs ranged from ~2 to ~48 GPa (average: 20 GPa) and the Young's modulus ranged from ~50 to ~1360 GPa (average: 790 GPa) (Yamamoto et al., 2010). It seems that the average tensile strength of the pristine MWCNT used in this research were somewhat lower than that of the arc-discharge grown MWCNTs (Yu et al., 2000).

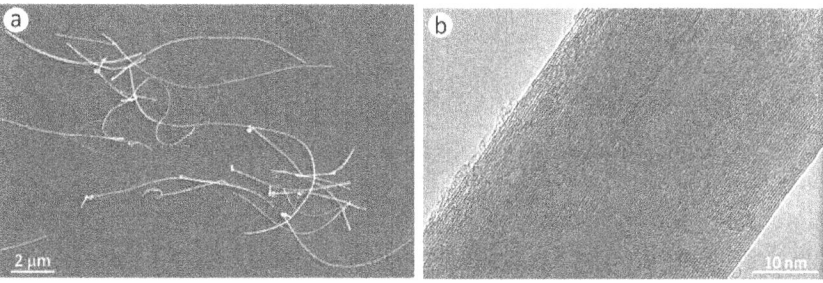

Figure 1: a) SEM and (b) TEM image of pristine MWCNTs used in this research.

Table 1: Measured geometrical and mechanical properties of pristine MWCNTs. Shown are the nanotube inner diameter (ID), outer diameter (OD), length (l), density (ρ), tensile strength (σ_f), Young's modulus (E), moment of inertia of cross sectional area (I) and flexural rigidity (EI), respectively

ID (nm)	OD (nm)	L (μm)	ρ (Mg/m^3)	σ_f (GPa)	E (GPa)	I (nm^4)	EI (N·nm^2)
7 (3~12)	70 (33~124)	8.7 (1.1~22.5)	2.1	20 (2~48)	790 (50~1360)	1.2×1^{0}6	9310×1^0-4

ACID Treatment of MWCNTS

The rationale behind the acid treatment is to introduce nanoscale defects and adsorb negatively charged functional groups at the MWCNT ends and along their lengths. The pristine MWCNTs were refluxed in 3:1 (volume ratio) concentrated H_2SO_4:HNO_3 mixture at a temperature of 70°C for 1 hour, 2 hours and 4 hours, washed thoroughly with distilled water to be acid-free, and then finally dried in an air oven at 60°C. Fig. 2 shows the typical TEM images of a series of the acid-treated MWCNTs and the corresponding distribution of nanodefect depths treated with the various conditions. It is demonstrated that with the acid treatment of the pristine MWCNTs, we have deliberately introduced nanoscale defects on the surface of the MWCNTs. The depth of the nanodefects is on the nanoscale and the average size is in the range of 4.4~7.0 nm for the acid treatment used in this study. We can see that the nanodefects density, i.e., the number of nanodefects per unit of a MWCNT surface area increases with the increasing treatment time. Hereafter, the number of the nanodefects per unit of the MWCNT surface area is referred to as the nanodefect density.

In addition to the nanodefects density, the average size of the nanodefect depths appears to vary with respect to the treatment time. When the treatment time increases from 1 hour to 2 hours, the average size of nanodefect depths increases from 4.4 nm to 6.5 nm. Furthermore, when the treatment time further increases to 4 hours, it increases to 7.0 nm. The aspect ratio (α) of the nanodefects on the MWCNT surface were estimated using the equation $\alpha = L_{width}/L_{depth}$, where L_{width} is the average size of nanodefect widths on the MWCNT surface, and L_{depth} is the average size of nanodefect depths. For the acid treated products with treatment time of 1 hour, 2 hours and 4 hours was 4.4, 4.9 and 3.9, respectively. The experimental results demonstrate that the present method, which uses the acid treatment, may provide an effective route for preparation of the nanodefects on the MWCNT surface, and it may be possible to adjust and control the average size and nanodefect density by varying the treatment time. According to the current TEM observations, peel-off of a few layers in the MWCNT structure was frequently observed for the MWCNT powders acid-treated for 4 hours. Thus, reduction of α may be due to the decrease in the MWCNT diameter by the peel-off of a few layers in the MWCNT structure and imply that the excessive acid treatment of the MWCNT resulted in degradation of the quality and mechanical properties of MWCNTs (Yamamoto et al., 2010). As previously reported (Liu et al., 1998), SWCNTs can be cut into shorter segments by acid treatment of 3:1 (volume ratio) concentrated H_2SO_4:HNO_3 mixture. In this study, however, when the acid treatment times are 1 hour and 2 hours, no such change in the length has been found in the acid-treated MWCNTs. The average lengths of the acid-treated MWCNTs were 8.7 μm and 8.3 μm, respectively. In contrast, average length was decreased slightly with a further increase in the treatment time up to 4 hours, and reached about 7.2 μm.

The zeta potential values of the pristine MWCNTs and the acid-treated MWCNTs at different pH values are shown in Fig. 3a. Here, the changes in zeta potentials were measured in 1.0 mM KCl aqueous solution of varying pH using a zeta potential analyzer (ZEECOM ZC2000, Microtec). The pH value of the aqueous solution was adjusted with HCl and NaOH. Zeta potential values were calculated using the Smoluchowski equation. The isoelectric point (pH_{iep}) for the pristine MWCNTs is located at about 3.0, whereas the acid treatment process makes the surface more negatively charged at tested pH values. The change in the zeta potential may be mainly due to the introduction of more functional groups after the acid treatment (Esumi et al., 1996; Liu et al., 1998). These functional groups make them easily dispersed in polar solvents, such as water and ethanol. Fig. 3b shows a photograph of the pristine MWCNTs and acid-treated MWCNTs suspensions at pH 6, respectively. It is clear that the pristine MWCNTs are not dispersed at pH 6. In contrast, the dispersion of the acid-treated MWCNTs is seen to improve dramatically. Furthermore, it can be expected that the larger electrical repulsive force between the acid-treated MWCNTs will facilitate their dispersion and prevent them from tangling and agglomeration. The zeta potential of the aluminum hydroxide, which is used as the starting material for synthesis of the alumina matrix, exhibited positive values over a wide pH range (pH = 3~9), while that of the acid-treated MWCNTs was negative in this pH range. On these two colloidal suspensions are mixed, particles of the aluminum hydroxide will bind onto the acid-treated MWCNTs because of the strong electrostatic attractive force between them, and this results in a homogeneous MWCNTs and aluminum hydroxide solution.

Preparation of MWCNT/Alumina Composites

A typical synthesis procedure for the composite preparation is as follows. The 50 mg MWCNTs acid-treated for 2 hours or pristine MWCNTs were dispersed in 400 ml ethanol with aid of ultrasonic agitation. 15.2 g aluminum hydroxide (Wako Pure Chemical Industries) was added to this solution and ultrasonically agitated. 73 mg magnesium hydroxide (Wako Pure Chemical Industries) was added to prevent excessive crystal growth. Here, the weight loss of the hydroxides caused by the dehydration process was accounted for in the calculation of the composite composition.

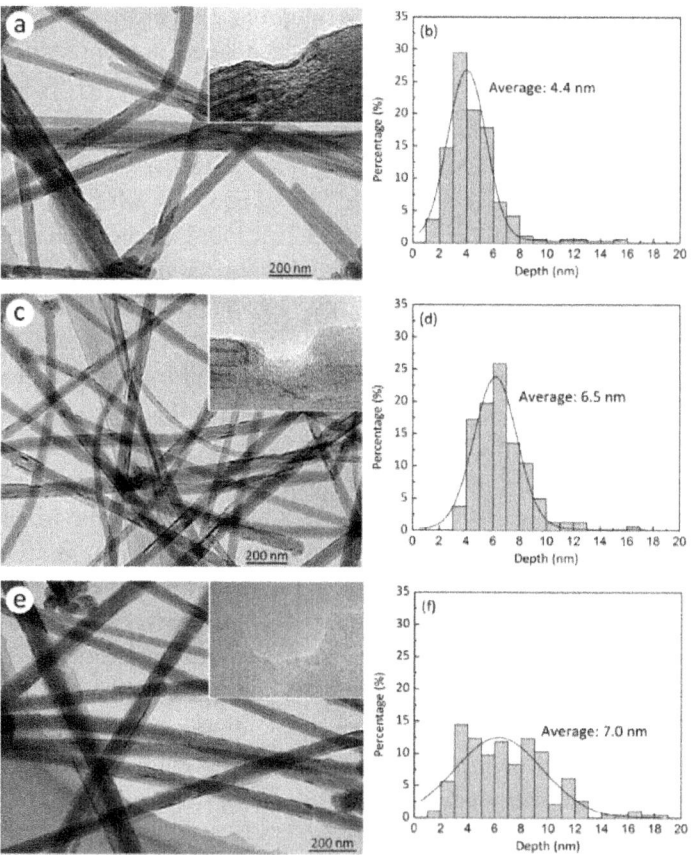

Figure 2: Typical low-magnification TEM images of acid-treated MWCNTs treated with the various conditions of (a) 1 hour, (c) 2 hours and (e) 4 hours. The insets show the high magnification images. (b), (d), (f) Corresponding depth distribution of the nanodefects in the sample (a), (c) and (e), respectively. The solid lines in (b), (d), (f) represent the Gaussian fitting curves. The observation was made for approximately 200 defects.

The weight loss of the aluminum hydroxide and the magnesium hydroxide was 34.7% and 31.9%, respectively. The resultant suspension was filtered and dried in an air oven at 60°C. Finally, the product obtained in the previous step was put into a half-quartz tube and was dehydrated at 600°C for 15 min in argon atmosphere. The composites were prepared by spark plasma sintering (SPS, SPS-1050 Sumitomo Coal Mining) (Omori, 2000) in a graphite die with an inner diameter of 30 mm at a temperature of 1500°C under a pressure of 20 MPa in vacuum for 10 min. For comparison, similar preparation processes were applied while using the pristine MWCNTs as the starting material. Fig. 4 shows X-ray diffraction patterns (M21Mac Science) of the (a) aluminum hydroxide–MWCNT mixture, (b) dehydrated product and (c) sintered body, respectively. It is difficult to distinguish the MWCNT peaks from all XRD patterns, probably due to the small quantity of MWCNTs. The diffraction peaks corresponding to the aluminum hydroxide and the intermediate were observed in the aluminum hydroxide–MWCNT mixture. However, the diffraction peaks corresponding to the aluminum hydroxide and the intermediate disappeared completely in the sintered body, suggesting the phase transformation of the aluminum hydroxide to form α-alumina via an amorphous phase (b). These results clearly indicate that alumina was successfully synthesized by SPS at 1500°C under 20 MPa in vacuum.

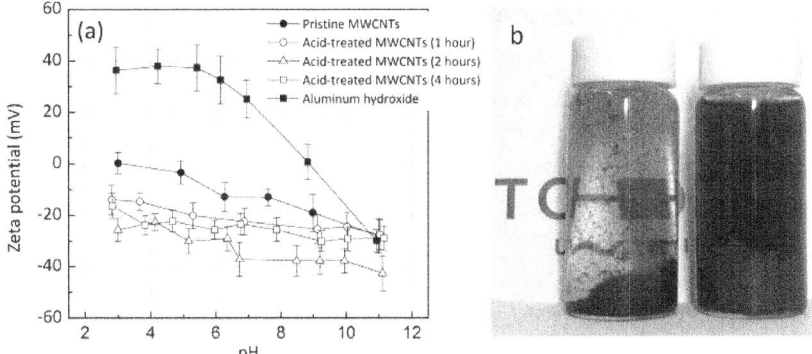

Figure 3: a) Zeta potential values of MWCNTs and acid-treated MWCNTs at different pH. (b) Undistributed one-day old aqueous suspensions of (left) pristine MWCNTs and (right) acid-treated MWCNTs.

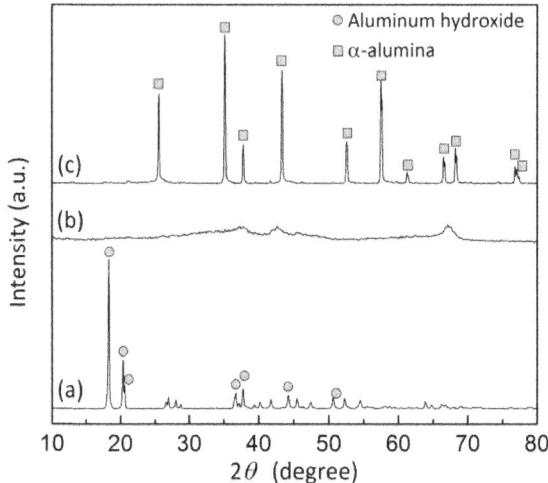

Figure 4: XRD patterns of the (a) aluminum hydroxide–MWCNT mixture, (b) dehydrated product and (c) sintered body, respectively.

Micro- And Nanostructures of MWCNT/Alumina Composites

We now discuss the micro- and nanostructures of the acid-treated MWCNT/alumina composites using SEM and TEM analysis. An interesting geometric structure was observed between the individual MWCNT and the alumina matrix, as shown in Fig. 5. It is revealed that a nanodefect on the acid-treated MWCNT is filled up with alumina crystal, which may be intruding into the nanodefect during grain growth. This nanostructure is novel in that its structure resembles a nanoscale anchor with an alumina crystal spiking the surface of the MWCNT.

From the SEM observations on the fracture surface, the following features can be noted. First, numerous individual MWCNTs protrude from the fracture surface, and the pullout of the MWCNTs can be clearly observed (Fig. 6a), which had not been obtained until now for conventional CNT/ceramic composites. Most of MWCNTs are located in the intergranular phase and their lengths are in the range 0~10 μm. The alumina grains have sizes in the micron range, around 1.5 μm (The grain size of the composite was obtained using SEM images, and the observation was made for 224 grains.). No clear difference in the grain size is observed between the acid-treated MWCNT/alumina composites and the pristine MWCNT/alumina composites, even though the incorporation of MWCNTs seems to suppress the grain growth of the alumina. Second, in the case of the smaller amount of the acid-treated MWCNTs, no severe phase segregation was observed, whereas the composites made with the pristine MWCNTs revealed an inhomogeneous structure even for MWCNT addition as low as 0.9 vol.%. In addition to the above features, some MWCNTs on the fracture surface showed a "clean break" near the crack plane, and that the diameter of MWCNT drastically slenderized toward their tip, as illustrated in Figs. 6 b and 6c, respectively. As SEM cannot clearly resolve the thickness of a single MWCNT, TEM was used to determine if the fracture phenomenon of MWCNTs was indeed occurring during crack opening.

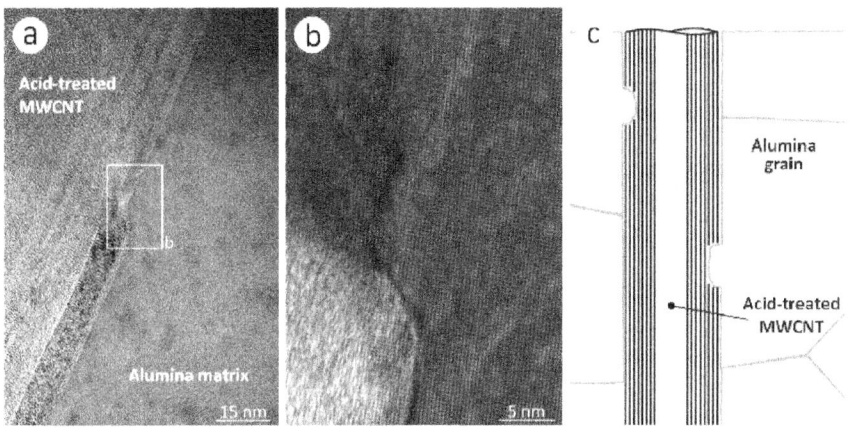

Figure 5: MWCNT morphology in the composites. (a) It is demonstrated that a nano-defect on the acid-treated MWCNT is filled up with alumina crystal. (b) Enlarged TEM image, taken from the square area. (c) Schematic description of MWCNT morphology in the composites.

TEM observations on the fracture surface demonstrated that a diameter change in the MWCNT structure was evidently observed for a certain percentage of the MWCNTs (Fig. 7a). At least, 25% MWCNT appear to have an apparent diameter change (The observation was made for 281 MWCNTs.). As shown in Fig. 7b, the high magnification TEM image clearly showed a change in diameter, and this morphology is quite similar to a "sword-in-sheath"-type failure (Yu et al., 2000; Peng et al., 2008; Yamamoto et al., 2010). Key features are illustrated in enlarged TEM image, taken from the square area in Fig. 7b. The inset showed that outer-walls having approximately 10 shells were observed to break up at the location where the MWCNT undergo failure, and that the edges of the broken outer shells were observed to be perpendicular to the cylinder axis. Since no apparent variation in the diameter of the MWCNTs has been observed along the axis in the as-received MWCNTs, these results imply that some MWCNTs underwent failure in the sword-in-sheath manner prior to pullout from the matrix. Note that MWCNT failure was also observed in fracture surfaces of alumina composites made with arc-discharge-grown and chemical vapor deposition-grown MWCNTs prepared under the same processing conditions (Yamamoto et al., 2011).

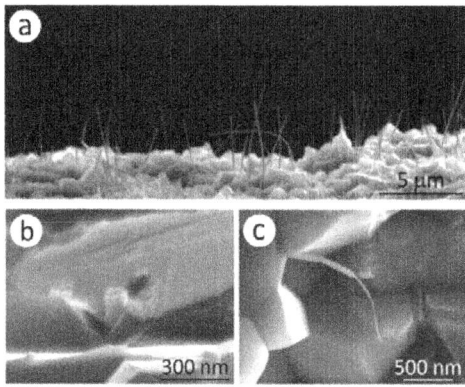

Figure 6: Fracture surface of acid-treated MWCNT/alumina composites. (a) Numerous individual MWCNTs protrude from the fracture surface. (b,c) Some MWCNTs have broken in the multi-wall failure.

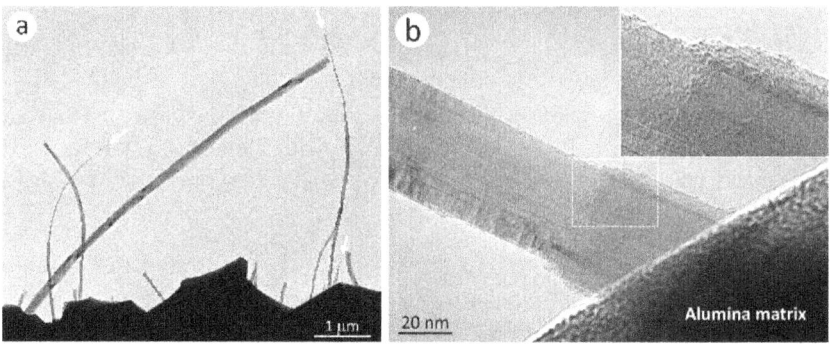

Figure 7: TEM images of the fracture surface of the composite acquired (a) low and (b) high magnification images.

Physical and Mechanical Properties of MWCNT/Alumina Composites

The bending strength of the composites was measured by the three-point bending method under ambient conditions, in which the size of the test specimens was 2.0 mm (width) × 3.0 mm (thickness) × 24.0 mm (length). The span length and crosshead speed for the strength tests were 20.0 mm and 0.83 μm/s, respectively. The fracture toughness was measured by the single-edge notched beam (SENB) method (Japanese Industrial Standards, 1995) under ambient conditions, in which the size of test specimens was 2.0 mm (width) × 3.0 mm (thickness) × 15.0 mm (length). A notch with depth and width of 0.3 mm and 0.1 mm was cut in the center part of the test specimens. A span length of 12.0 mm and crosshead speed of 0.83 μm/s were applied for the toughness test. The bending strength (σ_b) and fracture toughness (K_{Ic}) are given by the following equations:

$$\sigma_b = 3P_b L / 2bh^2 \tag{1}$$

$$K_{Ic} = \left(3P_b L / 2bh^2\right) \cdot a^{1/2} Y \tag{2}$$

where P_b is the maximum load, L is the span length, b is the specimen width, h is the specimen thickness, a is the notch depth and Y is the dimensional factor. All surfaces of the specimens were finely ground on a diamond wheel, and the edges were chamfered. The indentation tests were done on a hardness tester (AVK-A, Akashi) with a diamond Vickers indenter under ambient conditions. The 0.9 vol.% acid-treated MWCNT/alumina composite with surface roughness of 0.1 μm (Ra) was indented using a Vickers diamond pyramid with a load of 98.1 N (P) applied on the surface for 15 s. The diagonal (d) and the radial crack length (C) were measured by the SEM. The hardness (Hv) and indentation toughness values (K_{Ic}) were calculated by the following equations:

$$Hv = 0.1891 P / d^2 \tag{3}$$

$$K_{Ic} = 0.016 \left(E / H_v\right)^{1/2} \left(P / C^{3/2}\right) \tag{4}$$

where E is the Young's modulus of the composite (E = 362.8 GPa) measured by a pulse-echo method.

It was found that surface modification of the MWCNTs is effective in improvement of bending strength and fracture toughness of the MWCNT/alumina composites. Figs. 8 a and 8bshow the dependence of the bending strength and the fracture toughness on MWCNT content in the composites. There are few papers which report significant improvement in the mechanical properties such as toughness (Zhan et al., 2003), and the improvement by MWCNT addition has been limited so far in previous studies (Ma et al., 1998; Sun et al., 2002; Wang et al., 2004; Sun et al., 2005; Cho et al., 2009). In our composites, however, the bending strength and the fracture toughness simultaneously increased with the addition of a small amount of the acid-treated MWCNTs. The bending strength and the fracture toughness of the 0.9 vol.% acid-treated MWCNT/alumina composite reached 689.6 ± 29.1 MPa and 5.90 ± 0.27 MPa m$^{1/2}$, respectively. At the same time, the bending strength and the fracture toughness of the acid-treated MWCNT/

alumina composites were always higher than those of the pristine MWCNT/alumina composites with identical MWCNT content, indicating enhanced stress transfer capability from the alumina to the acid-treated MWCNTs. The Vickers indentation toughness calculated by using the Eq. (4) was 6.64 MPa m$^{1/2}$, which is a slightly larger value than that measured by using SENB method (5.90 MPa m$^{1/2}$). These observations revealed that the high structural homogeneity and enhanced frictional resistance of the structural components led to a simultaneous increase in the strength and the toughness of the acid-treated MWCNT/alumina composites. In contrast, for the larger amount of the MWCNTs, the degradation of mechanical properties of both the composites may be primarily attributed to the severe phase segregation. Because a bundle of segregated CNTs has poor load-carrying ability, the effect of this kind of CNT aggregate in the matrix may be similar to that of pores (Yamamoto et al., 2008a, 2008b).

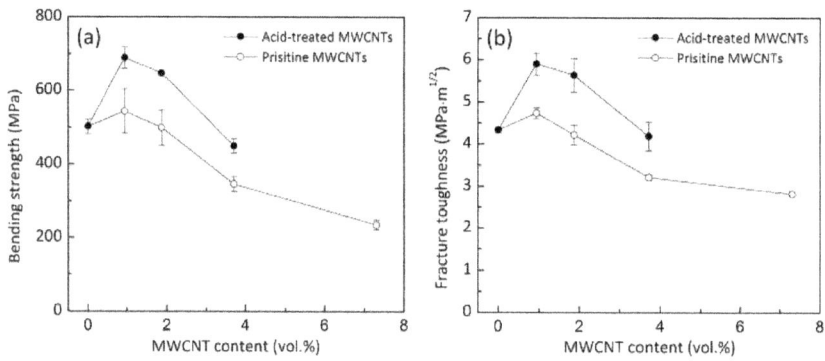

Figure 8: a) Bending strength and (b) fracture toughness as a function of MWCNT content.

EVALUATION OF CRACK BRIDGING CHARACTERISTICS

Ceramic-CNT interfacial behavior is another key factor in controlling the mechanical and physical properties of fiber reinforced composite materials (Evans, 1990; Hull & Clyne, 1996; Chen et al., 2011). In general, strong interfacial connectivity facilitates effective load transfer effect, but it prevents CNT pull-out toughening from occurring. Weak interfacial connectivity favors CNTs pull-out but fails to toughen the ceramic matrix. Thus, a balance must be maintained between CNT pull-out and toughening mechanics. It is well recognized that improved toughness of fiber-reinforced ceramic composites is obtained under moderate fiber-ceramic interfacial connectivity. In this regard, suitable (neither too strong nor too weak) ceramic-CNT interfacial connectivity is needed to ensure effective load transfer, and to enhance the toughness and strength of ceramic-CNT composites. Here, the failure mechanism of the MWCNTs during crack opening in a MWCNT/alumina composite is investigated through TEM observations and single nanotube pullout tests. Achieving tougher ceramic composites with MWCNTs is discussed based on these results.

Pullout Experiment Sample Preparation

The MWCNT failure during crack opening motivated our research of the crack bridging characteristics through the single nanotube pullout tests. The single nanotube pullout experiments were carried out using an in-situ SEM (Quanta 600 FEG; FEI) method with a nanomanipulator system (Yu et al., 2000;Yamamoto et al., 2010). An atomic force microscope (AFM) cantilever (PPP-ZEILR, nominal force constant 1.6 N/m; NANOSENSORS) was mounted at the end of a piezoelectric bender (ceramic plate bender CMBP01; Noliac) on an X–Y linear motion stage, and the composite with fracture surface (that was coated with platinum) was mounted on an opposing Z linear motion stage. The piezoelectric bender was used to measure the resonant frequency of each cantilever in vacuum. A single MWCNT on the fracture surface was clamped onto a cantilever tip by local electron-beam-induced deposition (EBID) of a carbonaceous material (Ding et al., 2005). As a precursor source for the EBID, we used n-docosane ($C_{22}H_{46}$, Alfa Aesar), which was dissolved in toluene to make a 3 mass% solution. A small amount of the solution was dropped on a cut-in-half copper TEM grid. After the solution evaporated, the TEM grid with paraffin source was mounted on the AFM chip, as shown in Fig. 9. The deposition rate of the EBID depends on several factors (Ding et al., 2005). Thus, the amount of the paraffin source, deposition time, and distance between the paraffin source and the cantilever tip were experimentally-optimized. The cantilevers serve as force-sensing elements and the spring constants of each were calculated in-situ prior to the pullout test using the resonance method (Sader et al., 1999). In brief, for the case of a rectangular cantilever, the force constant (k) is given by following equation,

$$k = M_e \rho_c bhL\omega_{vac}^2 \tag{5}$$

where ω_{vac} is the fundamental radial resonant frequency of the cantilever in vacuum, h, b, and L are the thickness, width, and length of the cantilever, respectively, ρ_c is the density of the cantilever (= 2.33 Mg/m³), and Me is the normalized effective mass which takes the value Me = 0.2427 for L/b > 5 (Sader et al., 1995). We measured ω_{vac}, h, b and L of each cantilever in the SEM and used the measured, not the nominal provided, values to calculate k.

The h, b and L are determined by counting the number of pixels in the acquired SEM images. The applied force is calculated from the angle of deflection at the cantilever tip in the acquired SEM images (Ding et al., 2006). The deflection (δ) and angle of deflection (θ) at the cantilever tip are given by

$$\delta = PL^3 / 3EI \tag{6}$$

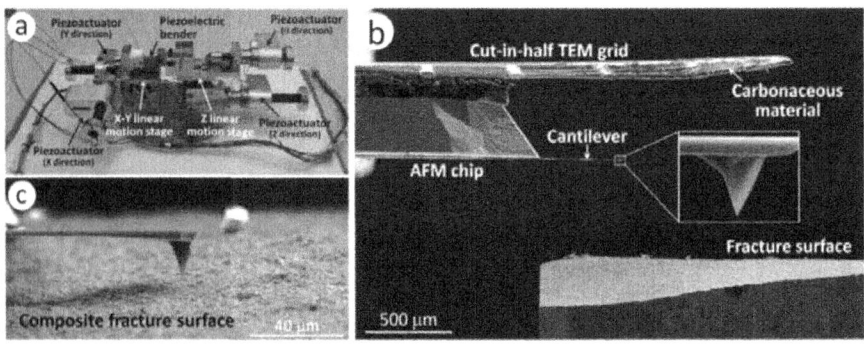

Figure 9: SEM image showing the experimental setup for pullout experiments.

$$\theta = PL^2 / 2EI \tag{7}$$

where P is the load applied at the cantilever tip, L is the cantilever length, E is the elastic modulus andI is the moment of inertia of the cantilever (Ding et al., 2006). Thus, the deflection at the cantilever tip can be represented by the angle of deflection with the following relationship (Ding et al., 2006):

$$\delta = 2\theta L / 3 \tag{8}$$

A crosshead speed – i.e., movement rate of the cantilever – of about 100 nm/s was applied for the pullout tests.

We fractured a composite specimen by conducting the fracture tests, which caused single MWCNT to project from the crack plane, as exemplified in Fig. 6a. This allows single MWCNT "pickup" with cantilever tip for subsequent tensile loading using the nanomanipulator. As mentioned above, however, the MWCNTs crossing the crack planes were strained during crack opening and possibly underwent failure, as shown in Figs. 6b, 6c and 7. Therefore, by observing the fracture surface on the composites, MWCNTs with no apparent damages were selected for the pullout tests. The physical and mechanical properties, and electrical conductivity of the composite used for the pullout testes are shown in Table 2.

Table 2: The properties of the composite with 0.9 vol.% pristine MWCNTs. The Young's modulus and Poisson's ratio were measured by the ultrasonic pulse echo method

Relativedensity (%)	Grainsize (μm)	Bendingstrength (MPa)	Fracturetoughness (MPa·m$^{1/2}$)	Hardness (GPa)	Young'smodulus (GPa)	Poisson'sratio
98.9	1.43±0.31	543.8±60.9	4.74±0.12	17.0 ± 0.4	358.0	0.20

Nanotube Fracture during the Failure of MWCNT/Alumina Composites

Results obtained from the pullout experiments revealed that strong load transfer was demonstrated, and no pullout behavior was observed for all 15 MWCNTs tested in this present research. Eight of these MWCNTs fractured at the composite surface and the remaining 7 MWCNTs underwent failure in the region between the fixed point on the cantilever and the crack plane, as illustrated in Fig. 10.

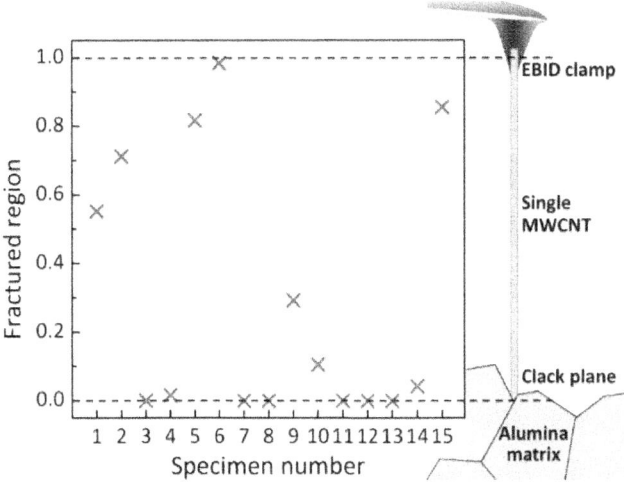

Figure 10: Fracture location of single MWCNTs under pullout loading. Of the 15 MWCNTs tested here, 8 MWCNTs fractured on the composite surface (sample numbers: 3, 4, 7, 8, 11–14) and remaining 7 MWCNTs fractured in the middle (sample numbers: 1, 2, 5, 6, 9, 10, and 15).

Two series of SEM and TEM images for each of two individual MWCNTs, captured before and after their breaking, are shown in Figs. 11 and 12. In the first series (Fig. 11; sample number 14), a MWCNT projecting 5.72 ± 0.01 μm from the fracture surface (Fig. 11a) was "welded" to a cantilever tip by local EBID, and then loaded in increments until failure. The resulting fragment attached on the cantilever tip was at least 10.9 μm long (Fig. 11b), whereas the other fragment remained lodged in a grain boundary of the alumina matrix (Fig. 11c), suggesting that MWCNT underwent failure in a sword-in-sheath manner. TEM images show a change in diameter at the location where the MWCNT underwent failure, and that the inner core protruding from the outer shells has a multi-walled closed-end structure, as shown in Figs. 11d and 11e, respectively. Given that uniformity of the interwall spacing of 0.34-nm-thick cylinder structure, approximately 11 shells underwent failure. There results strongly suggest that the MWCNTs broke in the outer shells and the inner core was then completely pulled away, leaving the companion fragment of the outer shells in the matrix. The sword-in-sheath failure did not always occur. Instead a few MWCNT failed leaving either a very short sword-in-sheath failure or a clean break. As for one example (Fig. 12; sample number 10), a MWCNT projecting 5.34 ± 0.01 μm from the crack plane (Fig. 12a) underwent failure on the composite fracture surface. The resulting fragment attached on the cantilever tip was at least 5.7 μm long (Fig. 12b), and no fragment was observed at the original position on the crack plane, suggesting that in this case the MWCNT failed by breaking inside the matrix, and did not pull out. Fig. 12c shows the TEM image of the tip of the same MWCNT which underwent very short sword-in-sheath failure or clean break during crack opening.

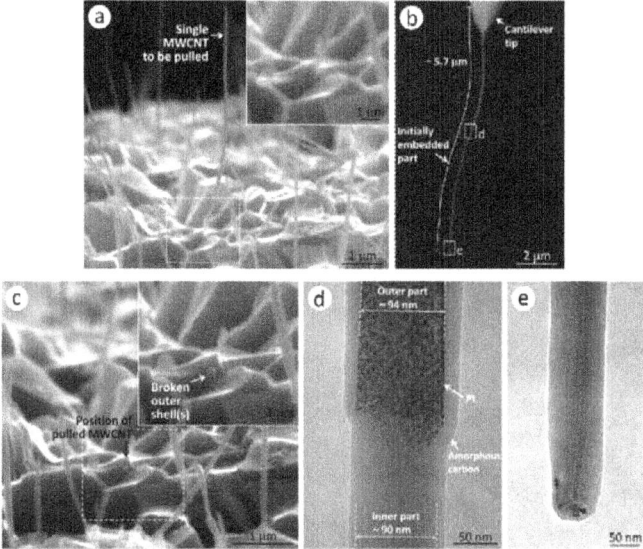

Figure 11: SEM images show (a) a free-standing MWCNT having a 5.72 ± 0.01 μm-long on the fracture surface of the composite. (b) After breaking, one fragment of the

same MWCNT attached on the cantilever tip had a length ~10.9 μm. (c) The other fragment remained in the matrix. (d,e) TEM images show a change in diameter at the location where the MWCNT underwent multi-wall failure, and that it clearly has a multi-walled closed-end structure.

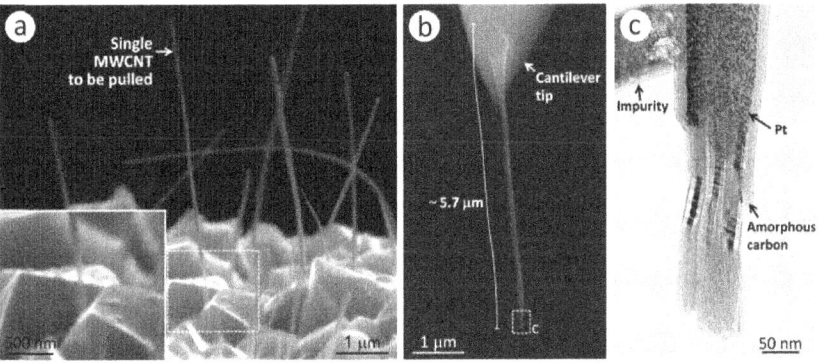

Figure 12: In the second series, (a) a tensile-loaded MWCNT with a length of 4.46 ± 0.01 μm fractured on the crack plane. (b) The resulting fragment on the cantilever tip had a length ~5.7 μm. (c) TEM image shows the MWCNT which underwent the very short sword-in-sheath failure or clean break.

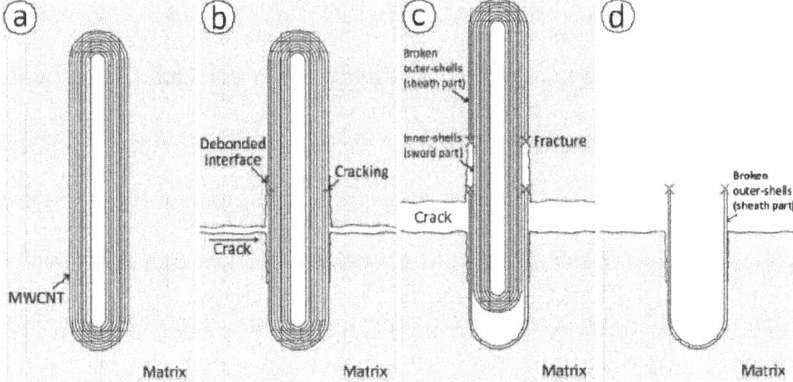

Figure 13: Schematic description of possible fracture mechanisms of the MWCNT (sample number 14). (a) Initial state of a MWCNT. (b) Tensile stresses lead to matrix crack and partial debonding formation. (c,d) As displacement increases, the MWCNTs, rather than pulling out from the alumina matrix, undergo failure in the outer shells and the inner core is pulled away, leaving the fragment of the outer shells in the matrix.

Next, we schematically describe possible processes and mechanics, explaining the MWCNT failure during crack opening (Fig. 13). As for one example, considering the sample number 14 (Fig. 11), the initial state of the

MWCNT in an ideal case is a completely impregnated and isolated embedded in the matrix (Fig. 13a). Tensile stresses parallel to the axis of MWCNT length lead to matrix crack formation. Subsequently, interfacial debonding between two phases may occur (Fig. 13b), perhaps over a limited distance (but this is unlikely to make a major contribution to the fracture energy.). Since there is variability in the MWCNT strength in the debonded region on either side of the crack plane, and it is possible for the MWCNT to break at a certain position, when the stress in the MWCNT reaches a critical value. As displacement increases, the MWCNTs, rather than pulling out from the alumina matrix, undergo failure in the outer shells and the inner core is pulled away, leaving the fragments of the outer shells in the matrix (Figs. 13c and 13d).

CONCLUSION

Creating tough, fracture-resistant ceramics has been a central focus of MWCNT/ceramic composites research. In this research, the MWCNT/alumina composite with enhanced mechanical properties of 689.6 ± 29.1 MPa for bending strength and 5.90 ± 0.27 MPa m$^{1/2}$ for fracture toughness have been successfully prepared by a novel processing method. A combination of the precursor method for synthesis of the alumina matrix, the acid treatment of the pristine MWCNTs and the spark plasma sintering method can diminish the phase segregation of MWCNTs, and render MWCNT/alumina composites highly homogeneous. The universality of the method developed here will be applicable to a wide range of functional materials such as tribomaterials, electromagnetic wave absorption materials, electrostrictive materials, and so on. Our present work may give a promising future for the application of MWCNTs in reinforcing structural ceramic components and other materials systems such as polymer- and metal-based composites.

We have also shown from TEM observations and single nanotube pullout experiments on the MWCNT/alumina composites that strong load transfer was revealed, and no MWCNT pullout behavior was observed. It is well recognized the fracture properties of fiber-reinforced composites are dominated by the fiber bridging force resulting from debonding and sliding resistance, which dictates the major contribution to the strength and toughness (Evans, 1990; Hull & Clyne, 1996). The results reported here suggest that modest improvements in toughness reported previously may be due to the way MWCNT's fail during crack opening in the MWCNT/ceramic composites. Our finding suggests important implications for the design of tougher ceramic composites with MWCNTs. The important factor for such tougher ceramic composites will thus be the use of MWCNT having a much higher load carrying capacity (as well as a good dispersion in the matrix).

ACKNOWLEDGEMENT

The authors thank our colleague, Dr. M. Omori, Mr. K. Shirasu, Mr. Y. Nozaka, Mr. Y. Aizawa and Ms. N. Suzuki of Fracture and Reliability Research Institute (FRRI), Tohoku University, for their helpful discussions, and Mr. T. Miyazaki of Technical Division, School of Engineering, Tohoku University, for technical assistance in the TEM analysis. The authors acknowledge Prof. R.S. Ruoff of The University of Texas at Austin for his useful guidance. This work is partially supported by Grand-in-Aids for Scientific Research (Nos. 23860004 and 21226004) from the Japanese Ministry of Education, Culture, Sports, Science and Technology. This work is performed under the inter-university cooperative research program of the Advanced Research Center of Metallic Glasses, Institute for materials Research, Tohoku University.

REFERENCES

1. J. Chen, M. A. Hamon, H. Hu, Y. Chen, A. M. Rao, P. C. Eklund, R. C. Haddon, 1998Solution properties of single-walled carbon nanotubes. Science, 2829598
2. Y. L. Chen, B. Liu, Y. Huang, K. C. Hwang, 2011Fracture toughness of carbon nanotube-reinforced metal- and ceramic-matrix composites. Journal of Nanomaterials, 2011Article ID 746029.
3. J. Cho, A. R. Boccaccini, M. S. P. Shaffer, 2009Ceramic matrix composites containing carbon nanotubes. Journal of Materials Science, 4419341951
4. H. J. Dai, E. W. Wong, C. M. Lieber, 1996Probing electrical transport in nanomaterials: conductivity of individual carbon nanotubes. Science, 272523526
5. M. J. De Andrade, M. D. Lima, C. P. Bergmann, G. D. O. Ramminger, N. M. Balzaretti, T. M. H. Costa, M. R. Gallas, 2008Carbon nanotube/silica composites obtained by sol-gel and high-pressure techniques. Nanotechnology, 19article 265607
6. W. Ding, D. A. Dikin, X. Chen, R. D. Piner, R. S. Ruoff, E. Zussman, X. Wang, X. Li, 2005Mechanics of hydrogenated amorphous carbon deposits from electron-beam-induced deposition of a paraffin precursor. Journal of Applied Physics, 98article 014905
7. W. Q. Ding, L. Calabri, X. Q. Chen, K. M. Kohhaas, R. S. Ruoff, 2006Mechanics of crystalline boron nanowires. Composites Science and Technology, 6611121124

8. T. W. Ebbesen, H. J. Lezec, H. Hiura, J. W. Bennett, H. F. Ghaemi, T. Thio, 1996Electrical conductivity of individual carbon nanotubes. Nature, 3825456
9. K. Esumi, M. Ishigami, A. Nakajima, K. Sawada, H. Honda, 1996Chemical treatment of carbon nanotubes. Carbon, 34279281
10. A. G. Evans, 1990Perspective on the development of high-toughness ceramics. Journal of the American Ceramic Society, 73187206
11. J. P. Fan, D. Q. Zhao, M. S. Wu, Z. Xu, J. Song, 2006Preparation and microstructure of multiwalled carbon nanotubes-toughened composite. Journal of the American Ceramic Society, 89750753
12. J. P. Fan, D. M. Zhuang, D. Q. Zhao, G. Zhang, M. S. Wu, F. Wei, Z. J. Fan, 2006Toughening and reinforcing alumina matrix composite with single-wall carbon nanotubes. Applied Physics Letters, 8912191012191 03
13. J. Gonzalez-Julian, P. Miranzo, M. I. Osendi, M. Belmonte, 2011Carbon nanotubes functionalization process for developing ceramic matrix nanocomposites. Journal of Materials Chemistry, 2160636071
14. J. Y. Huang, S. Chen, Z. Q. Wang, K. Kempa, Y. M. Wang, S. H. Jo, G. Chen, M. S. Dresselhaus, Z. F. Ren, 2006Superplastic carbon nanotubes-conditions have been discovered that allow extensive deformation of rigid singlewalled nanotubes. Nature, 439281
15. D. Hull, T. W. Clyne, 1996An Introduction to Composite Materials (Second edition). Cambridge University Press, 0521388554, The Edinburgh Building, Cambridge CB2 2RU, UK.
16. Japanese Industrial Standards (JIS).1995R 1607.
17. J. Kita, H. Suemasu, I. J. Davies, S. Koda, K. Itatani, 2010Fabrication of silicon carbide composites with carbon nanofiber addition and their fracture toughness. Journal of Materials Science, 4560526058
18. J. Liu, A. G. Rinzler, H. Dai, J. H. Hafner, R. K. Bradley, P. J. Boul, A. Lu, T. Iverson, K. Shelimov, C. B. Huffman, F. Rodriguez-Macias, Y. S. Shon, T. R. Lee, D. T. Colbert, R. E. Smalley, 1998Fullerene pipes. Science, 28012531256
19. R. Z. , J. Wu, B. Q. Wei, J. Liang, D. H. Wu, 1998Processing and properties of carbon nanotubes-nano-SiC ceramic. Journal of Materials Science, 3352435246
20. N. Miyahara, K. Yamaishi, Y. Mutoh, K. Uematsu, M. Inoue, 1994Effects of grain size on strength and fracture toughness in alumina, JSME International journal 37231237

21. J. Mukerji, 1993Ceramic matrix composites. Defence Science Journal, 43385395
22. M. Omori, 2000Sintering, consolidation, reaction and crystal growth by the spark plasma system (SPS). Materials Science and Engineering A, 287183188
23. A. Peigney, 2003Composite materials: tougher ceramics with nanotubes. Nature Materials, 21516
24. A. Peigney, E. Flahaut, Ch. Laurent, F. Chastel, A. Rousset, 2002Aligned carbon nanotubes in ceramics-matrix nanocomposites prepared by high-temperature extrusion. Chemical Physics Letters, 3522025
25. B. Peng, M. Locascio, P. Zapol, S. Y. Li, S. L. Mielke, G. C. Schatz, H. D. Espinosa, 2008Measurements of near-ultimate strength for multiwalled carbon nanotubes and irradiation-induced crosslinking improvements. Nature Nanotechnology, 3626631
26. R. W. Rice, 1996Grain size and porosity dependence of ceramic fracture energy and toughness at 22°C, Journal of Materials Science, 3119691983
27. J. E. Sader, J. W. M. Chon, P. Mulvaney, 1999Calibration of rectangular atomic force microscope cantilevers. Review of Scientific Instruments, 7039673969
28. J. E. Sader, I. Larson, P. Mulvaney, L. R. White, 1995Method for the calibration of atomic force microscope cantilevers. Review of Scientific Instruments, 6637893798
29. B. W. Sheldon, W. A. Curtin, 2004Nanoceramic composites: tough to test. Nature Materials, 3505506
30. L. Sun, L. Gao, X. Li, 2002Colloidal processing of carbon nanotube/alumina composites. Chemistry of Materials, 1451695172
31. J. Sun, L. Gao, Jin. X. Xihai, (2005, 2005Reinforcement of alumina matrix with multi-walled carbon nanotubes. Ceramics International, 31893896
32. M. M. J. Treacy, T. W. Ebbesen, J. M. Gibson, 1996Exceptionally high Young's modulus observed for individual carbon nanotubes. Nature, 381678680
33. X. Wang, N. P. Padture, H. Tanaka, 2004Contact-damage-resistant ceramic/single-wall carbon nanotubes and ceramic/graphite composites. Nature Materials, 3539544
34. G. Yamamoto, M. Omori, K. Yokomizo, T. Hashida, K. Adachi, 2008Structural characterization and frictional properties of carbon

nanotube/alumina composites prepared by precursor method. Materials Science and Engineering B, 148265269

35. G. Yamamoto, M. Omori, T. Hashida, H. Kimura, 2008A novel structure for carbon nanotube reinforced alumina composites with improved mechanical properties. Nanotechnology, 19article 315708

36. G. Yamamoto, M. Omori, K. Yokomizo, T. Hashida, 2008Mechanical properties and structural characterization of carbon nanotube/alumina composites prepared by precursor method. Diamond and Related Materials, 1715541557

37. G. Yamamoto, J. W. Suk, J. An, R. D. Piner, T. Hashida, T. Takagi, R. S. Ruoff, 2010The influence of nanoscale defects on the fracture of multi-walled carbon nanotubes under tensile loading. Diamond and Related Materials, 19748751

38. G. Yamamoto, K. Shirasu, T. Hashida, T. Takagi, J. W. Suk, J. An, R. D. Piner, R. S. Ruoff, 2011Nanotube fracture during the failure of carbon nanotube/alumina composites. Carbon, 4937093716

39. W. Yao, J. Liu, T. B. Holland, L. Huang, Y. Xiong, J. M. Schoenung, A. K. Mukherjee, 2011Grain size dependence of fracture toughness for fine grained alumina, Scripta Materialia, 65143146

40. M. F. Yu, O. Lourie, M. J. Dyer, K. Moloni, T. F. Kelly, R. S. Ruoff, 2000Strength and breaking mechanism of multiwalled carbon nanotubes under tensile load. Science, 287637640

41. G. D. Zhan, J. D. Kuntz, J. Wan, A. K. Mukherjee, 2003Single-wall carbon nanotubes as attractive toughening agents in alumina-based nanocomposites. Nature Materials, 23842

Chapter 11

MANUFACTURING AND PROPERTIES OF QUARTZ (SIO2) PARTICULATE REINFORCED AL-11.8%SI MATRIX COMPOSITES

M. Sayuti[1,2], S. Sulaiman[1], B.T.H.T. Baharudin[1], M.K.A Arifin[1] and T.R. Vijayaram[3]

[1]Department of Mechanical and Manufacturing Engineering, Faculty of Engineering, Universiti Putra Malaysia, Serdang, Selango, Malaysia

[2]Department of Industrial Engineering, Faculty of Engineering, Malikussaleh University, Lhokseumawe, Aceh, Indonesia

[3]Faculty of Engineering and Technology (FET) Multimedia University, Jalan Ayer Keroh Lama, Bukit Beruang, Melaka, Malaysia

INTRODUCTION

Metal matrix composites (MMC) are a class of composites that contains an element or alloy matrix in which a second phase is fixed firmly deeply and distributed evenly to achieve the required property improvement. The property of the composite varies based on the size, shape and amount of the second phase (Sayuti et al., 2010; Sulaiman et al., 2008). Discontinuously reinforced metal matrix composites, the other name for particulate reinforced composites, constitute 5 – 20 % of the new advanced materials (Gay et al., 2003). The mechanical properties of the processed composites are greatly influenced by their microstructure. An increased stiffness, yield strength and ultimate tensile strength are generally achieved by increasing the weight fraction of the reinforcement phase in the matrix. Inspite of these advantages, the usage of particulate reinforced MMCs as structural components in some applications is limited due to low ductility (Rizkalla and Abdulwahed, 1996). Owing to this and to overcome the draw-backs, a detailed investigation on the strengthening mechanism of composites has been carried out by composite experts (Humphreys, 1987). They have found that the particle size and its weight fraction in metal matrix composites influences the generation of dislocations due to thermal mismatch. The effect is also influenced by the developed residual and internal stresses too. The researchers have predicted that the dislocation density is directly proportional to the weight fraction and due to the amount of thermal mismatch. As a result, the strengthening effect is proportional to the square root of the dislocation density. This effect would be significant for fine particles and for higher weight fractions. The MMCs yield improved physical and mechanical properties and these outstanding benefits are due to the combined metallic and ceramic properties (Hashim et al., 2002). Though there are various types of MMCs, particulate-reinforced composites are the most versatile and economical (Sayuti, Sulaiman, Vijayaram, et al., 2011; Sayuti, Suraya, et al., 2011).

In the past 40 years, the researchers and design experts have perceived their research to emphasis on finding lightweight, environmental friendly, low-cost, high quality, and good performance materials (Feest, 1986). In accordance with this trend, MMCs have been attracting growing interest among researchers and industrialists. The attributes of MMCs include alterations in mechanical behavior (e.g., tensile and compressive properties, creep, notch resistance, and tribology) and physical properties (e.g., intermediate density, thermal expansion, and thermal diffusivity) a change, primarily induced by the reinforced filler phase (Sayuti, et al., 2011). Even though MMCs posses various advantages, they still have limitations of thermal fatigue, thermo-chemical compatibility, and posses lower transverse creep resistance. In order to overcome these limitations, fabrication of discontinuously reinforced Al-based MMCs was carried out by standard metallurgical processing methods such as powder metallurgy, direct casting, rolling, forging and extrusion. Subsequently, the products were shaped, machined and drilled by using conventional machining processes. Consequently, the MMCs would be available in suitable quantities with desirable properties, particularly for automotive applications (Sharma et al., 1997).

In general, composite materials posses good mechanical and thermal properties, sustainable over a wide range of temperatures (Vijayaram et al., 2006). The desirable factors such as property requirements, cost factor considerations and future application prospects would decide the choice of the processing method (Kaczmar et al., 2000). In practice, composite materials with a metal or an alloy matrix are fabricated either by casting or by powder metallurgy methods (Fridlyander, 1995). They are considered as potential material candidates for a wide variety of structural applications in the transportation, automobile and sport goods manufacturing industries due to the superior range of mechanical properties they exhibit (Hashim et al., 1999). MMCs represent a new generation of engineering materials in which a strong ceramic reinforcement is incorporated into a metal matrix to improve its properties such as specific strength, specific stiffness, wear resistance, corrosion resistance and elastic modulus (Baker et al., 1987; Chambers et al., 1996; Kok, 2005). As a virtue of their structure and bonding between the matrix and the reinforcement, MMCs combine metallic properties of matrix alloys (ductility and toughness) with ceramic properties of reinforcements (high strength and high modulus), therein leading to greater strength in shear and compression as well as higher service-temperature capabilities (Huda et al., 1993). Thus, they have scientific, technological and commercial significance. MMCs, because of their improved properties, are being used extensively for high performance applications such as in aircraft engines especially in the last decade. Recently, they also find application in automotive sectors (Surappa, 2003; Therén and Lundin, 1990).

Aluminum oxide (Al_2O_3) and silicon carbide (SiC) powders in the form of fibers and particulates are commonly used as reinforcements in MMCs. In the automotive and aircraft industries for example, production of engine pistons and cylinder heads, the tribological properties of the materials used are considered crucial. Hence, Aluminum oxide and silicon carbide reinforced aluminum alloy matrix composites are applied in these fields (Prasad and Asthana, 2004). Due to their high demand, the development of aluminum matrix composites is receiving considerable emphasis in modern application. Research reports ascertain that the incorporation of hard second phase particles in the alloy matrices to produce MMCs is beneficial and economical due to its high specific strength and corrosion resistance properties (Kok, 2005). Therefore, MMCs are those materials that have higher potential for a large range of engineering applications.

METAL MATRIX COMPOSITES (MMCS)

Metal matrix composites are a family of new materials which are attracting considerable industrial interest and investment worldwide. They are defined as materials whose microstructures compromise a continuous metallic phase (the matrix) into which a second phase, or phases, have been artificially introduced. This is in contrast to conventional alloys whose microstructures are produced during processing by naturally occurring phase transformations (Feest, 1986). Metal matrix composites are distinguished from the more extensively developed resin matrix composites by virtue of their metallic nature in terms of physical and mechanical properties and by their ability to lend themselves to conventional metallurgical processing operations. Electrical conductivity, thermal conductivity and non-inflammability, matrix shear strength, ductility (providing a crack blunting mechanism) and abrasion resistance, ability to be coated, joined, formed and heat treated are some of the properties that differentiate metal matrix composites from resin matrix composites. MMCs are a class of advanced materials which have been developed for weight-critical applications in the aerospace industry. Discontinuously reinforced aluminum composites, composed of high strength aluminum alloys reinforced with silicon carbide particles or whiskers, are a subclass of MMCs. Their combination of superior properties and fabricability makes them attractive candidates for many structural components requiring high stiffness, high strength and low weight. Since the reinforcement is discontinuous, discontinuously reinforced composites can be made with properties that are isotropic in three dimensions or in a plane. Conventional secondary fabrication methods can be used to produce a wide range of composites products, making them relatively inexpensive compared to the other advanced composites reinforced with continuous filaments. The benefit of using composite materials and the cause of their increasing adoption is to be looked for in the advantage of attaining property combinations that can result in a number of service benefits. Among these are increased strength, decreased weight, higher service temperature, improved wear resistance and higher elastic module. The main advantage of composites lies in the tailorability of their mechanical and physical properties to meet specific design criteria. Composite materials are continuously displacing traditional engineering materials because of their advantages of high stiffness and strength over homogeneous material formulations. The type, shape and spatial arrangement of the reinforcing phase in metal matrix composites are key parameters in determining their mechanical behavior. The hard ceramic component which increases the mechanical characteristics of metal matrix composites causes quick wear and premature tool failure in the machining operations. Metal matrix composites have been investigated since the early 1960s with the impetus at that time, being the high potential structural properties that would be achievable with materials engineered to specific applications (Mortensen et al., 1989).

In the processing of metal matrix composites, one of the subjects of interest is to choose a suitable matrix and a reinforcement material (Ashby and Jones, 1980). In some cases, chemical reactions that occur at the interface between the matrix and its reinforcement materials have been considered harmful to the final mechanical properties and are usually avoided. Sometimes, the interfacial reactions are intentionally induced, because the new layer formed at the interface acts as a strong bond between the phases (Gregolin et al., 2002).

During the production of metal matrix composites, several oxides have been used as reinforcements, in the form of particulates, fibers or as whiskers (Zhu and Iizuka, 2003). For example, alumina, zirconium oxide and thorium oxide particulates are used as reinforcements in aluminum, magnesium and other metallic matrices (Upadhyaya, 1990). Very few researchers have reported on the use of quartz as a secondary phase reinforcement particulate in an aluminum or aluminum alloy matrix, due to its aggressive reactivity between these materials (Sahin, 2003). Preliminary studies showed that the contact between molten aluminum and silica-based ceramic particulates have destroyed completely the second phase microstructure, due to the reduction reaction which provokes the infiltration of liquid metal phase into the ceramic material (Mazumdar, 2002). Previous works carried out by using continuous silica fibers as reinforcement phases in aluminum matrix showed that even at temperatures nearer to 400 ^{0}C, silica and aluminum can react and produce a transformed layer on the original fiber surface as a result of solid diffusion between the phases and due to the aluminum-silicon liquid phase formation (Seah et al., 2003). The organizations and companies that are very active in the usage of MMCs in Canada and United States include the following (Rohatgi, 1993):

1. Aluminium Company of Canada, Dural Corporation, Kaiser Aluminium, Alcoa, American Matrix, Lanxide, American Refractory Corporation
2. Northrup Corporation, McDonald Douglas, Allied Signal, Advanced Composite Materials Corporation, Textron Specialty Materials
3. DWA Associates, MCI Corporation, Novamet
4. Martin Marietta Aerospace, Oakridge National Laboratory, North American Rockwell, General Dynamics Corporation, Lockheed Aeronautical Systems
5. Dupont, General Motors Corporation, Ford Motor Company, Chrysler Corporation, Boeing Aerospace Company, General Electric, Westinghouse
6. Wright Patterson Air Force Base, (Dayton, Ohio), and
7. Naval Surface Warefare Centre, (Silver Spring, Maryland)

India also has substantial activity in PM and cast MMCs. It has had world class R&D in cast aluminium particulate composites which was sought even by western countries.

Classification of Composites

Among the major developments in materials in recent years are composite materials. In fact, composites are now one of the most important classes of engineered materials, because they offer several outstanding properties as compared to conventional materials. The matrix material in a composite may be ceramic based, polymer or metal. Depending on the matrix, composite materials are classified as follows:

- Metal matrix composites (MMCs)
- Polymer matrix composites (PMCs)
- Ceramic matrix composites (CMCs)

Majority of the composites used commercially are polymer-based matrices. However, metal matrix composites and ceramic matrix composites are attracting great interest in high temperature applications (Feest, 1986). Another class of composite material is based on the cement matrix. Because of their importance in civil engineering structures, considerable effort is being made to develop cement matrix composites with high resistance to cracking (Schey, 2000). Metal matrix composites (MMCs) are composites with a metal or alloy matrix. It has resistance to elevated temperatures, higher elastic modulus, ductility and higher toughness. The limitations are higher density and greater difficulty in processing parts. Matrix materials used in these composites are usually aluminum, magnesium, aluminum-lithium, titanium, copper and super alloys. Fiber materials used in MMCs are aluminum oxide, graphite, titanium carbide, silicon carbide, boron, tungsten and molybdenum. The tensile strengths of non metallic fibers range between 2000 MPa to 3000 MPa, with elastic modulus being in the range of 200 GPa to 400 GPa. Because of their lightweight, high specific stiffness and high thermal conductivity, boron fibers in an aluminum matrix have been used for structural tubular supports in the space shuttle orbiter. Metal matrix composites having silicon carbide fibers and a titanium matrix, are being used for the skin, stiffeners, beams and frames of the hypersonic aircrafts under development. Other applications are in bicycle frames and sporting goods (Wang et al., 2006). Graphite fibers reinforced in aluminum and magnesium matrices are applied in satellites, missiles and in helicopter structures. Lead matrix composites having graphite fibers are used to make storage-battery plates. Graphite fibers embedded in copper matrix are used to fabricate electrical contacts and bearings. Boron fibers in aluminum are used as compressor blades and structural supports. The same fibers in magnesium are used to make antenna structures. Titanium-boron fiber composites are used as jet-engine fan blades. Molybdenum and tungsten fibers are dispersed in cobalt-base super alloy matrices to make high temperature engine components. Squeeze cast MMCs generally have much better reinforcement distribution than compocast materials. This is due to the fact that a ceramic perform which is used to contain the desired weight fraction of reinforcement rigidly attached to one another so that movement is inhibited. Consequently, clumping and dendritic segregation are eliminated. Porosity is also minimized, since pressure is used to force the metal into interfiber channels, displacing the gases. Grain size and shape can vary throughout the infiltrated preform because of heat flow patterns. Secondary phases typically form at the fiber-matrix interface, since the lower freezing solute-rich regions diffuse toward the fiber ahead of the solidifying matrix (Surappa, 2003).

Significance of Composites

Composites technology and science requires interaction of various disciplines such as structural analysis and design, mechanics of materials, materials science and process engineering. The tasks of composites research are to investigate the basic characteristics of the constituents and composite materials, develop effective and efficient fabrication procedures, optimize the material for service conditions and understanding their effect on material properties and to determine material properties and predict the structural behavior by analytical procedures and hence to develop effective experimental techniques for material characterization, failure analysis and stress analysis (Daniel and Ishai, 1994). An important task is the non-destructive evaluation of material integrity, durability assessment, structural reliability, flaw criticality and life prediction. The structural designs and systems capable of operating at elevated temperatures has spurred intensive research in high temperature composites, such as ceramic/matrix, metal/ceramic and carbon/carbon composites. The utilization of conventional and new composite materials is intimately related to the development of fabrication methods. The manufacturing process is one of the most important stages in controlling the properties and ensuring the quality of the finished product. The technology of composites, although still developing, has reached a state of maturity. Nevertheless, prospects for the future are bright for a variety of reasons. Newer high volume applications, such as in the automotive industry, will expand the use of composites greatly.

Matrix

Matrix is the percolating alloy/metal/polymer/plastic/resin/ceramic forming the constituent of a composite in which other constituents are embedded. If the matrix is a metal, then it is called as a metal matrix and consecutively polymer matrix, if the matrix is a polymer and so on. In composites, the matrix or matrices have two important functions (Weeton et al., 1988). Firstly, it holds the reinforcement phase in the place. Then, under an applied force, it deforms and distributes the stress to the reinforcement constituents. Sometimes the matrix itself is a key strengthening element. This occurs in certain metal matrix composites. In other cases, a matrix may have to stand up to heat and cold. It may conduct or resist electricity, keep out moisture, or protect against corrosion. It may be chosen for its weight, ease of handling, or any of many other applications. Any solid that can be processed to embed and adherently grip a reinforcing phase is a potential matrix material.

In a composite, matrix is an important phase, which is defined as a continuous one. The important function of a matrix is to hold the reinforcement phase in its embedded place, which act as stress transfer points between the reinforcement and matrix and protect the reinforcement from adverse conditions (Clyne, 1996). It influences the mechanical properties, shear modulus and shear strength and its processing characteristics. Reinforcement phase is the principal load-carrying member in a composite. Therefore, the orientation, of the reinforcement phase decides the properties of the composite.

Reinforcing Phase / Materials

Reinforcement materials must be available in sufficient quantities and at an economical rate. Recent researches are directed towards a wider variety of reinforcements for the range of matrix materials being considered, since different reinforcement types and shapes have specific advantages in different matrices (Basavarajappa et al., 2004). It is to be noted that the composite properties depend not only on the properties of the constituents, but also on the chemical interaction between them and on the difference in their thermal expansion coefficients, which both depend on the processing route. In high temperature composites, the problem is more complicated due to enhanced chemical reactions and phase instability at both processing and application temperatures. Reinforcement phases in MMCs are embedded in the form of continuous reinforcement or discontinuous reinforcement in the matrix material. The reinforcing phase may be a particulate or a fiber, continuous type or discontinuous type. Some of the important particulates normally reinforced in composite materials are titanium carbide, tungsten carbide, silicon nitride, aluminum silicate, quartz, silicon carbide, graphite, fly ash, alumina, glass fibers, titanium boride etc. The reinforcement second phase material is selected depending on the application during the processing of composites (Clyne, 1996). The reinforcement phase is in the form of particulates and fibers generally. The size of the particulate is expressed in microns, micrometer. However, the discontinuous fiber is defined by a term called as 'Aspect Ratio'. It is expressed as the ratio of length to the diameter of the fiber. To improve the wettabilty with the liquid alloy or metal matrix material, the reinforcement phase is always preheated (Adams et al., 2003).

Factors Affecting Reinforcement

The interface between the matrix and the reinforcement plays an important role for deciding and explaining the toughening mechanism in the metal matrix composites. The interface between the matrix and the reinforcement should be organized in such a way that the bond in between the interface should not be either strong or weak (Singh et al., 2001).

Matrix Interface / Interphase of Matrices

Interfaces are considered particularly important in the mechanical behavior of MMCs since they control the load transfer between the matrix and the reinforcement. Their nature depends on the matrix composition, the nature of the reinforcement, the fabrication method and the thermal treatments of the composite. For particular matrix/reinforcement associations and especially with liquid processing routes, reactions can occur which change the composition of the matrix and lead to interfacial reaction products, thus changing the mechanical behavior of the composites. The interfacial phenomena in MMCs have been surveyed by several authors. Considering physical and chemical properties of both the matrix and the reinforcing material, the actual strength and toughness desired for the final MMCs, a compromise has to be achieved balancing often several conflicting requirements. A weak interface will lead to crack propagation following the interface, while a strong matrix associated with a strong interface will reveal cracks across both the matrix and the reinforcements. If however the matrix is weak in comparison with the interface and the particle strength, the failure will propagate through the matrix itself. The wettability of the reinforcement material by the liquid metallic matrix plays a major role in the bond formation. It mainly depends on heat of formation, electronic structure of the reinforcement and the molten metal temperature, time, atmosphere, roughness and crystallography of the reinforcement. Similarity between metallic bond and covalent bond is reflected in some metal, like titanium carbide and zirconium carbide which are more easily wetted than strong ionic bonds found in ceramics such as alumina that remains poorly wetted. Surface roughness of the reinforced material improves the mechanical interlocking at the interface, though the contribution of the resulting interfacial shear strength is secondary compared to chemical bonding. Large differences in thermal expansion coefficient between the matrix and the reinforcement should be avoided as they can include internal matrix stresses and ultimately give rise to interfacial failures. From a purely thermo dynamical point of view, a comparison of free enthalpy of formation at various temperatures shows that many metals in the liquid state are reactive toward the reinforcing materials in particular oxides or carbides. Though thermodynamically favored, some reactions are however not observed and practically the kinetics of these reactions has to be considered in conjunction with thermodynamic data in order to evaluate the real potential of the reactions. The consequences of such interfacial reactions are the chemical degradation of the reinforcing material associated with a decrease of its mechanical properties, the formation of brittle reaction products at the interface, as well as the release of elements initially part of the reinforcing material toward the matrix may generate inopportune metallurgical phases at the vicinity of the reinforcing materials. Moreover in the case of alloyed matrices, the selective reactivity and depletion of given elements from the alloy can generate compositional gradients in the matrix and may therefore alter its properties close to the interface. Though a moderate reaction may improve the composite bonding, extended reactions usually ruin the reinforcing material. The relation between interfacial reactions and interface strength depends on the materials. The elaboration of MMC requires often a very short solidification time to avoid excess interfacial reaction. During the cooling process, differences in thermal capacity and thermal conductivity between the reinforcing material and the matrix induce localized temperature gradients. Solidification of the metallic matrix is believed to be generally a directional outward process, starting from the inside of the metallic matrix while ending at the reinforcing material surface. Finally, the processing type and the parameters have to be selected and adjusted to a particular MMC system. Metals are generally more reactive in the liquid rather than in the solid state. Consequently, shorter processing time, that is, short contact time between the liquid metal and the reinforcement can limit the extent of interfacial reactions. The study of reinforcement and matrix bonding is important in composite matrix structure, which has been described by Gregolin (2002). While the load is acting on the composite, it has been distributed to the matrix and the reinforcement phase through the matrix interface. The reinforcement is effective in strengthening the matrix only if a strong interfacial bond exists between them. The interfacial properties also influence the resistance to crack propagation in a composite and therefore its fracture toughness (Dusza and Sajgalik, 1995). The two most important energy-absorbing failure mechanisms in a composite are debonding and particle pull-out at the particle matrix interface. If the interface between the matrix and reinforcement debonds, then the crack propagation is interrupted by the debonding process and instead of moving through the particle, the crack moves along the particle surface allowing the particle to carry a higher load (El-Mahallawy and Taha, 1993).

Physical Phenomena of Wettability and Application

Wettability is defined as the extent to which a liquid will spread over a solid surface. Interfacial bonding is due to the adhesion between the reinforcement phase and the matrix. For adhesion to occur during the manufacturing of a composite, the reinforcement and the matrix must be brought into an intimate contact. During a stage in composite manufacture, the matrix is often in a condition where it is capable of flowing towards the reinforcement and this behavior approximates to that of the flow of a liquid. A key concept in this contact is wettability. Once the matrix wets the reinforcement particle, and thus the matrix being in intimate contact with the reinforcement, causes the bonding to occur (Hashim et al., 2001; Oh et al., 1987). Different types of bonding will occur and the type of bonding varies from system to system and it entirely depends on the details such as the presence of surface contaminants. The different types of bonding observed are mechanical bonding, electrostatic bonding, chemical bonding, and inter diffusion bonding (Burr et al., 1995). The bonding strength can be measured by conducting the tests like single particle test, bulk specimen test, and micro-indention test (Dusza and Sajgalik, 1995).

Poor wettability of most ceramic particulates with the molten metals is a major barrier to processing of these particulate reinforced MMCs by liquid metallurgy route. The characterization and enhancement of wettability is therefore, of central importance to successful composite processing (Asthana and Rohatgi, 1993). Wettability is shown in the Figure 1 below and it is customarily represented in terms of a contact angle defined from the Young-Dupre equation which is expressed as follows:

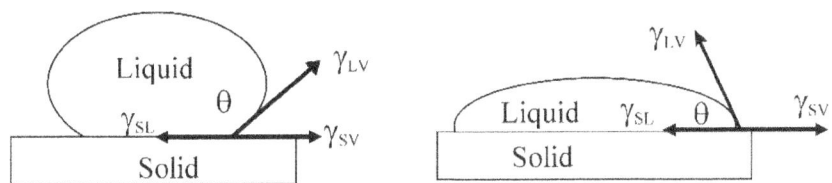

Figure 1: A sessile drop to the left is an example of poor wetting (>90) and the sessile drop to the right is an example of good wetting (<90) (Rajan et al., 1998).

$$\gamma_{lv} \cos \theta = \gamma_{sv} - \gamma_{sl} \quad (1)$$

Where γ_{SV} = Solid/Vapor surface energy, γ_{SL} = Solid/Liquid surface energy and γ_{LV} = Liquid/Vapor surface energy.

The wetting behavior of a liquid on a solid can be characterized by the wetting or contact angle that is formed between the liquid and the solid substrate. A "sessile drop" is a continuous drop of liquid on a flat, solid surface under steady-state conditions. To neglect the effects of gravity, the gravitational forces should be small compared to the surface tension of the drop. If this condition is satisfied, the drop will approach a hemispherical shape which represents its smallest area and lowest surface free energy. The sessile drop is placed on the solid substrate and the angle between the solid surface and the tangent to the liquid surface at the contact point is measured. This is known as the contact angle or wetting angle. The contact angle can vary between 0 and

180 and is a measure of the extent of wetting. The conditions of good wetting (<90) and partial wetting (>90) are illustrated in Figure 1. Complete wetting (also referred to as spreading) is obtained at an angle of 0 and complete non-wetting occurs at an angle of 180. The contact angle is the vector sum of the interfacial surface energies between the solid/liquid (γ_{sl}), liquid/vapor (γ_{lv}), and solid/vapor (γ_{sv}) phases. Young's equation represents a steady-state condition for a solid/liquid interface in stable or metastable thermodynamic equilibrium. Temperature changes have been shown to affect the contact angle of many different systems. The temperature effect, in most cases, can be explained by a reaction at the liquid/solid interface. Thermally activated reactions can occur because many systems are not at chemical equilibrium. The reactions that contribute to wetting (decrease of the contact angle) are those that increase the driving force for wetting ($\gamma_{SV} - \gamma_{SL}$), which is acting at the surface of the liquid drop and the solid substrate. The reactions that contribute to the driving force for wetting are the ones in which the composition of the substrate changes by dissolution of a component of the liquid. On the contrary, if the reaction results in a change of the liquid's composition by dissolution of the solid substrate, but with no change in the composition of the substrate, there is no contribution to the driving force for wetting.

As mentioned above, if the solid substrate is an active participant in the reaction, the free energy of the outer surface of the liquid drop will contribute to the driving force for wetting. As the drop expands on the substrate, the perimeter remains in contact with the unreacted solid and thus the reaction continues to contribute to the driving force for wetting. Examination of phase diagrams representing the interaction between the constituents of the liquid and solid surfaces can help to predict the wetting behavior of a system.

Moreover, measurement of wettability of powders consisting of irregular and polysized particles is extremely difficult. Several techniques have been proposed in the thermodynamic literature to measure wettability. However, these techniques have been applied mostly to non-metallic liquids and their application to metal ceramic systems with reference to pressure casting of composites has been quite limited. The engineering approaches to increasing wettability can be broadly classified into two categories. One method is the surface modification of the reinforcement phase and the other technique is melt treatment. Surface modifications of reinforcements include heat treatment of the particulates to determine surface gas desorbtion, surface oxidation and coating of particles with materials that react with the matrix. Melt treatment is usually done to promote reactivity between the metal and the particulate surface. The wetting reaction must be constrained to prevent reinforcement degradation during the fabrication of subsequent utilization (Ho and Wu, 1998).

Particulate Reinforcement

The improvement in toughness due to the particulate reinforcement depends on the residual stresses surrounding the particles, the weight fraction of the particles, size and shape of the particles (Suery and Esperance, 1993). Particles can be spherical, disk-shaped, rod shaped, and plate shaped. Each particle forces the crack to go out of plane, and can force the crack to deflect in more than one direction and thus increase the fracture surface energy (Gogopsi, 1994). Plate and rod shaped particles can increase the composite toughness by another mechanism called as 'pullout' and 'bridging'. The residual stress around the particles results from thermal expansion mismatch between the particles and the matrix, which helps to resist the crack propagation. The term 'particulates' is used to distinguish these materials from particle and referred as a large, diverse group of materials that consist of minute particles. The second phase particle can produce small but significant increase in toughness and consequently increases its strength through crack deflection processes. The particles, sometimes given a proprietary coating can be used for improving strength. When compared to whiskers-reinforcement systems, particle reinforcement systems have less processing difficulties and should permit to add higher weight fractions of the reinforcing phase. The orientation of particles appears as flat plates (Matthew and Rawlings, 1999; Pardo et al., 2005).

EXPERIMENTAL PROCEDURE

Materials Selected for Processing Composites

Aluminum. – 11.8% silicon (LM6)

The main materials used in this project are LM6 aluminum alloy as a matrix material and SiO_2-quartz as a particulate reinforced added in different percentages. Pure (99.99%) aluminum has a specific gravity of 2.70 and its density is equals 2685kg/m^3. The details of the LM6 alloy properties and composition is shown in Table 1 and Table 2.

Table 1: Composition of LM6(Sayuti, Sulaiman, Baharudin, et al., 2011)

Composition LM6	
Al	85.95
Cu	0.2
Mg	0.1

Si	11.8
Fe	0.5
Mn	0.5
Ni	0.1
Zn	0.1
Lead	0.1
Tin	0.05
Titanium	0.2
Other	0.2

Table 2: Physical, Mechanical and thermal properties of LM6 (Sulaiman, et al., 2008)

PHYSICAL PROPERTIES	VALUES
Density (g/cc)	2.66
MECHANICAL PROPERTIES	**VALUES**
Tensile strength, Ultimate (MPa)	290
Tensile Strength, Yield (MPa)	131
Elongation %; break (%)	3.5
Poisson's ratio	0.33
Fatigue Strength (MPa)	130
THERMAL PROPERTIES	**VALUES**
CTE, linear 20°C (μm/m-°C)	20.4
CTE, linear 250°C (μm/m-°C)	22.4
Heat Capacity (J/g- °C)	0.963
Thermal Conductivity (W/m-K)	155
Melting Point (°C)	574

Quartz

Pure and fused silica is commonly called quartz. Quartz is a hard mineral which is abundantly available as a natural resource. It has a rhombohedra crystal structure with a hardness of 7 on the Mohs scale and has a low specific gravity ranging from 2.50 to 2.66. It provides excellent hardness when incorporated into the soft lead-alloy, thereby making it better suited for applications where hardness is desirable. It also imparts good corrosion resistance and high chemical stability. It is a mineral having a composition SiO_2, which is the most common among all the materials, and occurs in the combined and uncombined states. It is estimated that 60% of the earth's crust contain SiO_2. Sand, clays, and rocks are largely composed of small quartz crystals. SiO_2 is white in color in the purest form. The properties of pure quartz are listed in the Table 3.

Table 3: Properties of quartz

Properties of quartz	
Molecular weight	60.08
Melting Point °C	1713
Boiling Point °C	2230
Density gm/cc	2.32
Thermal Conductivity	0.01 W/cm K (bulk)
Thermal Diffusivity	0.009 cm2/sec (bulk)
Mohs Hardness @ 20 °C	7 Modified Mohs
Si %	46.75
O %	53.25
Crystal Structure	Cubic
Mesh size	230
Size	65 microns (65 μm)

Preparation. of Materials

The materials used in this work were Aluminum LM6 alloy as the matrix and SiO_2 as reinforcement particulates with different weight percentages. The tensile test specimens were prepared according to ASTM standards B 557 M-94 (ASTM, 1991). Sodium silicate and CO_2 gas was used to produce CO_2 sand mould for processing composite casting. The aluminum alloy, LM6, was based on British standards that conform to BS 1490-1988 LM6. Alloy of LM6 is actually a eutectic alloy having the lowest melting point that can be seen from the Al-Si phase diagram. The main composition of LM6 is about 85.95% of aluminum and 11.8% of silicon.

The SiO_2 particulate used as a second phase reinforcement in the alloy matrix was added on the molten LM6 by different weights fraction such as 5%, 10%, 15%, 20%, 25%, and 30%. The mesh size of Silicon Dioxide particulate is 230 microns and the average particle size equal to 65 microns (65μm).

Fabrication. of composites

Only one type of pattern was used in this project and the procedure for making the pattern involves the preparation of drawing, selection of pattern material and surface finishing. Carbon dioxide moulding process was used to prepare the specimens as per the standard moulding procedure. Quartz-particulate reinforced MMCs were fabricated by casting technique. Six different weight fractions of SiO_2 particle in the range from 5%, 10%, 15%, 20%, 25%, and 30% by weight were used. In this research work, the particulates were preheated to 200 °C in a heat treatment muffle furnace for 2 hours and it was transferred immediately in the crucible containing liquid LM6 alloy.

Testing Methods

Tensile. Testing

Tensile test was conducted to determine the mechanical properties of the processed SiO_2 particulate reinforced LM6 alloy composites. Test specimens were made in accordance to ASTM standard B557 M-94. A 250 KN servo hydraulic INSTRON 8500 UTM was used to conduct the tensile test. The tensile testing of the samples was performed based on the following specifications and procedures according to the ASTM standards, which of one crosshead speed of 2.00 mm/minute, grip distance 50.0 mm, specimen distance 50.0 mm and temperature 24 °C.

Hardness. Measurement

The hardness testing was done on a Rockwell Hardness Tester. The hardness of composites was tested by using MITUTOYO ATK-600 MODEL hardness tester. For each sample, ten hardness readings were taken randomly from the surface of the samples. Hardness values of different types of the processed composites are determined for different weight fraction % of titanium carbide particulate containing aluminum-11.8% silicon alloy and graphs were plotted between the hardness value and the corresponding type of particulate addition on weight fraction basis.

Impact. testing

The impact test was conducted in accordance with ASTM E 23-05 standards at room temperature using izod impact tester. The casting processing steps and testing shows are shown in Figure 2.

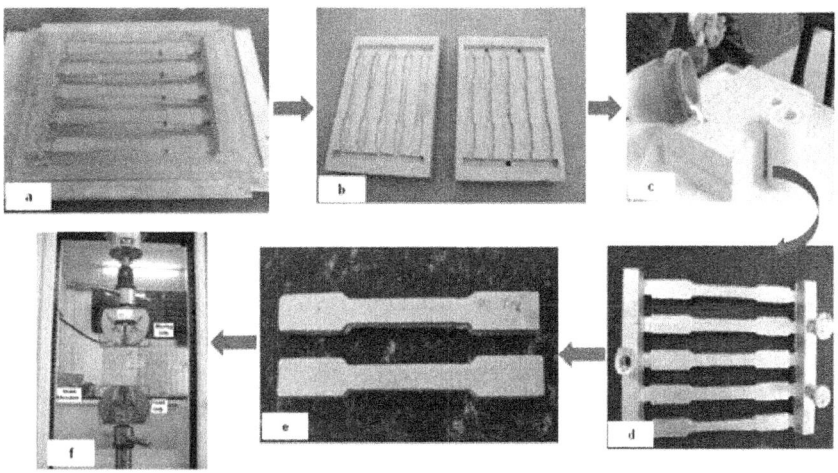

Figure 2: The casting processing steps; (a) Pattern of mould (b) sand mould : drag and copper (c) melting and pouring in the sand mould (d) tensile specimens with gating system (e) tensile specimen after removing of gating systems (f) tensile testing

Density. Measurement

The density of a material is defined as its mass per unit volume. A&D-GR 200 – Analytical Balance was used to conduct the density measurement. The theoretical density of each set of composites was calculated using the rule of mixtures (Rizkalla and Abdulwahed, 1996). Each pellet was weighed in air (W_a), then suspended in Xylene and weighed again (W). The density of the pellet was calculated according to the formula:

$$Density = \frac{Wa}{(Wa - Ww)} \times density\ of\ Xylene$$

Thermal. Diffusivity Measurement

Thermal diffusivity of composite materials is measured using the photo flash method. The photoflash detection system consists of a light source, sample holder, thermocouple, low noise pre amplifier, oscilloscope, photodiode and a personal computer. The temperature rise at the back surface of the sample is detected by the thermocouple. The detected signal is amplified by a low-noise preamplifier and processed by a digital oscilloscope (Carter and Norton, 2007; Yu et al., 2002).

The voltage supplied to the camera flash is always maintained below 6 Volts before switching on the main power supply. The sample is machined to acquire flat surface to obtain better quality result and it is attached directly to the thermocouple. The camera flash is located at 2 cm in front of the sample holder. Before starting the equipment, the set up was tested using a standard material such as aluminium. Measurement was carried out every 10 minutes to allow the sample to thermally equilibrate at room temperature. The data was analyzed before running the next measurement.

Photoflash detection system is not an expensive method and the standard thermal diffusivity value for aluminum is equal to 0.83 cm^2/sec for thickness greater than 0.366 cm (Muta et al., 2003). In the photo flash system, the excitation source consists of a high intensity camera flash. This method is well suitable for aluminum, aluminum alloys and aluminum-silicon particulate metal matrix composites (Collieu and Powney, 1973). The thermal diffusivity values can be obtained for different thicknesses of the test samples. The thermal diffusivity α determines the speed of propagation of heat waves by conduction during changes of temperature with time. It can be related to α, the thermal conductivity through the following equation (Michot et al., 2008; Taylor, 1980).

The photo flash technique was originally described by Parker and it is one of the most common ways to measure the thermal diffusivity of the solid samples. The computer is programmed to calculate the thermal diffusivity, α, using the equation:

$$\alpha = \frac{(1.37 \times L^2)}{[(3.14)^2 \times t_{0.5}]}$$

Where L = thickness in mm and $t_{0.5}$ = half rise time in seconds.

Scanning. Electron Microscopy (SEM)

LEO 1455 variable pressure scanning electron microscope with Inca 300 Energy Dispersive X-ray (EDX) was used to investigate the morphological features. Results and data obtained from the tensile tested samples were correlated with the reported mechanical properties for each volume fraction of silicon dioxide percentage addition to the LM6 alloy matrix.

RESULTS

Tensile. Properties

The average value of tensile strength (MPa) and Young's Modulus (MPa) versus weight fraction of SiO_2 is shown in the Figures 3 and 4.

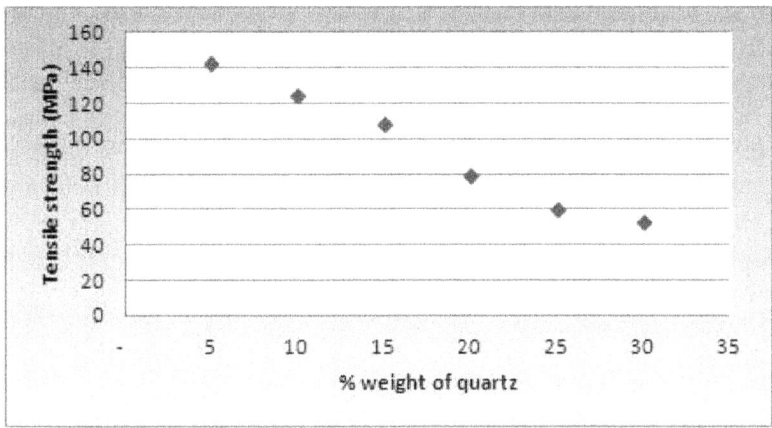

Figure 3: Tensile strength Vs % weight.

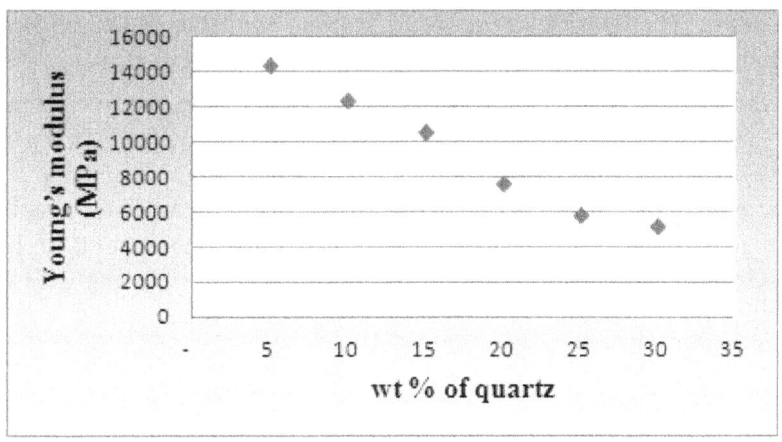

Figure 4: Young"s modulus Vs % weight.

The graph plotted between the average tensile strength and modulus or elasticity values versus variation in weight fraction of quartz particulate addition to LM6 alloy indicates that both the properties decreases with increase in the addition of the quartz particulate. The increase of closed pores content with increasing quartz particulate content would create more sites for crack initiation and hence lower down the load bearing capacity of the composite. The fluctuation maybe due to the non-uniform distribution of quartz particulates, due to experimental errors and also depends on the cooling rate of the castings (ASTM, 1991; Seah, et al., 2003). When particulates increase, particles are no longer isolated by the ductile aluminum alloy matrix, therefore cracks will be not arrested by the ductile matrix and gaps would propagate easily between the quartz particulates. This residual stress affects the material properties around it and the crack tips and the fracture toughness values would be altered. Consequently, these residual stresses would probably contribute for the brittle nature of the composites. It should be noted that the compressive strength of the quartz particulate dominates which is more than the tensile strength of the LM6 alloy matrix and hence the tensile strength decreases with more amount of addition of quartz particulate afact which is well supported is well supported and evidenced from the literature citation (Rizkalla and Abdulwahed, 1996; Seah, et al., 2003).

Hardness

Similarly, for a given S_iO_2 reinforcement content, some differences in the hardness values were observed depending upon the particle size of the constituents. From the Table 4, data on hardness of quartz particulate reinforced composites made in sand mold is listed. It was found that the hardness value increased gradually with the increased addition of quartz particulate by weight fraction percentage as shown in Figure 5.

The maximum hardness value obtained based on the Rockwell superficial 15N-S scale was 67.85 for 30% weight fraction addition. The EDS spectrums for 30% wt of SiO_2 are shown in Figure 5. Their respective elemental analysis is shown in Table 4. It was observed that the grain-refined composite casting has higher weight percentage of Si compared with the original LM6 casting. These results indicate the interrelationship between the thermal properties and hardness.

Impact. Strength

Impact strength data of quartz particulate reinforced composite castings processed was determined and it is listed in the Table 4. From the plotted graph shown in the Figure 6, it is found that the impact strength values were gradually increased with the increased addition of quartz particulate in the alloy matrix. The maximum value of impact strength was 24.80 N-m for 30% weight fraction addition of quartz particulate to the alloy matrix. A reason for the increased volume impact–abrasive wear of the SiO2 particle reinforced composites lies in the propensity of the carbides to fracture and spall as a result of the repeated impact from the quartzite. In the monolithic ferrous-based alloys, the matrix can absorb substantial damage in the form of plastic deformation. This plastic deformation is in fact beneficial in that, the matrix will get harder as a result, and wear, fatigue type processes ending as a material removal mechanism. In the SiO_2 particle reinforced composites, however, the high weight fraction of SiO_2 limits the amount of plastic deformation that the matrix can absorb.

This leads more quickly to SiO_2 reinforcement fracture, matrix– SiO_2 particle delamination, and S_1O_2 particle spalling. As a consequence, volume impact–abrasive wear increases at a more rapid rate for the composite materials as the hardness increases. However, for the very 'hardest' S_1O_2 particle reinforced composites, impact–abrasion resistance is very good. The summary of mechanical properties of quartz particulate reinforced composite castings processed was determined and it is listed in the Table 4.

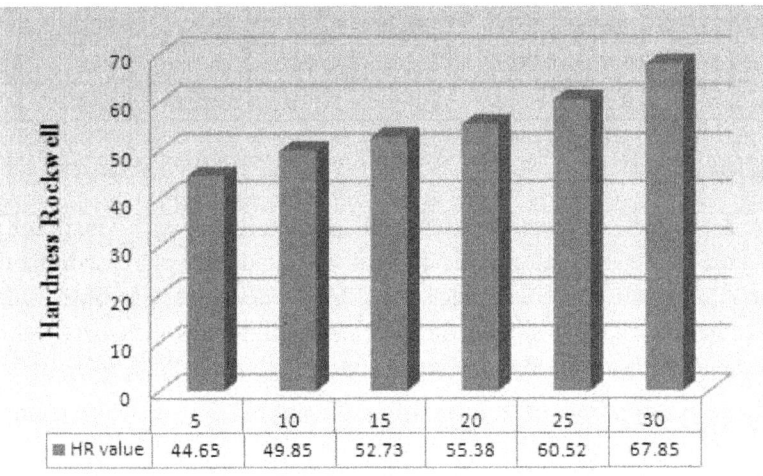

Figure 5: Hardness Vs wt % of quartz

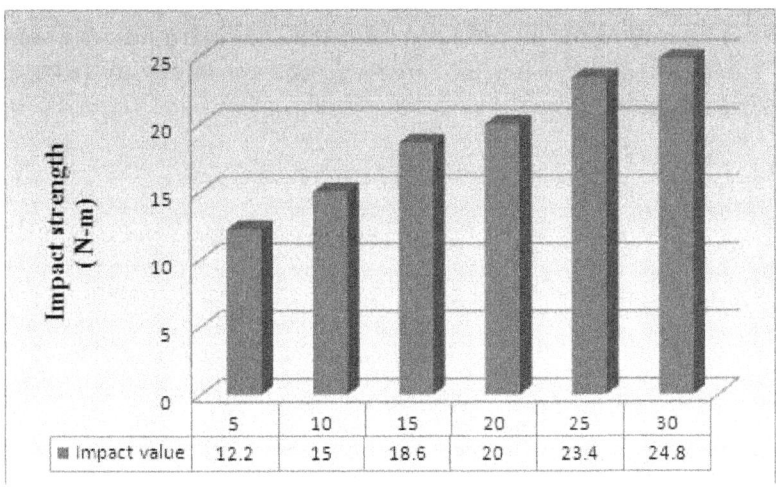

Figure 6: Impact strength Vs Weight fraction % of quartz.

Table 4: Mechanical properties of quartz particulate composites

Wt % of quartz	UTS (MPa)	Yield (MPa)	Young's modulus MPa	Fracture stress MPa	Ductility %	Reduction in area %	Rockwell Hardness	Impact (N-m)
5 %	142.99	132.00	14351	189.50	1.214	2.863	44.65	12.20
10 %	124.74	129.60	12350	164.60	1.412	2.864	49.85	15.00
15 %	108.47	118.50	10635	142.20	1.422	3.042	52.73	18.60
20 %	78.97	109.60	7621	128.40	1.632	3.264	55.38	20.00
25%	59.53	100.50	5853	115.30	1.824	3.625	60.52	23.40
30%	52.64	92.65	5242	104.60	1.741	3.482	67.85	24.80

Density

Figure 8 gives the influence of quartz addition on the density. The graph shows that as the quartz-silicon dioxide content was gradually increased, the density of the Aluminum composite decreased. Slight decrease was observed in the density because quartz-silicon dioxide has a slight lower density value than LM6 (the density of LM6 is 2.65grs/cc and of quartz is 2.23grs/cc).

The investigation of the aluminum composite was well documented. The percentage of the closed pores in the sintered composites increased with increasing quartz content. This can be attributed to silica being harder than aluminum and non deformation at all under the applied compaction load. The morphological features of quartz particles were significantly different from those of Aluminum and as a result, the interparticle friction effects were different. Therefore, the increase in the amount of closed pores with increasing quartz content would justify the observed decrease in density (Rizkalla and Abdulwahed, 1996).

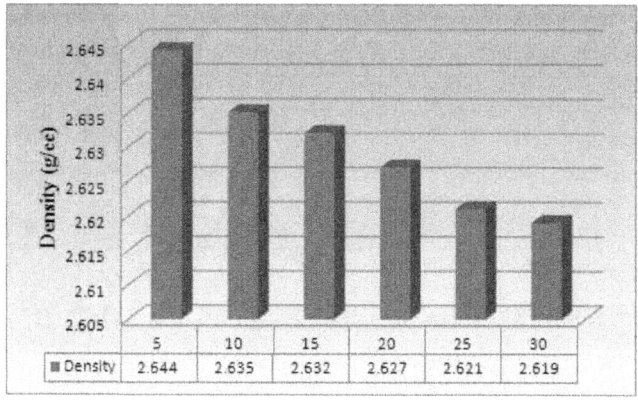

Figure 7: Graph plotted on density versus %wt fraction of S_1O_2

Thermal. Properties

Quartz particulate reinforced composite castings made in grey cast iron mold were tested and analyzed for thermal properties. Graphs are plotted between the weight fraction % addition of quartz and thermal diffusivity and thermal conductivity values. It is found that the thermal diffusivity of the quartz composites decreased with the increased addition in the alloy matrix. Reversely, the thermal conductivity of the quartz composites decreased with the increased addition of quartz particulate in the alloy matrix. Quartz particulates are a ceramic reinforcement phase and on addition of this in the alloy matrix reduces the thermal conductivity. The data for thermal diffusivity and thermal conductivity of the quartz particulate reinforced composites made in sand mold is given in the Table 4. These are illustrated in the plotted graphs and are shown in Figure 8 and 9. The thermal diffusivity and thermal conductivity for 30% weight fraction addition of quartz are 0.2306 cm^2/sec and 52.9543 W/mK respectively and it is well supported from the literature citation (Collieu and Powney, 1973). The summary of physical properties of quartz particulate reinforced composite castings processed was determined and it is listed in the Table 5.

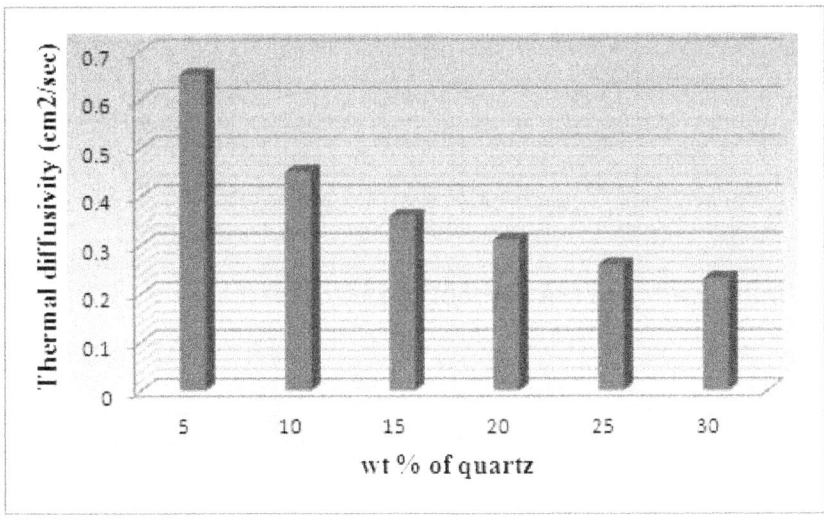

Figure 8: Thermal diffusivity V_s Wt Fraction % of quartz

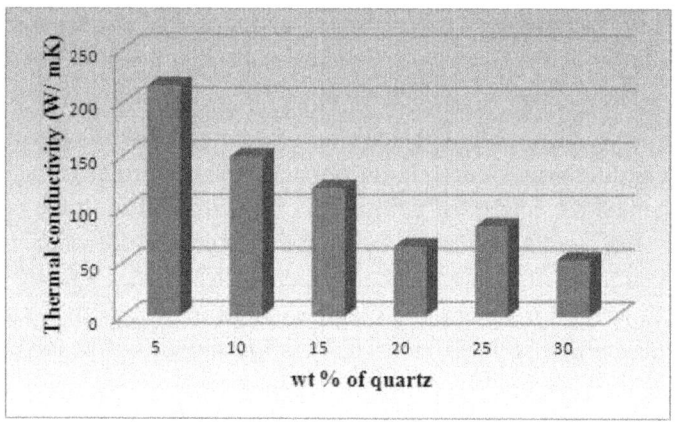

Figure 9: Thermal conductivity Vs W_t Fraction % of quartz.

Table 5: Physical properties of quartz particulate composites

Wt % of quartz	Density (g/cc)	Thermal diffusivity cm²/sec	Thermal conductivity (W/ mK)
5 %	2.644	0.6513	215.826
10 %	2.635	0.4514	149.584
15 %	2.632	0.3595	119.933
20 %	2.627	0.3102	65.6860
25%	2.621	0.2590	84.6830
30%	2.619	0.2306	52.9543

Scanning. Electron Microscopy (SEM)

Scanning Electron Microscopy and energy dispersive spectroscopy was employed to obtain some qualitative evidences on the particle distribution in the matrix and bonding quality between the particulate and the matrix. Besides this the fracture surface of the composite was analyzed by using SEM to show the detail of chemically reacted interfaces. Thus, in order to increase the potential application of MMCs, it is necessary to concentrate on the major aspects, like particle size of quartz and quartz distribution concentration.

The fracture surfaces or fractographs are shown in the Figures 10-15 after tensile testing the specimens having different weight fraction of quartz particulate. It was observed that the increase of SiO_2 content would create more sites for crack initiation and would lower the load bearing capacity of MMCs. In addition the number of contacts between quartz particles would increase and more particles were no longer isolated by the ductile aluminum alloy matrix. Therefore, cracks were not arrested by the ductile matrix and they would propagate easily between quartz particulates. Decrease of SiO_2 content to less than 30% in the matrix and a particle size of 230 micron could increase the tensile strength. Hence cracking on the surface is not too dominant. This phenomenon is shown in Figure 10. The problem on interfacial bonding between the particulate quartz and the matrix during the solidification of composites can be ignored because the phenomena of cracking occurs only in a small part of the surface (Seah, et al., 2003). In contrast, when the content of quartz was increased (30%), interfacial bonding concept would be an important phenomenon because the surface cracking will be distributed on the surface of the parts. The other problem caused by the interaction between Aluminum alloy and quartz particle is not a significant one and it is removed while solidification during the pouring process and due to slip inter bonding/ inter granular movement which is illustrated with the aid of Figure 11.

Figure 10: EDX Spectrum and Fractograph of 5wt% quartz particulate reinforced in quartz -LM6 alloy matrix composite at 250X magnification by SEM after tensile testing.

Figure 11: EDX Spectrum and Fractograph of 10wt% quartz particulate reinforced in quartz -LM6 alloy matrix composite at 100X magnification by SEM after tensile testing.

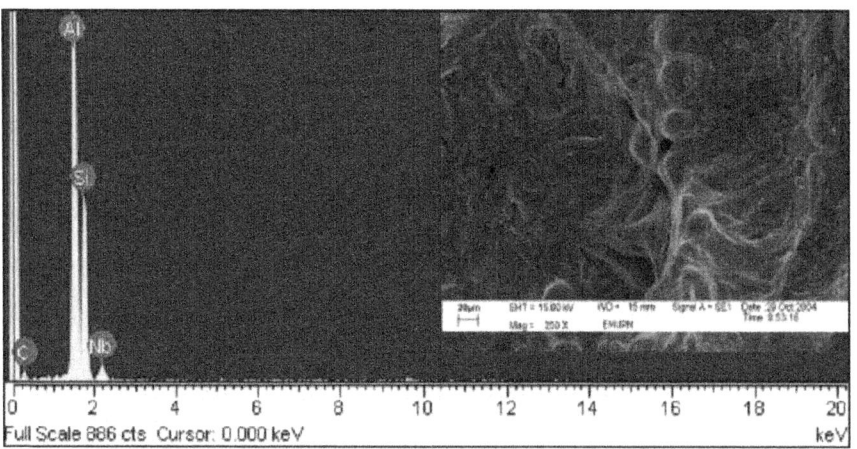

Figure 12: EDX Spectrum and Fractograph of 15wt% quartz particulate reinforced in quartz -LM6 alloy matrix composite at 250X magnification by SEM after tensile testing.

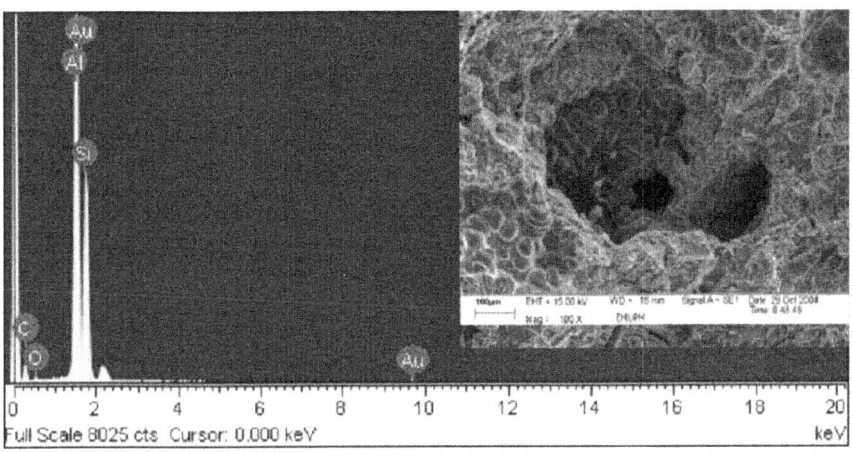

Figure 13: EDX Spectrum and Fractograph of 20wt% quartz particulate reinforced in quartz -LM6 alloy matrix composite at 100X magnification by SEM after tensile testing.

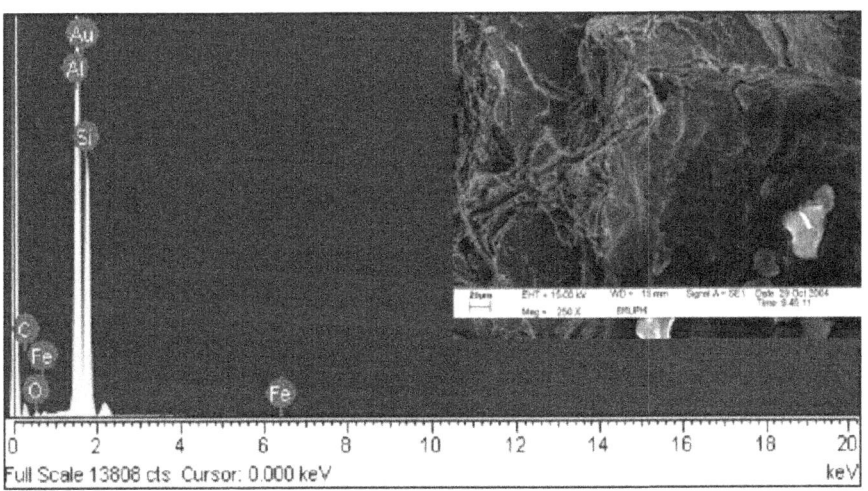

Figure 14: EDX Spectrum and Fractograph of 25wt% quartz particulate reinforced in quartz -LM6 alloy matrix composite at 250X magnification by SEM after tensile testing.

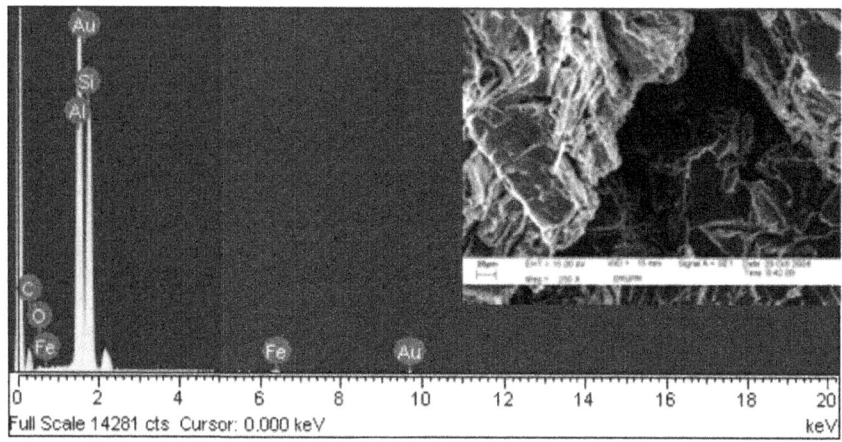

Figure 15: EDX Spectrum and Fractograph of 30wt% quartz articulate reinforced in quartz -LM6 alloy matrix composite at 250X magnification by SEM after tensile testing

CONCLUSIONS

In this study, the compressive strength of the silicon dioxide particulate reinforcement dominates and influences more effectively than the tensile strength of the LM6 alloy matrix phase. Hence the values of tensile strength and modulus of elasticity are decreased with the increased addition of S_iO_2 particulate from 5 to 30% by volume fraction basis. This fact from the present experimental research is well supported and validated from the literature. The mechanical behaviour of the processed composite had a strong dependence on the volume fraction addition of the second phase reinforcement particulate on the alloy matrix. On the other hand, decreasing the S_iO_2 particulate content less than 30% by weight along with the particle size constraint as 230 mesh-65 microns would increase the tensile strength and cracking on the surface might not be too dominant. The hardness value of the silicon reinforced aluminum silicon alloy matrix composite is increased with the addition of quartz particulate in the matrix.

The density of these composites decreased slightly with increasing quartz content. Slight decrease was observed in the density because quartz-silicon dioxide has a slightly lower density value than LM6. For a given particle size combination, the thermal diffusivity and thermal conductivity decreases as S_iO_2 wt % of the composite increases. The particle size ratio of the constituents becomes an important factor for thermal properties, especially above 10wt. % S_iO_2. A higher Al/ S_iO_2 particle size ratio results in segregation of S_iO_2 particles along the LM6 boundaries. This yields lower thermal conductivity with respect to the homogeneously distributed reinforcement. Therefore, a thermal conductivity value that is less than the expected one might be attributed to the micro-porosity in the segregated structure. Similar tendencies were also observed for the results of hardness tests.

In future, it is strongly recommended that tensile tests be performed by reinforcing the second phase quartz particulate addition to the LM6 alloy matrix by limiting it up to 15 wt%. In addition, compressive strengths testing of the processed composite samples can be done to highlight the benefits, advantages and applications of these composites. It is also worthwhile to conduct heat treatment studies of these processed composites and this will be in the scope of future research work.

ACKNOWLEDGEMENT

The authors would like to express their deep gratitude and sincere thanks to the Department of Mechanical and Manufacturing Engineering, Universiti Putra Malaysia for their help to complete this work.

REFERENCES

1. D. F. Adams, L. A. carlsson, R. B. Pipes, 2003Expeirmental Characterization of Advanced Composite Materials (3rd ed.). Florida, USA: CRC Press LLC.
2. M. F. Ashby, D. R. H. Jones, 1980Engineering Materials: An Introduction to their properties and applications, International series on materials science and technology. UK: Elsevier Science and Technology.
3. R. Asthana, P. K. Rohatgi, 1993A Study of Metal-Ceramic Wettability in Sic-Al Using Dynamic Melt Infiltration of SiCKey Engineering Materials47 EOF62 EOF
4. ASTM.1991American Society for Testing and Material, Anual Books of ASTM Standards. USA.
5. A. R. Baker, D. J. Dawson, D. C. Evans, 1987Ceramics and Composite, Materials for Precision Engine Components. Journal of Materials & Design, 8(6), 315-323.
6. S. Basavarajappa, G. Chandramohan, A. Dinesh, 2004Mechanical Properties of MMC'S-an Experimental Inversitgation. Paper presented at the International Symposium of research Students on Materials and Engineering, Chennai.

7. A. Burr, J. Y. Yang, C. G. Levi, F. A. Leckie, 1995The strength of metal-matrix composite jointsActa Metallurgica et Materialia3361 EOF3373 EOF
8. C. B. Carter, M. G. Norton, 2007Ceramic Material: Science and Engineering New York: Springer.
9. B. V. Chambers, M. L. Seleznev, A. Cornie, J. , S. Zhang, M. A. Rye, 1996The Strength and Toughness of Cast Aluminium Composite as a Function of Composition, Heat Treatment and Particulate. SAE International Journal, 164169
10. T. W. Clyne, 1996Interfacial Effects in Particulate, Fibrous and Layered Composite MaterialsKey Engineering Materials, 116-117, 133-152.
11. A. M. B. Collieu, D. J. Powney, 1973The Mechanical and Thermal Properties of Materials
12. UK, London: Edward ArnoldPublishers) Ltd.
13. I. M. Daniel, O. Ishai, 1994Engineering Mechanics of Composite MaterialsUSA: Oxford University Press.
14. J. Dusza, P. Sajgalik, 1995Fracture Toughness and Strength Testing of Ceramic Composites. In N. P, Cheremisinoff and P. N.Cheremisinoff (Eds.), Handbook of Advanced Materials Testing (399435New York, USA: Marcel Dekker, Inc.
15. N. A. El-Mahallawy, M. A. Taha, 1 EOF14 EOF
16. E. A. Feest, 1986Metal matrix composites for industrial applicationMaterials & Design58 EOF64 EOF
17. J. N. Fridlyander, 1995Metal Matrix CompositesUK, London: Chapman & Hall.
18. D. Gay, V. Hoa, S. , W. Tsai, S. , 2003Composite Materials: Design and Applications.USA: CRC Press LLC.
19. Y. G. Gogopsi, 1994Particulate Silicon Nitride- Based Composite. Materials science, 29(4), 2541-2556.
20. Goldenstein. H. Gregolin, M. d. C. Gonçalves, Santos, R. G. d. , 2002Aluminium Matrix Composites Reinforced with Co-continuous Interfaced Phases Aluminium-alumina Needles. Materials Research, 5(3), 337-342.
21. J. Hashim, L. Looney, M. S. J. Hashmi, 1 EOF1999Metal matrix composites: production by the stir casting methodJournal of Materials Processing Technology

22. J. Hashim, L. Looney, M. S. J. Hashmi, 2001The enhancement of wettability of SiC particles in cast aluminium matrix compositesJournal of Materials Processing Technology329 EOF
23. J. Hashim, L. Looney, M. S. J. Hashmi, 2002Particle distribution in cast metal matrix composites--PartI. Journal of Materials Processing Technology251 EOF
24. H. Ho, N. , S. Wu, T. , 1998The wettability of molten aluminum on sintered aluminum nitride substrateMaterials Science and Engineering: A, 248(1-2), 120 EOF
25. D. Huda, M. A. El Baradie, M. S. J. Hashmi, 1993Metal-matrix composites: Manufacturing aspects. Part IJournal of Materials Processing Technology513 EOF528 EOF
26. J. Humphreys, 1987Composites for automotive on-engine applicationsMaterials & Design147 EOF151 EOF
27. J. W. Kaczmar, K. Pietrzak, W. Wlosinski, 2000The production and application of metal matrix composite materialsJournal of Materials Processing Technology58 EOF
28. M. Kok, 2005Production and mechanical properties of Al2O3 particle-reinforced 2024 aluminium alloy compositesJournal of Materials Processing Technology381 EOF387 EOF
29. F. L. Matthew, R. D. Rawlings, 1999Composite Material; Engineering and Science. UK: Imperial College of Science.
30. S. K. Mazumdar, 2002Composites Manufacturing: Materials, Product and Process Engineering.USA: CRC Press Inc.
31. A. e. Michot, D. S. Smith, S. Degot, C. Gault, 2008Thermal Conductivity and Specific Heat of Kaolinite: Evolution with Thermal TreatmentJournal of the European Ceramic Society2826392644
32. A. Mortensen, J. A. Cornie, Flemings, M. C. , 1989Solidification Processing of Metal-Matrix CompositesMaterials & Design, 10(2), 68-76.
33. H. Muta, K. Kurosaki, M. Uno, S. Yamanaka, 2003Thermoelectric properties of constantan/spherical SiO2 and Al2O3 particles compositeJournal of Alloys and Compounds326 EOF
34. S. Oh, Y. , J. A. Cornie, Russell, K. C. , 1987Particulate Wetting and Metal: Ceramic Interface Phenomena. Paper presented at the Ceramic Engineering Science Proceedings.
35. A. Pardo, M. C. Merino, S. Merino, F. Viejo, M. Carboneras, R. Arrabal, 2005Influence of reinforcement proportion and matrix composition

on pitting corrosion behaviour of cast aluminium matrix composites (A3xx.x/SiCp)Corrosion Science1750 EOF1764 EOF

36. S. V. Prasad, R. Asthana, 2004Aluminium Metal Matrix Composite for Automotive Applications: Tribological Considerations. Tribology Letters, 17(3), 445-453.

37. T. P. D. Rajan, R. M. Pillai, B. C. Pai, 1998Reinforcement coatings and interfaces in aluminium metal matrix compositesJournal of Materials Science3491 EOF

38. H. L. Rizkalla, A. Abdulwahed, 1996Some mechanical properties of metal-nonmetal Al---SiO2 particulate compositesJournal of Materials Processing Technology398 EOF

39. P. K. Rohatgi, 1993Metal Matrix Composite, Casting Processes. Science Journal, 43(4), 323-349.

40. Y. Sahin, 2003Preparation and some properties of SiC particle reinforced aluminium alloy compositesMaterials & Design, 24(8), 671 EOF679 EOF

41. M. Sayuti, S. Sulaiman, B. T. H. T. Baharudin, M. K. A. Arifin, S. Suraya, T. R. Vijayaram, 2010Mechanical properties of particulate reinforced aluminium alloy matrix composite

42. M. Sayuti, S. Sulaiman, B. T. H. T. Baharudin, M. K. A. Arifin, T. R. Vijayaram, S. Suraya, (2011) Influence of mechanical vibration moulding process on the tensile properties of TiC reinforced LM6 alloy composite castings. 66-68 (12071212).

43. M. Sayuti, S. Sulaiman, T. R. Vijayaram, B. T. H. T. Baharudin, M. K. A. Arifin, 2011The influence of mechanical vibration moulding process on thermal conductivity and diffusivity of Al-TiC particulate reinforced composites311-31338

44. M. Sayuti, S. Suraya, S. Sulaiman, T. R. Vijayaram, M. K. H. Arifin, B. T. H. T. Baharudin, 2011Thermal investigation of aluminium- 11.8% silicon (LM6) reinforced SiO2- Particles. 264-265620625

45. J. A. Schey, 2000Introduction to Manufacturing Processes3USA: McGraw Hill.

46. K. H. W. Seah, J. Hemanth, S. C. Sharma, 2003Mechanical properties of aluminum/quartz particulate composites cast using metallic and non-metallic chillsMaterials & Design87 EOF93 EOF

47. S. Sharma, K. H. W. Seah, B. M. Girish, R. Kamath, B. M. Satish, 1997Mechanical properties and fractography of cast lead-alloy/quartz particulate compositesMaterials & Design, 18(3), 149 EOF

48. M. Singh, D. P. Mondal, A. K. Jha, S. Das, A. H. Yegneswaran, 2001Preparation and properties of cast aluminium alloy-sillimanite particle compositeComposites Part AApplied Science and Manufacturing, 32(6), 787 EOF795 EOF

49. M. Suery, G. L. Esperance, 1993Interfacial Reactions and Mechanical Behaviour of Aluminium Matrix Composites Reinforced with Ceramic ParticlesKey Engineering Materials33 EOF

50. S. Sulaiman, M. Sayuti, R. Samin, 2008Mechanical properties of the as-cast quartz particulate reinforced LM6 alloy matrix compositesJournal of Materials Processing Technology731 EOF735 EOF

51. M. Surappa, 2003Aluminium matrix composites: Challenges and opportunitiesSadhana, 28(1), 319-334.

52. R. Taylor, 1980Construction of apparatus for heat pulse thermal diffusivity measurements from 3003000 KJournal of physics E : Scientific Instruments, 13(11).

53. K. Therén, A. Lundin, 1990Advanced composite materials for road vehiclesMaterials & Design71 EOF75 EOF

54. G. S. Upadhyaya, 1990Trends in advanced materials and processes. Materials & Design, 11(4), 171-179.

55. T. R. Vijayaram, S. Sulaiman, A. M. S. Hamouda, M. H. M. Ahmad, 2006Fabrication of fiber reinforced metal matrix composites by squeeze casting technologyJournal of Materials Processing Technology, 178(1-3), 34 EOF38 EOF

56. J.-j. Wang, J.-h. Guo, L.-q. Chen, 2006TiC/AZ91D composites fabricated by in situ reactive infiltration process and its tensile deformationTransactions of Nonferrous Metals Society of China, 16(4), 892 EOF896 EOF

57. J. W. Weeton, D. M. Peters, K. L. Thomas, 1988Engineers' Guide to Composite Materials. Metals Park, Ohio 44073, USA: American Society for Metals.

58. S. Yu, P. Hing, X. Hu, 2002Thermal conductivity of polystyrene-aluminum nitride compositeCompositesPart A: Applied Science and Manufacturing, 33(2), 289 EOF

59. S. J. Zhu, T. Iizuka, 2003Fabrication and mechanical behavior of Al matrix composites reinforced with porous ceramic of in situ grown whisker frameworkMaterials Science and Engineering A, 354(1-2), 306 EOF

CITATION

CHAPTER 1
Jang, K.-I. et al. Soft network composite materials with deterministic and bio-inspired designs. Nat. Commun. 6:6566 doi: 10.1038/ncomms7566 (2015).

CHAPTER 2
David Alejandro Arellano Escárpita, Diego Cárdenas, Hugo Elizalde, Ricardo Ramirez and Oliver Probst (2012). Biaxial Tensile Strength Characterization of Textile Composite Materials, Composites and Their Properties, Prof. Ning Hu (Ed.), ISBN: 978-953-51-0711-8, InTech, DOI: 10.5772/48105.

CHAPTER 3
Yuan Li, Sen Liu, Ning Hu, Weifeng Yuan and Bin Gu (2012). Molecular Simulations on Interfacial Sliding of Carbon Nanotube Reinforced Alumina Composites, Composites and Their Properties, Prof. Ning Hu (Ed.), ISBN: 978-953-51-0711-8, InTech, DOI: 10.5772/48816.

CHAPTER 4
Yun Lu, Liang Hao and Hiroyuki Yoshida (2012). Mechanical Coating Technique for Composite Films and Composite Photocatalyst Films, Composites and Their Applications, Prof. Ning Hu (Ed.), ISBN: 978-953-51-0706-4, InTech, DOI: 10.5772/48794.

CHAPTER 5

V. Alfred Franklin and T. Christopher, "Generation of R-Curve from 4ENF Specimens: An Experimental Study," Journal of Composites, vol. 2014, Article ID 956268, 10 pages, 2014. doi:10.1155/2014/956268

CHAPTER 6

Teimouri, H. , Milani, A. , Seethaler, R. and Heidarzadeh, A. (2016) On the Impact of Manufacturing Uncertainty in Structural Health Monitoring of Composite Structures: A Signal to Noise Weighted Neural Network Process. *Open Journal of Composite Materials*, **6**, 28-39. doi: 10.4236/ojcm.2016.61004.

CHAPTER 7

Sebastian, J. , Thachil, E. , Mathen, J. , Madhavan, J. , Thomas, P. , Philip, J. , Jayalakshmy, M. , Mahmud, S. and Joseph, G. (2015) Enhancement in the Electrical and Thermal Properties of Ethylene Vinyl Acetate (EVA) Co-Polymer by Zinc Oxide Nanoparticles. *Open Journal of Composite Materials*, **5**, 79-91. doi:10.4236/ojcm.2015.53011.

CHAPTER 8

Sumanta Bhandary and Biplab Sanyal (2012). Graphene-Boron Nitride Composite: A Material with Advanced Functionalities, Composites and Their Properties, Prof. Ning Hu (Ed.), ISBN: 978-953-51-0711-8, InTech, DOI: 10.5772/50729.

CHAPTER 9

Ilya Mazov, Vladimir Kuznetsov, Anatoly Romanenko and Valentin Suslyaev (2012). Properties of MWNT-Containing Polymer Composite Materials Depending on Their Structure, Composites and Their Properties, Prof. Ning Hu (Ed.), ISBN: 978-953-51-0711-8, InTech, DOI: 10.5772/48245.

CHAPTER 10

Go Yamamoto and Toshiyuki Hashida (2012). Carbon Nanotube Reinforced Alumina Composite Materials, Composites and Their Properties, Prof. Ning Hu (Ed.), ISBN: 978-953-51-0711-8, InTech, DOI: 10.5772/48667.

CHAPTER 11

M. Sayuti, S. Sulaiman, T.R. Vijayaram, B.T.H.T Baharudin and M.K.A. Arifin (2012). Manufacturing and Properties of Quartz (SiO2) Particulate Reinforced Al-11.8%Si Matrix Composites, Composites and Their Properties, Prof. Ning Hu (Ed.), ISBN: 978-953-51-0711-8, InTech, DOI: 10.5772/48095.

INDEX

A

Artificial Neural Networks (ANN) 137
Artificial Neural Networks (ANNs) 142

C

Carbon fibers 193
Carbon nanotubes (CNTs) 193
Central-notched flexure (CNF) 117
Ceramic matrix composites (CMCs) 248
Chemical vapor deposition (CVD) 79
Composite materials 195, 196, 198, 199, 205, 206, 207, 209, 211, 213, 214

D

Damage signature database (DSD) 138, 141, 146
Data acquisition (DAQ) 18

E

Electrical properties 193, 214
Electromagnetic irradiation (EMI) 209

F

Finite element (FE) 141

H

Hexagonal BNC (h-BNC) 176

I

Internal-notched flexure (INF) 117

M

Mechanical coating technique (MCT) 82, 83, 112
Metal matrix composites (MMC) 243
Metal matrix composites (MMCs) 248
Multiwall carbon nanotubes (MWNTs) 194

N

Nanotechnology applications 193, 194

P

Physical vapor deposition (PVD) 79
Polymer matrix composites (PMCs) 248
Polymethylmethacrylate (PMMA) 202

S

Scanning electron microscope (SEM) 8, 10

Signal-to-noise (SN) 137, 138, 145
Single-walled carbon nanotube (SWC-
  NT) 55
Single-wall nanotubes (SWNTs) 193
Structural Health Monitoring (SHM)
  137

T

Tension-tension (T-T) 26
Textile composites (TC 25, 32, 42
Transmission electron microscope
  (TEM) 222

Two-dimensional (2D) 3

U

Unidirectional composites (UDC) 25

W

World Wide Failure Exercises (WWFE)
  26

Z

Zigzag nanoribbons (ZGNRs) 189